# APPLIED ELECTRICITY AND ELECTRONICS FOR THE TECHNOLOGIES

# APPLIED ELECTRICITY AND ELECTRONICS FOR THE TECHNOLOGIES

Do Page 51 50

*Edward F. Mahoney*

PRENTICE HALL, *Englewood Cliffs, New Jersey 07632*

*Library of Congress Cataloging-in-Publication Data*

Mahoney, Edward F.
Applied electricity and electronics for the technologies / Edward F. Mahoney.
p. cm.
Includes index.
ISBN 0-13-040221-4
1. Electricity. 2. Electronics. 3. Electricity. I. Title.
TK7816.M18 1989
621.3—dc19 88-31740
CIP

Cover design: *20/20 Services, Inc.*
Manufacturing buyer: *Robert Anderson*



Printed in the United States of America

10 9 8 7 6 5 4 3 2 1

ISBN 0-13-040221-4

PRENTICE-HALL INTERNATIONAL (UK) LIMITED, *London*
PRENTICE-HALL OF AUSTRALIA PTY. LIMITED, *Sydney*
PRENTICE-HALL CANADA INC., *Toronto*
PRENTICE-HALL HISPANOAMERICANA, S.A., *Mexico*
PRENTICE-HALL OF INDIA PRIVATE LIMITED, *New Delhi*
PRENTICE-HALL OF JAPAN, INC., *Tokyo*
SIMON & SCHUSTER ASIA PTE. LTD., *Singapore*
EDITORA PRENTICE-HALL DO BRASIL, LTDA., *Rio de Janeiro*

*To my wife, Gloria, who finds more in me than I can.*
*Without her, this book would not be.*

# CONTENTS

# *PREFACE*

The increased use of electricity and electronics for control is evident in almost every industry. Production equipment is rapidly becoming automated through the use of digital computers tied through analog systems to handle repetitive processes. The American automobile is becoming increasingly controlled by the use of electronic sensors. Electronics is used in the blending of engine and transmission controls. Advanced power train controls, spark timing, manifold boost, and fuel mixture are all coming under increased electronic control. All wheel steering and advanced braking are other areas where the use of electronics is evident. In today's world if it is mechanical and requires positioning control, it probably includes electronics.

Students in all the technologies must become more familiar with electricity and electronics if they are to be productive in today's market. *Applied Electricity and Electronics for the Technologies* has been prepared to fit the need for general coverage of electrical and electronic phenomena. The text has 21 units, covering basic electricity to basic digital circuits. The material is presented in a simple, readable fashion that is easily understood. Questions are included at the end of each unit. The questions are not intended as tests but as a review and self-analysis.

The text will also be useful to individuals presently working in the technologies whose background in electronics is in need of expansion. Earlier programs did not include instruction in electricity and electronics.

Upon completion of this text, the reader will have a good exposure to basic electricity and electronics. Knowledge in this area is a *requirement* for everyone working in today's technological world.

*Edward F. Mahoney*

# APPLIED ELECTRICITY AND ELECTRONICS FOR THE TECHNOLOGIES

# UNIT ONE

# *ELECTRON THEORY*

The information presented in this unit is theoretical. Theory is a line of reasoning that is assumed to be correct. Theory may change from time to time as new information is gathered and new lines of reasoning are formed.

Electrical theory is highly developed at the present time and requires a knowledge of higher mathematics for complete understanding. However, mathematical explanations of electricity are not the concern of this text; instead, an elementary approach to those electrical concepts, theories, and formulas that are essential to the technician will be provided.

It is suggested that this unit be read as a story. The material presented is good background information that will assist the reader in understanding the nature of electricity. With a good grasp of the nature of electricity, a technician will find troubleshooting an easier task.

## *THE STRUCTURE OF MATTER*

Everything in the known universe is made up from the 96 stable elements along with the few unstable short-term elements. (See Appendix B.) These stable elements in their pure forms or in a combination with other elements (a combination of elements is called a *compound*) comprise the things we see around us. In their pure forms, the elements are made up of electrons, protons, and neutrons in a special arrangement for each element. The negatively charged, tiny object called an *electron* must be considered in order to understand the behavior of electricity. This consideration requires an investigation into the structure of matter—of what are things really made?

Imagine looking at a piece of copper under a microscope as the magnification is increased to an extremely high magnitude. The construction of copper as it really is might be seen.

With the microscope at 10× power, the copper looks very much like copper (Figure 1–1). As the magnification is increased to 100×, the rough crystalline structure

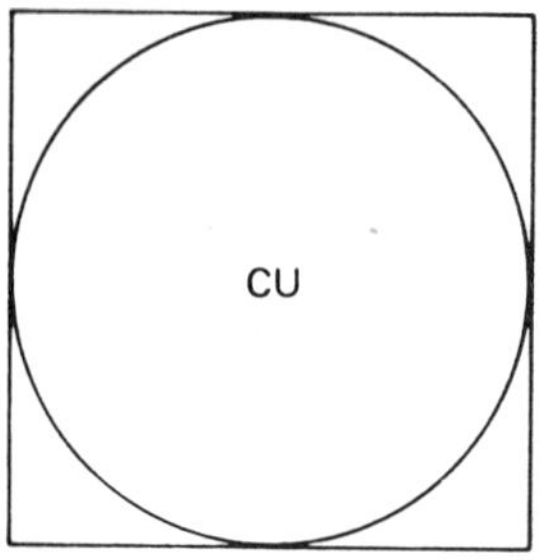


x 10

**Figure 1–1**

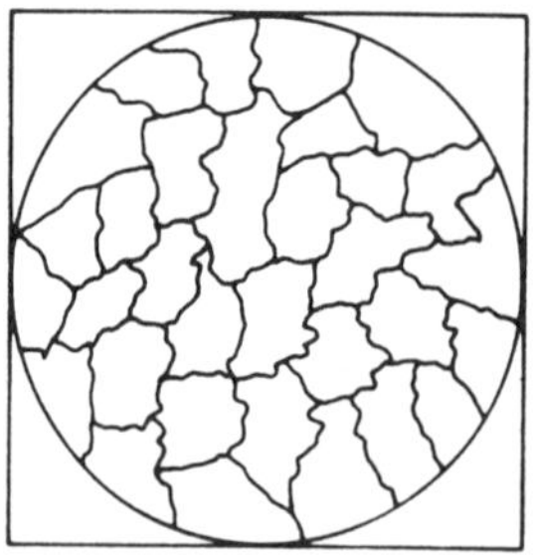

x 100

**Figure 1–2**

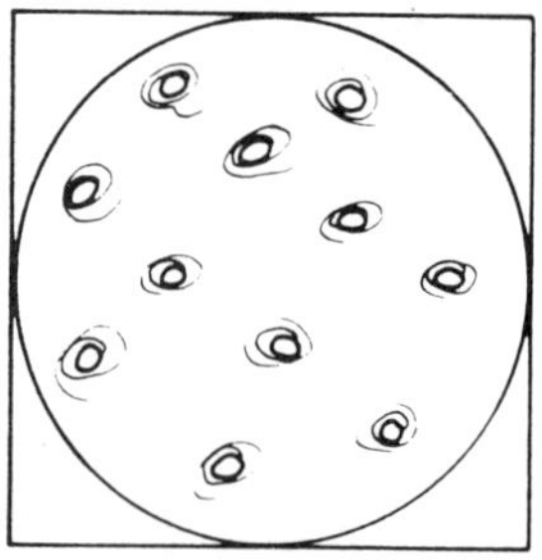

x 10,000,000

**Figure 1–3**

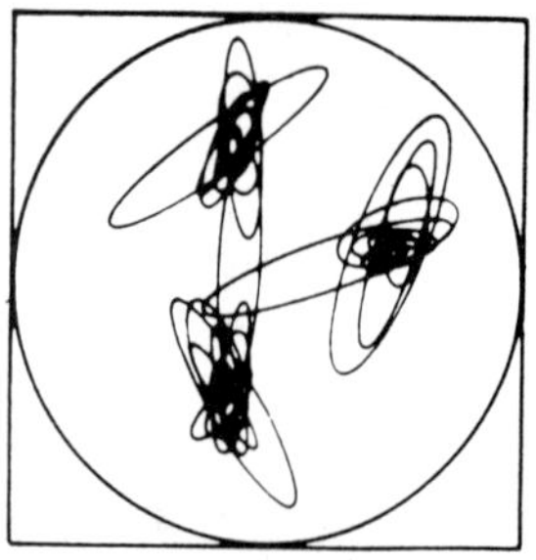

x 100,000,000

**Figure 1–4**

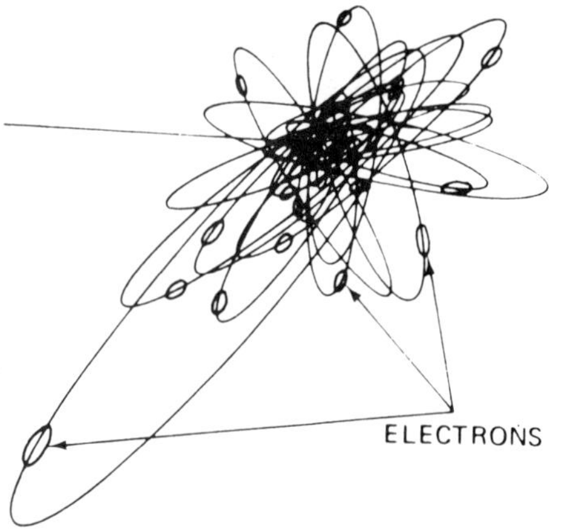


COPPER ATOM

**Figure 1–5**

of copper is seen (Figure 1–2). It takes a large jump in magnification to 10,000,000 × before the beginning of atomic structure becomes evident, as seen in the bumpy surface (Figure 1–3). Finally, at a magnification of 100,000,000 ×, individual atoms are seen (Figure 1–4).

What is there to see? The center of a copper atom, the *nucleus*, consists of positively charged and neutral particles, *protons* and *neutrons*. Negatively charged particles, *electrons*, whirl in four orbits around the center. It is on these negatively charged particles that the electron theory is based. An accepted representation of a copper atom is given in Figure 1–5.

In the study of *electricity*, the electrons whirling about the nucleus are the center of interest. Moving these electrons on command produces electricity that can be used to do work.

## *BOHR'S LAW*

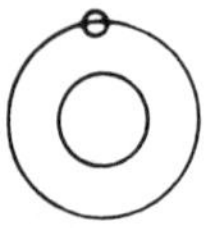

HYDROGEN ATOM

**Figure 1–6**

As scientists experimented with things of an electrical nature, they learned about the characteristics of electricity and things that related to electrical phenomena. Physicist Niels H. D. Bohr (1885–1962) developed a model of electron, proton, and neutron arrangements to represent everything in the universe; this became known as Bohr's quantum theory of atomic structure. According to Bohr, an atom of hydrogen, the lightest known element, consists of one proton in the nucleus and one electron rotating about it (Figure 1–6). Bohr's theory of the hydrogen atom is regarded as the basis of modern atomic nuclear physics.

Heavier materials move up the atomic scale, having a greater number of particles in the nucleus and a greater number of electrons whirling about that center. As the theory developed and is now understood, the arrangement of electrons in rings (orbits) around the nucleus is fixed.

1. In the first orbit around the nucleus, there can be no more than 2 electrons.
2. In the second orbit, the maximum is 8 electrons.
3. The third orbit has a maximum of 18 electrons.
4. The fourth orbit has a maximum of 32 electrons.

## *CONDUCTORS*

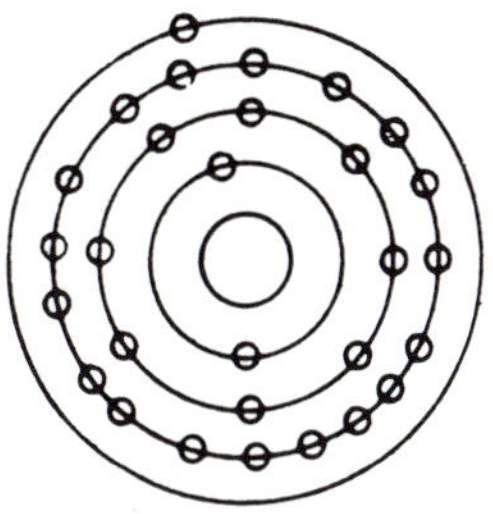

COPPER ATOM

**Figure 1–7**

As indicated above, the number of electrons in each orbit around the nucleus of an atom is fixed. The number of electrons in orbit as well as the number of protons at the center determines the type of material.

Copper, for example, has a nucleus of 29 protons and 34 neutrons and 29 electrons rotating around it. According to the fixed arrangement of electrons, there are 2 in the first orbit, 8 in the second orbit, and 18 in the third orbit. This accounts for only 28 electrons. The remaining electron orbits all by itself in the fourth ring (see Figure 1–7). Since this electron is all by itself in the fourth ring, it is not held as strongly to the nucleus as are the other electrons. The electron is "free" to move whenever an outside force acts on it. Copper is a good conductor of electricity because each of the billions of tiny atoms in a copper conductor has one free electron ready to move when a force is applied to it. A reasonably small copper wire will pass a relatively large amount of electricity.

Another example of a good conductor is aluminum. The aluminum atom has a total of 13 electrons in three orbits. Two are in the first orbit, eight in the second orbit, and three in the third orbit (Figure 1–8). There are positions for 18 electrons in the third orbit. The three electrons in the third orbit are not tightly bound to the nucleus, and the third orbit can momentarily accommodate extra electrons. Aluminum, therefore, is a good conductor of electricity, but *not as good* as copper. If the same

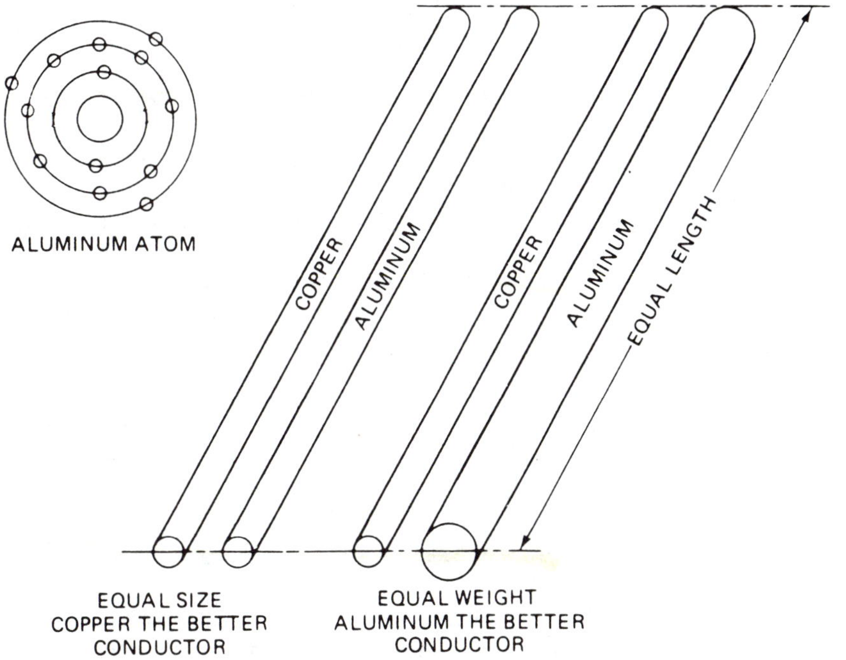


**Figure 1–8**

size of copper and aluminum wire were used, copper would conduct better. If weight and cost are factors to be considered, aluminum—a light-weight metal—is an excellent choice. With wire of equal length and weight (not size), aluminum will conduct electricity *twice as well* as copper. With equal weights, the aluminum wire will, of course, be much larger.

## THE BEHAVIOR OF ELECTRONS

After walking across a rug or sliding out of a car, you sometimes have the strange ability to cause an electric spark when you touch some other object. When you have that ability, you are ''electrically charged,'' or you carry an electric charge. This same type of electrical charge can be demonstrated by running a plastic comb through dry hair. The comb will attract light-weight objects, such as small scraps of paper or thread.

### Static Electricity

The electricity that is obtained by rubbing some things together is called ''static electricity.'' It is called static because electrons are picked up on an item and the electrical charge remains on the item until it touches some other object that does not have a like charge.

As experiments were performed on static electricity, it was found that two types of charges could be obtained, depending on the nature of the materials rubbed together. If two glass rods are rubbed with a silk cloth, they are charged alike and repel each other; if two hard rubber rods are rubbed with wool, they are also charged alike and also repel each other.

*Conclusion: Like charges repel each other.*

If a charged glass rod is approached by a charged rubber rod, the glass rod and rubber rod will be attracted to each other. The charges on the rods are unlike each other, so they attract each other.

*Conclusion: Unlike charges attract each other.*

In order to classify the charges, they were given names. Benjamin N. Franklin (1706–1790), an American scientist, suggested ''positive'' and ''negative'' charges. The glass rods have a positive charge; the rubber rods have a negative charge.

*Static electricity is electricity at rest.*

### Dynamic Electricity

Static electricity is not a practical form of electricity. A practical form of electricity, the form that is used to provide power and energy all over the world, is dynamic electricity.

*Dynamic electricity is electricity in motion.*

In an automobile, the battery provides dynamic electricity, as does the alternator (generator). For the power company, water—at high pressure behind dams—moves large alternators that provide dynamic electricity. In some areas where there are no

**Figure 1–9** Turkey Point power plants.

rivers for dams, large steam plants are used to turn alternators that generate dynamic electricity.

A new kind of power plant, which uses atomic energy to provide the heat needed for steam turbine operation, has been constructed in many areas of the country. One example of an atomic plant of this type that is in operation is located at Turkey Point, Florida; it produces much of the electricity used in south Florida (Figure 1–9).

Another energy source that may be of greater importance in the future, geothermal, is presently being investigated. In order to tap this energy source, a deep hole must be drilled into the earth to an area of intense heat. Water would be introduced into the hole where it would be converted to superheated steam at the lower level. The steam would return to the earth's surface to operate turbines that would rotate the electric alternators that provide electricity.

Regardless of the energy source, dynamic electricity is produced by generators or alternators converting mechanical energy into electrical energy.

## *VOLTAGE AND CURRENT*

A well-developed understanding of the basic terms relating to dynamic electricity and their relationship with each other is necessary in order to troubleshoot electrical systems effectively.

*Voltage* and *current* are terms associated with dynamic electricity. Voltage and current will be covered together because it is not possible to obtain one without some of the other.

*Voltage is the pressure that causes electrons to move.*
*Current is the movement of electrons.*

### Analogy of Electricity

As is often the case, it is easier to understand a new subject when it is related to some past experience. The electrical phenomenon will be related to air. More exactly, it will be related to an air-moving system.

Almost everyone has stood in front of a fan and felt the pressure of air moved

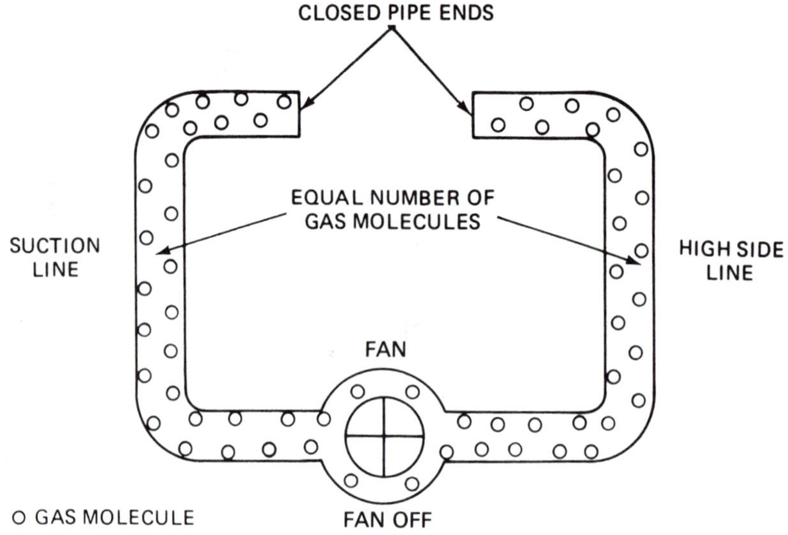


**Figure 1–10**

by the rotating fan blade. This can be compared to an electrical system. The fan or air mover must be placed in a closed system (see Figure 1–10) in order to relate to electricity as electricity is mainly confined to flow in wires.

Figure 1–11 shows the fan in operation. Some of the gas molecules are moved by the fan from the suction line (on the left) to the discharge side (on the right). There is a relative excess of gas molecules in the discharge line (compression), causing a pressure difference between the two lines.

If a small tube, as shown in Figure 1–12, were connected from the discharge line to the suction line, some of the gas molecules would move through the tube. There would be a continuous movement of gas molecules through the system as long as the fan was operating. The suction side of the system will continue to be under a vacuum, and the discharge side will continue at high pressure as long as the small tubing offers some restriction (resistance) to the movement (flow) of gas molecules through the system.

In Figure 1–13, an electric generator has been substituted for the fan, and the air lines have been replaced with copper wire. The circuit is open because there is

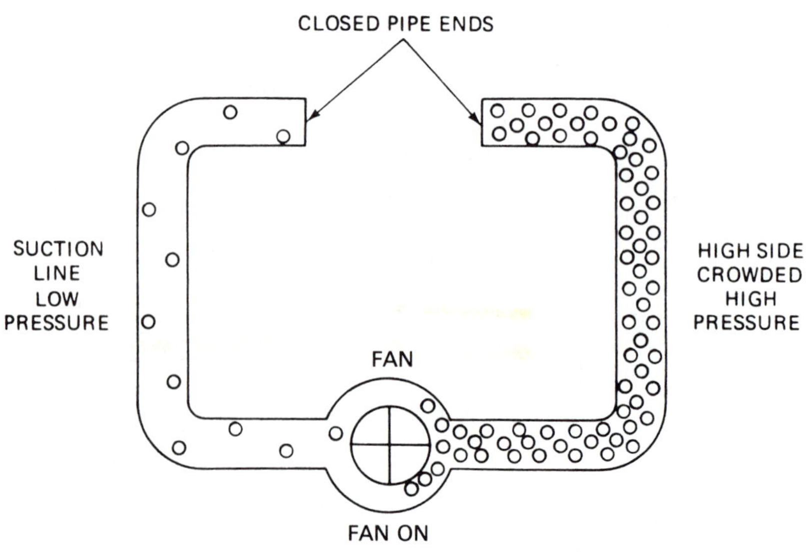


**Figure 1–11**

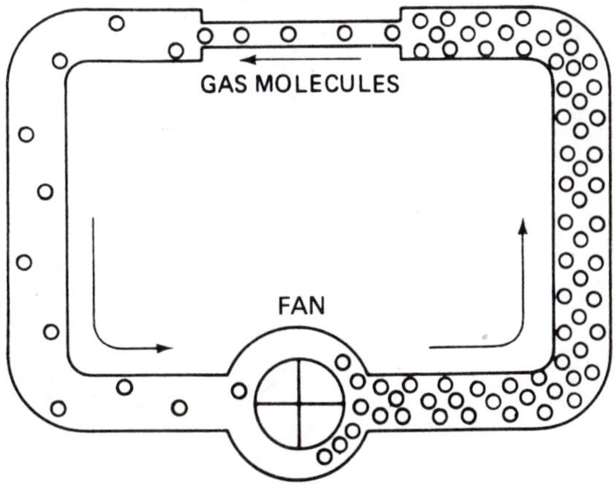


**Figure 1–12**

no connection between the wire from the left side of the generator and the wire from the right side. Within the copper wire, there are a large number of copper atoms. Each of the atoms has one free electron (easily moved) in its outer orbit.

The electrical generator, when running, is capable of moving electrons. This is similar to the fan's ability to move gas molecules. How the generator moves electrons is covered in later units.

When the generator is not running, there is an equal number of free electrons in the wires on either side of the generator: The system is in balance. When turned on, the generator draws some of the free electrons from the left side and pushes them to the right side, as in Figure 1–14. The movement of electrons from one side to the other creates an *unbalance*. As long as there is an unbalance, there will be a *pressure* called *voltage*. The pressure is measured in volts and exerted in an effort to try to balance the system. Another term for voltage is electromotive force, abbreviated "EMF."

Because there is an excess of electrons on the right side, this side of the generator is negative. Similarly, there is an absence of electrons on the left side. As the nucleus of the copper atoms has not changed, the protons remain the same on both sides of the generator. The situation now is that there are more protons on the left side than electrons. Thus, the left side of the generator is positive. As long as the generator is

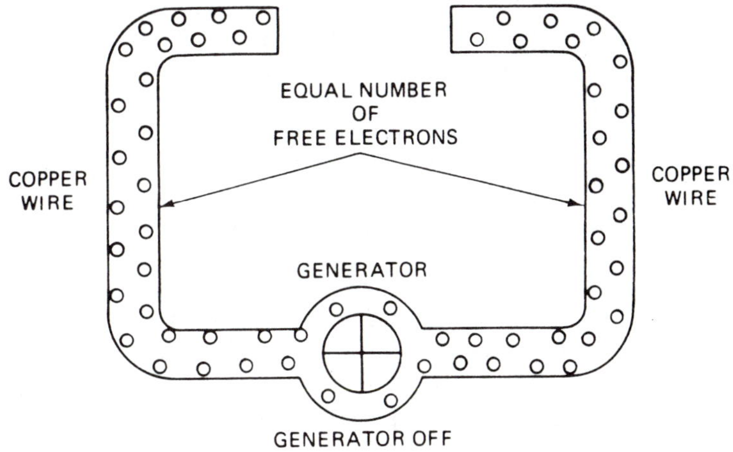


**Figure 1–13**

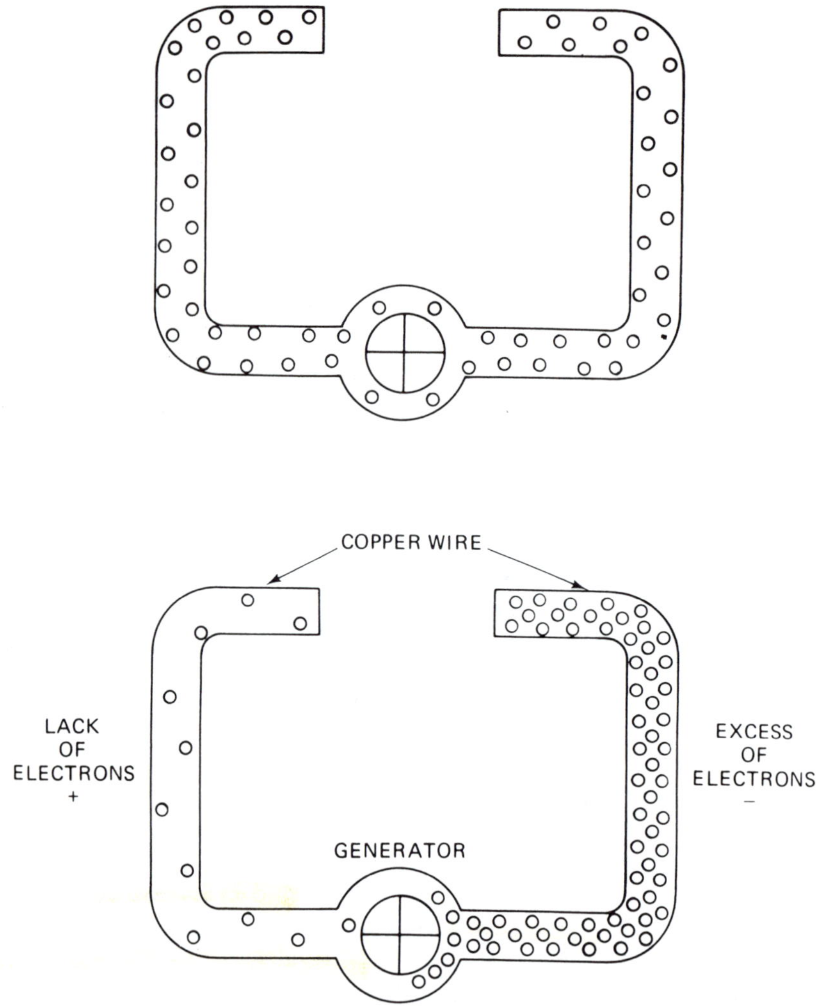


**Figure 1–14**

operating, the system will remain unbalanced; voltage will exist *between* any point on the right side and any point on the left side.

If a voltage-measuring device (called a *voltmeter*) were connected across the open wire ends, the meter needle would move (Figure 1–15). This indicates the unbalance in electrical pressure that exists *between the two points*. Voltage is always measured between two points in a circuit. (See Unit 3.)

***WARNING:*** Never connect a wire across the output terminals of a generator. The wire will short out the generator. It is used here only as a means of explanation.

If a small wire were connected between the left side of the generator and the right side, some of the excess electrons would move from the right side through the wire to the left side (Figure 1–16). The generator will continue to move electrons, just as the fan moves air. There will be an electric current flowing in the circuit as long as the circuit is complete and the generator is running. As long as there is an unbalance and a circuit, current will flow. Current, electrons in motion, is measured in *amperes*, commonly referred to as *amps*.

*The quantity of electrons ($6.25 \times 10^{18}$) is called a *coulomb*. One ampere is equal to 1 coulomb per second.

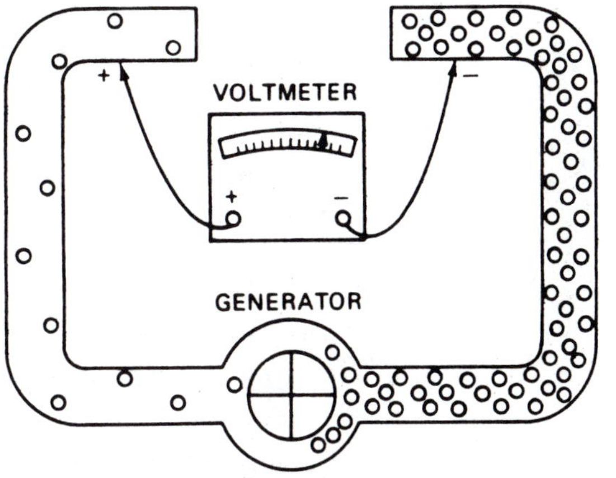


**Figure 1–15**

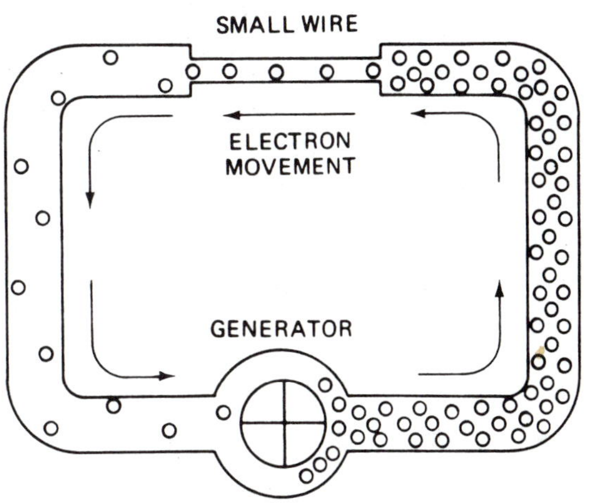


**Figure 1–16**

Voltage is caused by an unbalance in electrons. It is *not* the electrons but *the electrons which have moved* that create the unbalance. Similarly, the current in the circuit is the electrons in motion.

## REVIEW QUESTIONS

T__✓__ F_____ 1. Static electricity is electricity at rest.
T__✓__ F_____ 2. Dynamic electricity is electricity in motion.
T__✓__ F_____ 3. Like charges repel each other.
T_____ F__✓__ 4. Unlike charges repel each other.
T__✓__ F_____ 5. Voltage is a pressure that causes electrons to move.
T__✓__ F_____ 6. Current is the movement of electrons.
T__✓__ F_____ 7. The natural state is for everything to be in a balanced condition.
T__✓__ F_____ 8. Current is measured in amperes.

T______ F______ 9. Electromotive force, EMF, is measured in volts.

T______ F______ 10. With equal size and length of conductors, copper is a better conductor than aluminum.

T______ F______ 11. The number of electrons in a coulomb is 62,000,000.

T______ F______ 12. There are 2 electrons in orbit about the nucleus of many atoms.

T______ F______ 13. Copper weighs more than aluminum.

T______ F______ 14. There are 4 electrons in the second orbit of a hydrogen atom.

T______ F______ 15. Heavier materials have a greater number of electrons in each atom than do lighter materials.

T______ F______ 16. The heavier the material the better the conductor.

T______ F______ 17. The number of electrons in the third and higher orbits is a random number.

T______ F______ 18. One electron per second equals 1 ampere.

T______ F______ 19. Once an object is charged with static electricity, it can not be discharged.

T______ F______ 20. The maximum number of electrons in the first orbit of an atom is 2.

# UNIT TWO

# *MAGNETISM*

Magnets and magnetism are involved in many components of the air conditioning and refrigeration electrical circuits. The compressors are driven by electric motors that function through magnetism. Many of the electrical controls operate on magnetic principles. A good basic understanding of magnetic principles and magnetic circuits will be helpful to all technicians.

## *NATURAL MAGNETS*

Ancient peoples knew of the special properties of certain stones, notably those found in Magnesia, Asia Minor, which attracted small bits of iron. The stones are called *magnets*, after the location in which they were found. Because most ancient peoples had little scientific knowledge, the action of the magnets was attributed to magic. Natural magnets are stone with a high iron ore content. Because of its special characteristics, this stone was given the name *magnetite*.

Around the year A.D. 1000, the property of the natural magnets to point in a fixed direction was discovered. As these stones were now useful in navigation, they were given a new name, *loadstone* (also spelled lodestone). The name loadstone means ''leading stone'' and is still used to indicate special iron ore with magnetic characteristics.

## *ARTIFICIAL MAGNETS*

A piece of iron or steel may be magnetized by placing it in contact with a natural magnet or loadstone. Another method of producing an artificial magnet is to wrap a coil of wire around the iron or steel core and then to pass an electrical current through the wire.

If the material being magnetized is hardened steel, it will retain its magnetism

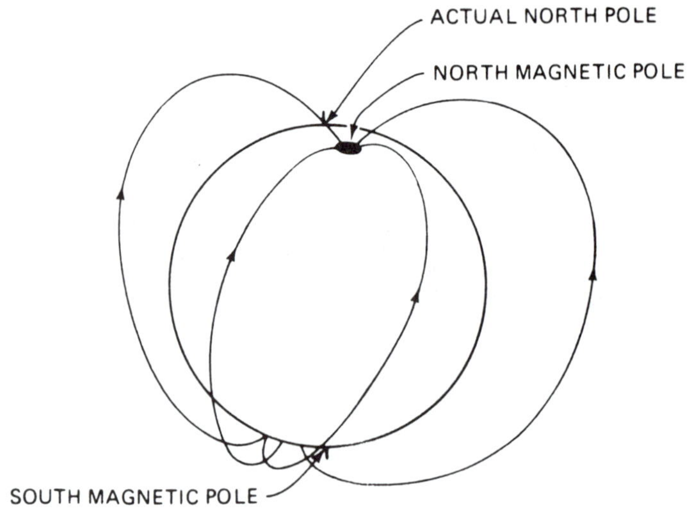

**Figure 2–1**

for a considerable length of time. If soft iron is used, a strong magnetism will appear during the magnetizing process, but little magnetism will remain after the magnetizing force is removed. Because of their different characteristics, permanent magnets are made from hardened steel and other special alloys. Soft iron or steel is used as the core of relays and solenoids where strong magnetism must be present only when current is flowing in the coils.

## *MAGNETIC POLARITY*

If a piece of metal, such as a steel bar, is magnetized, the magnetic effects concentrate at the ends of the bar. These points of magnetic concentration are called the *poles* of the magnet. Away from the ends and around the bar an invisible force is present, the effects of which can be seen if small bits of iron or steel are placed near the magnet. The invisible force around the magnet is called a *magnetic field*.

The planet Earth is a permanent magnet, with one end near the north pole and the other near the south pole (Figure 2–1). If a bar magnet were balanced near the center by a small thread, it would tend to line up with the Earth's magnetic field. One end of the bar magnet would point north. It is, therefore, labeled north-seeking pole and usually is marked with the letter N. The other end of the bar would be pointing south and is marked S, for south-seeking pole (Figure 2–2).

By definition, the magnetic field of a bar magnet is said to emerge from the

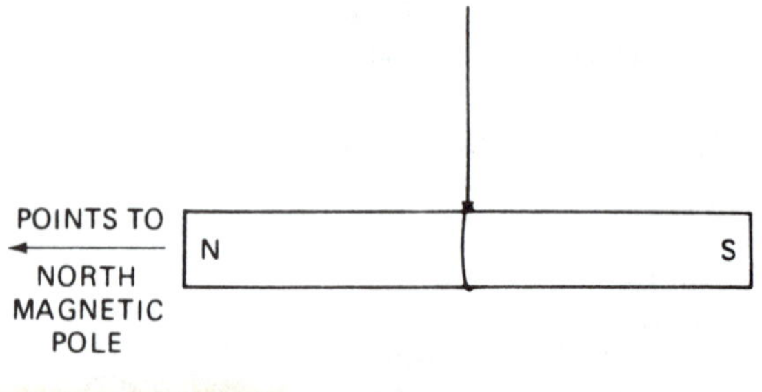


A BAR MAGNET BALANCED AND HUNG BY A STRING WILL ALIGN ITSELF WITH THE EARTH'S MAGNETIC FIELD
**Figure 2–2**

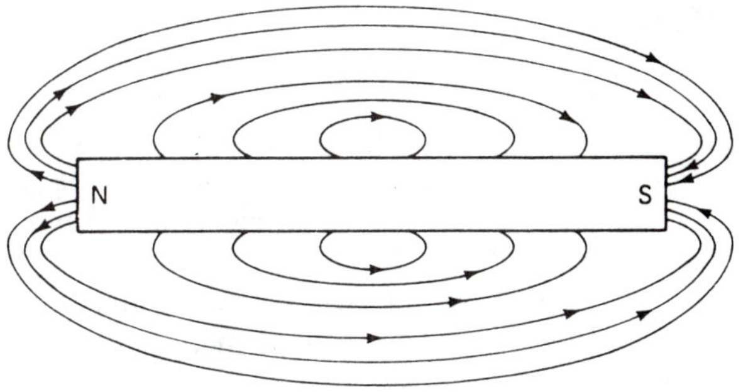


Figure 2–3

north pole and to reenter the magnet at the south pole. Some lines will emerge from the sides of the magnet, but the major concentration is at the poles (Figure 2–3).

If two bar magnets were hung at their balance points and then brought together, the repelling and attracting forces could be demonstrated. In Figure 2–4, two magnets are hung by strings, and the repulsion of two like poles is shown. As the magnetic bars approach each other, the two north poles swing away from each other.

*Like poles repel each other.*

In Figure 2–5, the attraction of two unlike poles is shown. As the two unlike poles approach, they are attracted to each other, as the pull on the strings indicates.

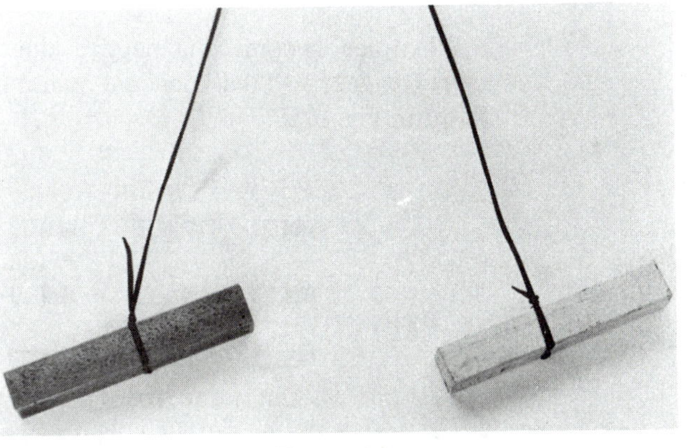

Figure 2–4

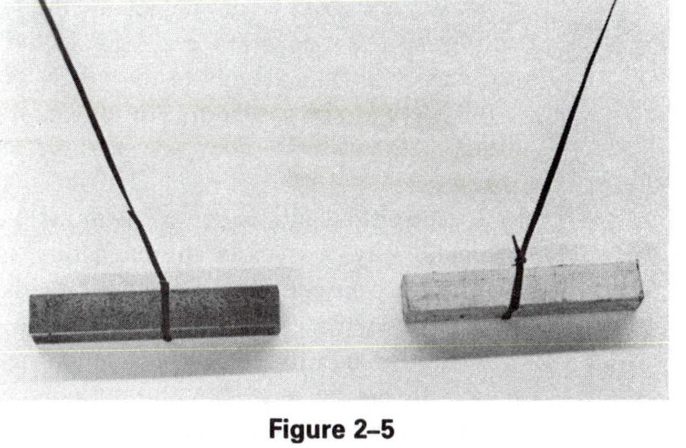

Figure 2–5

## MAGNETIC MATERIALS

Most materials are not magnetic. Copper, aluminum, brass, wood, paper, and glass are examples of nonmagnetic materials. Magnetic fields will pass through these materials, but the material will not be magnetized. Some materials are magnetic; iron, steel, cobalt, and nickel are examples of magnetic materials. There are also many alloys of iron, cobalt, and nickel that provide excellent magnetic properties. Some of these alloys contain nonmagnetic materials, such as aluminum, but the final alloy is magnetic and, in most cases, very highly so.

## MOLECULAR MAGNETS

In ordinary steel, each molecule of steel is a tiny, permanent magnet. In unmagnetized steel, the molecular magnets are arranged haphazardly throughout the metal (Figure 2–6). As each molecular magnet is pointing in a different direction, the total magnetic effect is zero. The steel is not a magnet and will not attract magnetic materials.

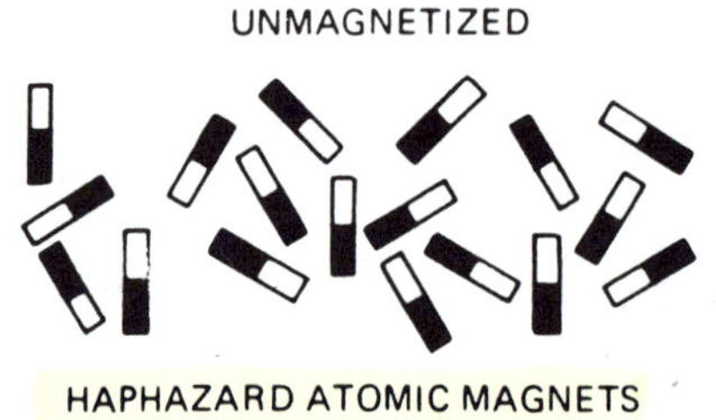


**Figure 2–6**

When steel becomes magnetized, the magnetizing process rotates the molecular structure (magnets) so that most are pointing in the same direction. This arrangement is shown in Figure 2–7.

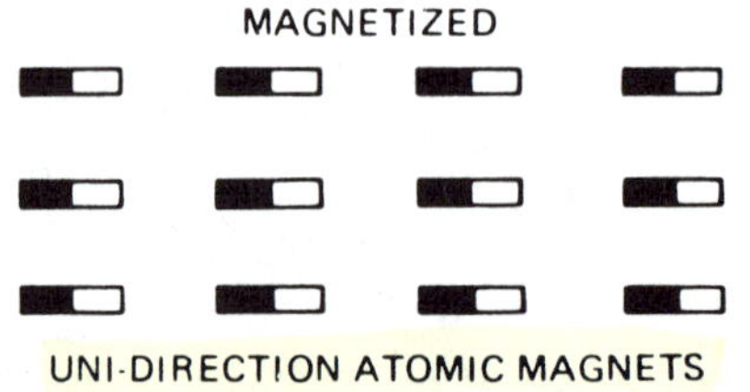


**Figure 2–7**

## MAGNETIC FIELD

Although the magnetic field is invisible, its effect can be demonstrated. If a piece of paper is placed over a bar magnet and the paper is sprinkled with iron filings, a pattern will form. The magnetic force, or field, will arrange the filings in lines running from one end of the bar magnet to the other, as shown in Figure 2–8.

There are no insulators from magnetic fields. Magnetic lines of force can pass through all materials. They will be deflected or bent by other magnetic fields but not stopped or blocked.

The term ''antimagnetic'' is in most cases a play on words. A wrist watch, for example, may be marked antimagnetic. As the delicate mechanisms within a watch may be very susceptible to magnetic fields, these small, delicate parts are protected by the frame of the watch, which is made of soft iron (Figure 2–9). Magnetic fields will pass through the iron frame rather than the hard steel parts. The small parts are protected, not because they are antimagnetic, but because the magnetic fields take the best path through the soft iron.

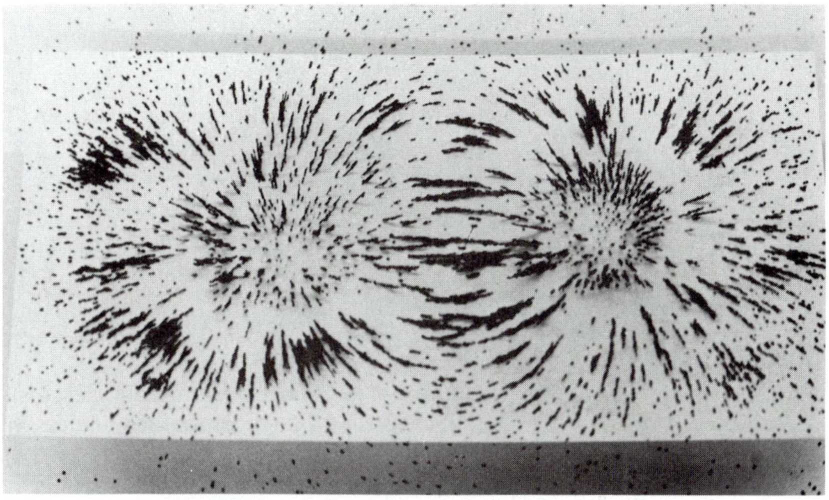

**Figure 2–8** Iron filings show field around bar magnet.

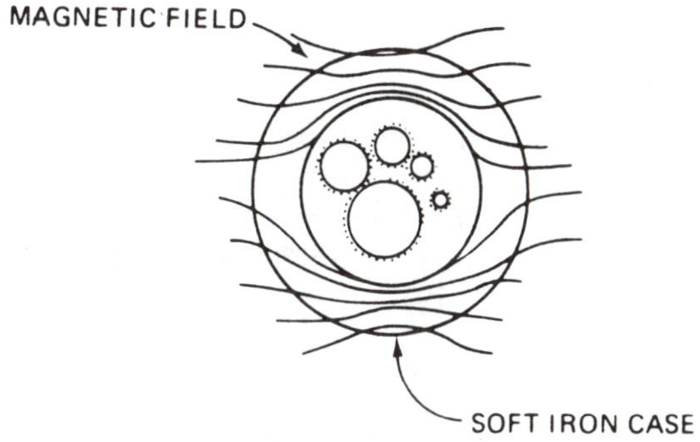


**Figure 2–9**

## *ELECTROMAGNETISM*

As indicated previously, a magnet may be created by passing an electric current through a coil of wire. This may be explained by one of the basic tenets of electricity, namely, a magnetic field exists about each electron in motion. When electrons move through a wire, the magnetic field set up around the wire is proportional to the amount of current in the wire. This is a basic relation between electricity and magnetism. The magnetic field around a wire carrying current is in the form of concentric circles (Figure 2–10).

The method for demonstrating the magnetic field about a wire is shown in Figure 2–11. A wire carrying current is passed through a sheet of paper. Compasses placed

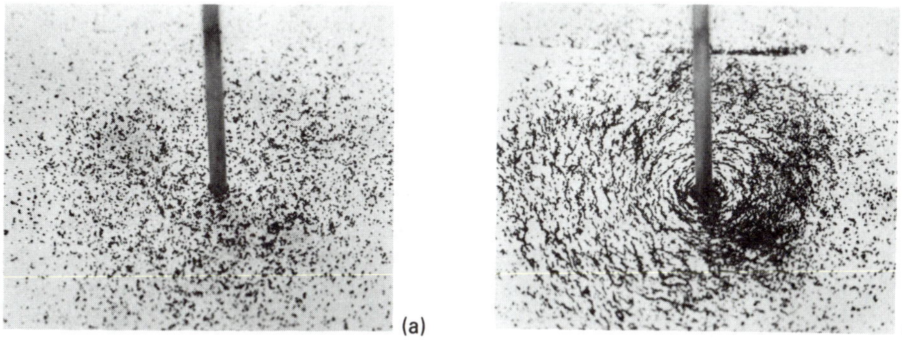


**Figure 2–10** (a) No current. (b) Current flow.

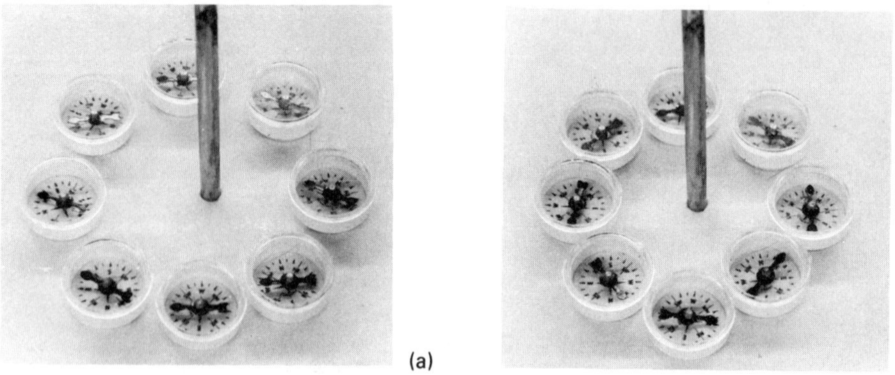

(a) (b)

**Figure 2–11** (a) No current; compasses all point north. (b) Electrons traveling down compasses line up with field.

on the paper demonstrate the existence of a magnetic field as well as the direction of the field. The compass needles align themselves with the magnetic field.

## *LEFT-HAND RULE*

If the direction of electron flow in a wire is known, the direction of the magnetic field about the wire can be determined by following the *left-hand rule*: Grasp the current-carrying wire with the left hand with the thumb pointing in the direction of electron flow. The fingers will point in the direction of the magnetic lines of force (Figure 2–12).

If the direction of the field is known but not the direction of electron flow, the rule can be used to determine the current direction. Grasp the wire with the left hand with fingers pointing in the direction of the magnetic lines. The thumb will point in the direction of electron flow.

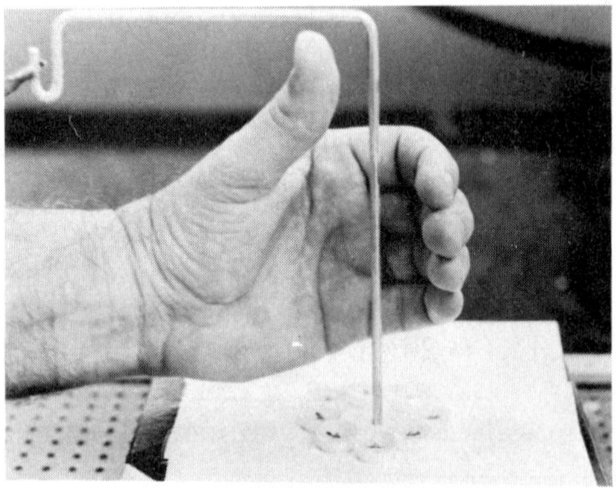

**Figure 2–12** Left-hand rule.

## *ELECTROMAGNETS*

If a current-carrying wire is formed into a loop or coil, the concentric lines of force will all be in the same direction through the center of the loop (Figure 2–13). If loops are placed close together, there will be further concentration of lines of force. Some lines will merge and go around the combined loops, as shown in Figure 2–14.

If several turns of insulated wire are formed into a coil, lines of force will enter

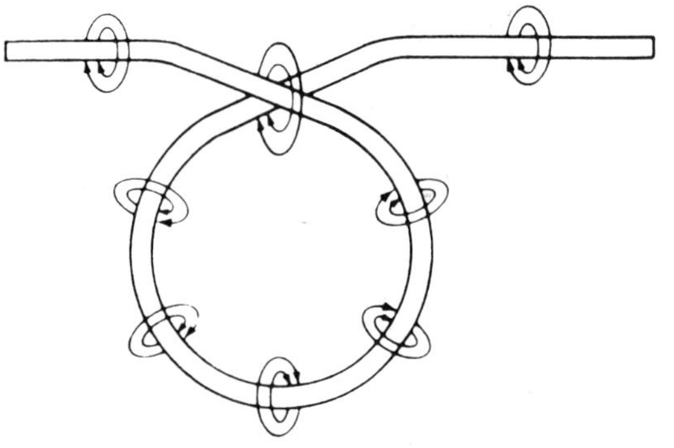

Figure 2–13

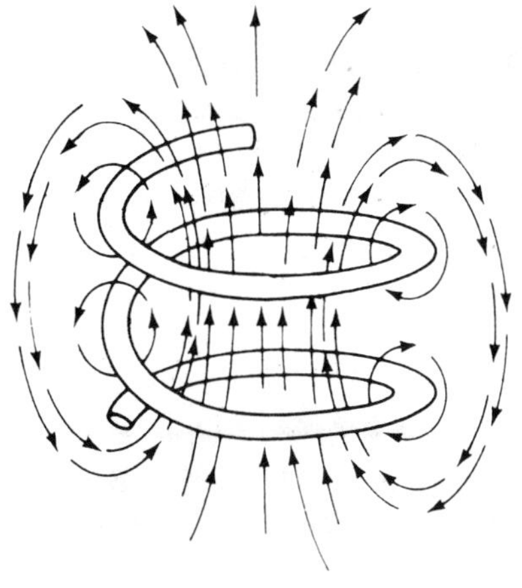

Figure 2–14

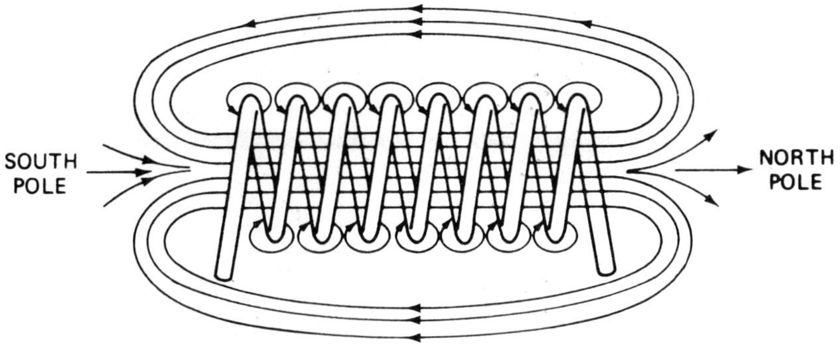


Figure 2–15

one end of the coil, pass through it, and emerge at the other end. The lines of force will be completed outside of the coil, as shown in Figure 2–15.

## *MAGNETIC POLARITY OF ELECTROMAGNETS*

The polarity of an electromagnet may also be determined by using the left-hand rule. The direction of electron current must be known in order to use the rule. Grasp the coil in the left hand with the fingers pointing in the direction of electron flow. The thumb will point in the direction of lines of force through the center of the coil and to the north pole of the electromagnet (Figure 2–16).

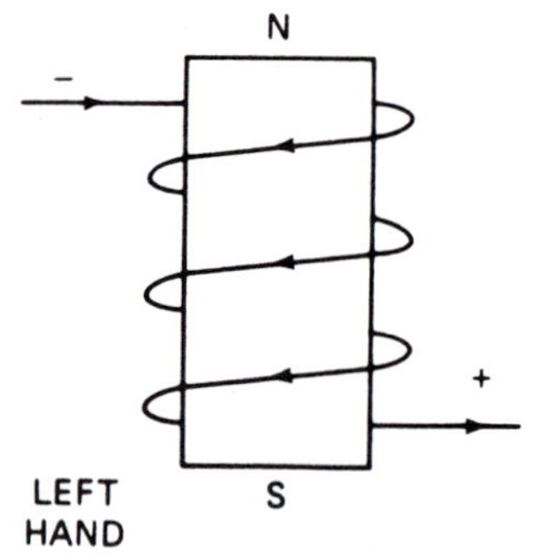


Figure 2–16

### *MAGNETIC STRENGTH*

The strength of an electromagnet is dependent on the size, the length, the material of the core, the number of turns in the coil, and the amount of current flowing through it. Other conditions being the same, an electromagnet with a soft iron core will be stronger than one with a core of steel.

### *AMPERE-TURNS (NI)*

The magnetizing force of a coil is based on the ampere-turns of the coil. The ampere-turns of a coil are equal to the amount of current flowing through the coil (I) multiplied by the number of turns (N). In Figure 2–17, the ampere-turn rating of different coil-current combinations is given.

#### Field Intensity (H)

The ampere-turns of a coil indicate the coil's magnetic force. The magnetic field that results from the magnetizing force is also dependent on the length of the coil. The formula is

$$H = \frac{NI}{l}$$

where H is the field intensity, NI is the ampere-turns, and l is the length of the coil in meters.

#### Iron Core Coils

Many materials are magnetic. If the material is used as the core for a current carrying coil, the strength of the produced magnetic field will be greater than if air were used. Iron is one of the main core materials.

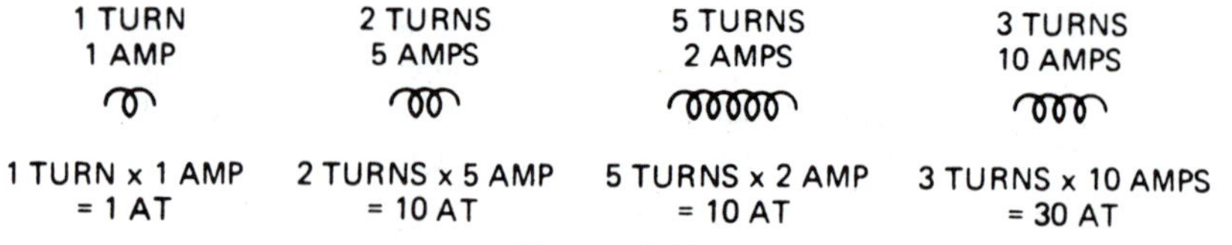


Figure 2–17

### Permeability

The ratio of the number of lines of force in the material to the number of lines if the material were replaced by air is called the *permeability* of the material. The Greek letter $\mu$ is used as the symbol for permeability.

---

**EXAMPLE 1**

A certain air-core coil has a flux density (lines per square centimeter) of 200. When an iron core is used in place of the air core, the flux density increases to 40,000. What is the permeability of the iron?

***Solution***

$$\mu = \frac{40{,}000}{200} = 200$$

The formula provides for the calculation of the magnetic strength of an iron core coil.

---

## *MAGNETIZATION CURVE*

The *magnetization curve* shows the relation between the magnetizing force (H) and the total number of flux lines (B) of a magnetic circuit (iron core).

The curve in Figure 2–18 shows that as the magnetizing force (H) increases, the flux density (B) increases. As the magnetizing force (H) reaches about 4,000, the curve starts to level off. The curve shows that for this particular core material, a magnetizing force of 4,000 starts to saturate the core. Increasing the magnetizing force beyond 4,000 does not increase the flux density very much. The core is saturated.

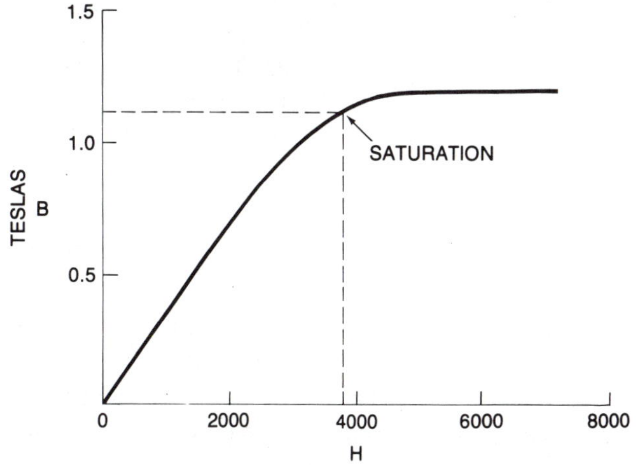


**Figure 2–18**

## *HYSTERESIS*

*Hysteresis* is the lagging behind of the magnetic flux lines (B) to the magnetizing force H in a magnetic circuit. A certain amount of energy is needed to rearrange the magnetic dipoles in the core material. If the direction of the magnetizing force is changed repeatedly, the magnetic flux lines (B) will always lag behind the magnetizing force (H).

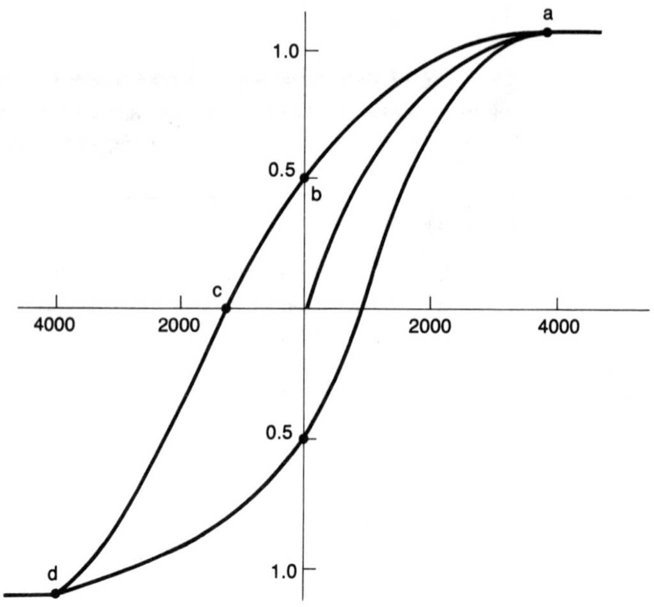


Figure 2–19

## Hysteresis Loop

The characteristics of a given core material may be shown by plotting the flux density (B) against the magnetizing force (H) varying in magnitude and direction periodically. The curve of Figure 2–19 shows this relationship.

The curve of Figure 2–19 also shows an increasing flux density (B) as the magnetizing force H is increased from 0 to +4,000. At this point (*a*), the core is approaching saturation. As the magnetizing force is reduced to 0, the curve indicates a residual flux density of 0.5 teslas (point *b*).

The direction of the magnetizing force must be changed and increased to approximately −1,800 before the flux density is reduced to zero at point *c*. Increasing the magnetization force to −4,000 provides for saturation in the negative direction at point *d*. Notice the continuous lagging behind of B to H with H having to rise to +1,800 before B returns to 0 from its negative excursion.

## SUMMARY

1. Permanent magnets are made with hardened steel.
2. Soft iron is used in temporary magnets.
3. Like poles of magnets repel each other.
4. Unlike poles of magnets attract each other.
5. There are no insulators for magnetism or magnetic lines of force.
6. Soft iron is often used to shield items from magnetism by providing a better path for the magnetic lines of force.
7. When an electric current flows in a wire, a magnetic field will exist around the wire.
8. When a current-carrying wire is formed into a loop or coil, the magnetic field will be concentrated in the center of the loop.
9. When an iron core is used as the core of a coil carrying an electric current, the magnetic field will be greatly increased.

## REVIEW QUESTIONS

1. With magnets, like poles (repel, attract) _______ each other.
2. The concentration of a magnetic field in a bar magnet is at the (side, ends) _______ of the bar.
3. Two adjacent wires carrying current in the same direction will (attract, repel) _______ each other.
4. Two adjacent wires carrying current in the opposite direction will (attract, repel) _______ each other.
5. If a wire carrying current is formed into a loop or coil, the magnetic field will (increase, decrease) _______.
6. If a magnetic material, such as soft iron, is used as the core of the electromagnet, the strength of the magnet will (increase, decrease) _______.
7. Soft iron is used as the core of (permanent, temporary) _______ magnets.
8. Hard steel is used as the core of (permanent, temporary) _______ magnets.
9. (Nothing, glass) _______ will isolate from magnetic lines of force.
10. Electric motors containing more iron and windings are usually (cheaper, more expensive) _______ to operate.
11. What is the field intensity (H) of a coil of 10 cm length containing 100 turns? The current flowing through the coil is 2 amperes.
12. When a magnetic material is used as the core material of a coil, the flux density increases to 84,000. The flux density was 300 with an air core. What is the permeability of the magnetic material?
13. In the curve of Figure 2–18, what is the flux density for a magnetizing force of 6,000?
14. In the curve of Figure 2–18, what is the flux density for a magnetizing force of 2,000?
15. In Figure 2–19, what is the flux density for an increasing magnetizing force of 2,000 in the positive direction?
16. In Figure 2–19, what is the flux density for a magnetizing force of 2,000 decreasing from point *a*?

# UNIT THREE

# *OHM'S LAW*

In Units 1 and 2, voltage and current were covered. There is another factor that is of equal importance in the study of electricity and electrical circuits: *resistance*. In the discussion of voltage and current, the observation was made that when a generator is operating, electrons are pulled from the wire on one side and forced out on the wire on the other side; this creates an unbalance called voltage. It is shown in Figure 3–1. As long as the generator operates, it will keep the unbalance, and a voltage will exist between the two wires.

Note that in the circuit of Figure 3–1 there is no wire connected between the output wires of the generator. The resistance of air is extremely high and may be considered infinite. With such a high resistance, no current will flow.

If a small wire were connected between the ends of the original wires, a current would flow in the circuit, as shown in Figure 3–2. The amount of current that would flow in the circuit would depend on the type of material of which the wire is made,

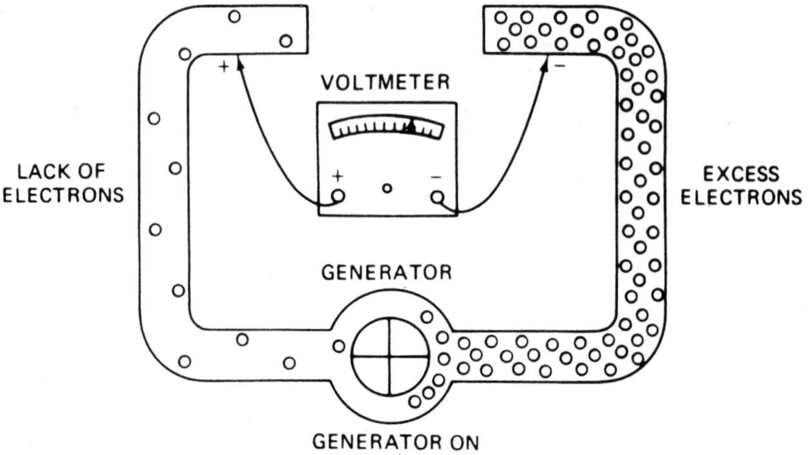


**Figure 3–1**

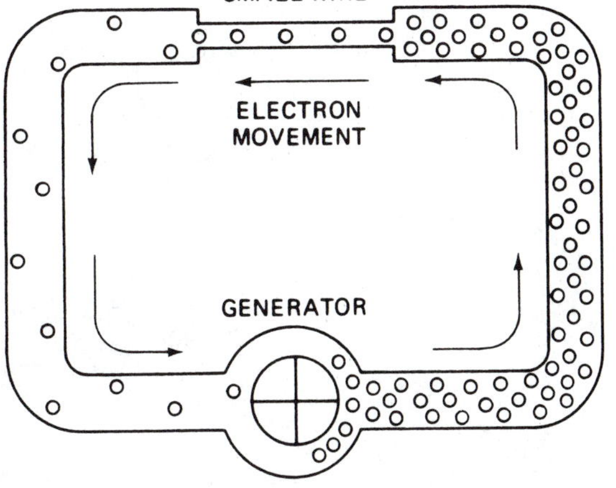


**Figure 3–2**

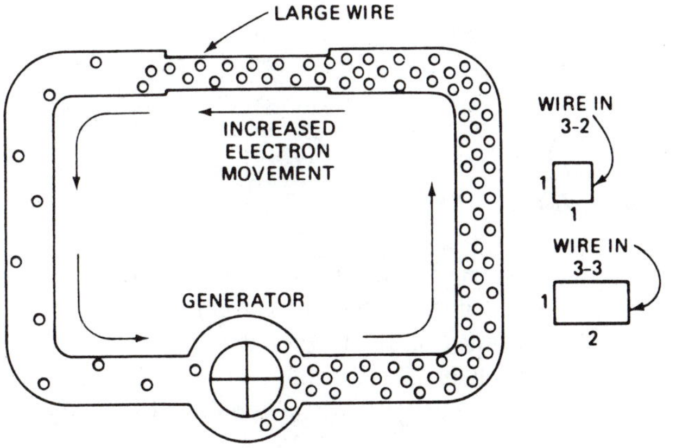


**Figure 3–3**

the size (diameter) of the wire, and its length. In the circuit in Figure 3–2, the small wire is said to offer resistance to the current flow. Although resistance is offered, current will flow.

The connecting wire in the circuit in Figure 3–3 is the same length and of the same material. However, the cross-sectional area is twice the size of the wire used in Figure 3–2.

More current flows in the circuit in Figure 3–3 than in the circuit in Figure 3–2 because resistance is reduced. Actually, *twice* as much current flows when the larger wire is used.

The resistance of a wire is dependent on four things:

1. Material used to make the wire (available free electrons).
2. Cross-sectional area of the wire (size).
3. Length of the wire.
4. Temperature (normally as temperature increases, resistance increases).

Resistors come in many shapes and sizes (Figure 3–4). Pictured here are four resistors: Three are different types of wire-wound power resistors, and one is a carbon resistor.

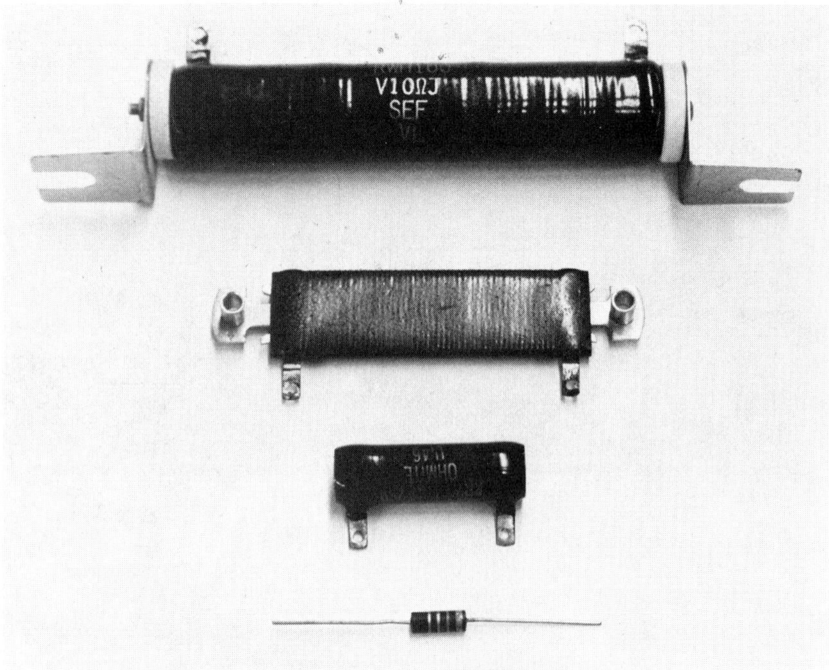


**Figure 3–4** Resistors.

## RESISTANCE VALUES

Resistors used in electrical and electronic circuits vary in value from less than 1 ohm to values in the million ohms (megohm). Some of the resistors will have their value printed on the face of the resistor. For example, a 200-ohm, 25-watt resistor would probably have printed right on the resistor, 200 # 25W. Other resistors of smaller wattage (W) may have a coded indication of the resistor value.

## RESISTOR COLOR CODE

The color-coding of resistors has been in use for many years. The color-code system uses bands of color about the resistor to indicate the value of the resistor and its tolerance. The color-code numbering system is as follows:

| *Color* | *1st band* | *2nd band* | *3rd band times or add zeros* | | | *4th band* |
|---|---|---|---|---|---|---|
| black | 0 | 0 | $10^0$ | 1 | none | 1% |
| brown | 1 | 1 | $10^1$ | 10 | 1 | 2% |
| red | 2 | 2 | $10^2$ | 100 | 2 | |
| orange | 3 | 3 | $10^3$ | 1,000 | 3 | |
| yellow | 4 | 4 | $10^4$ | 10,000 | 4 | |
| green | 5 | 5 | $10^5$ | · | 5 | |
| blue | 6 | 6 | $10^6$ | · | 6 | |
| violet | 7 | 7 | $10^7$ | · | 7 | |
| gray | 8 | 8 | $10^8$ | · | 8 | |
| white | 9 | 9 | $10^9$ | · | 9 | |
| gold | | | $10^{-1}$ | 0.1 | | 5% |
| silver | | | $10^{-2}$ | 0.01 | | 10% |
| no color | | | | | | 20% |

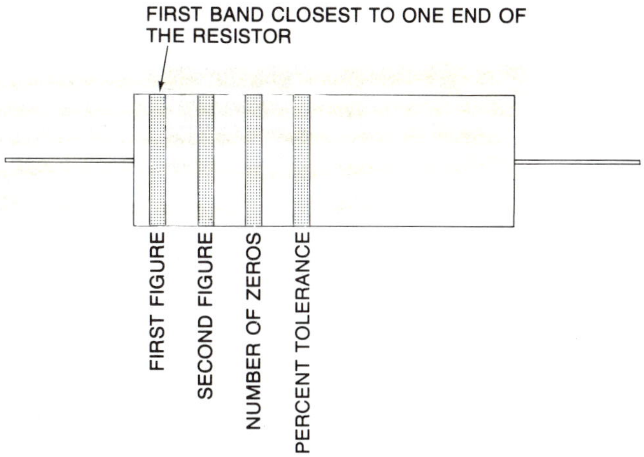


**Figure 3–5**

Standard color-coding calls for a minimum of three color bands on a resistor and a maximum of six color bands. The fifth and sixth bands are of engineering and statistical significance and will not be of concern in this book. They provide information as to temperature sensitivity and failure rate.

The first four color bands are of particular significance to technicians. The coded bands are called out in Figure 3–5.

---

**EXAMPLE 1**

What is the resistance value of a resistor coded yellow, violet, orange, gold?

***Solution***

1. The first figure is 4.
2. The second figure is 7.
3. The number of zeros is 3.
4. The tolerance is 5%.

The resistance value is 47,000 ohms ± 2,350 ohms. Another way of stating the resistance value is to say that the manufacturer indicates that the resistor is between 44,650 ohms and 49,350 ohms.

---

**EXAMPLE 2**

What is the value of a resistor coded brown, black, blue?

***Solution***

1. The first figure is 1.
2. The second figure is 0.
3. The number of zeros is 6.
4. The tolerance is 20%.

The resistor is 10 megohm.

---

Standard resistors are color-coded in wattage sizes from $\frac{1}{8}$ watts to 2-watt values.

### VARIABLE RESISTORS

*Variable resistors* are used in circuits whenever there is a requirement for changing the magnitude of voltage or current fed to other components in the circuit. If the variable resistor is connected to provide a variable voltage, it is called a *potentiometer* or, as in general use, a "pot." If the variable resistor is connected in the circuit to provide a variable current, it is called a *rheostat*. See Figure 3–6. There is a tendency to use the term "pot" in both situations.

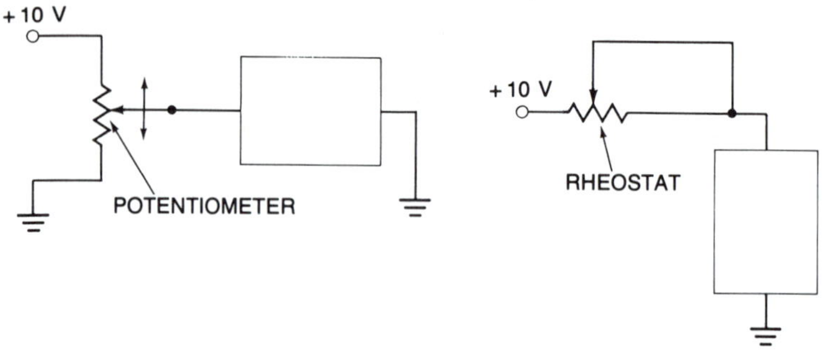


**Figure 3–6**

### OHM'S LAW

The relationship of voltage, current, and resistance may be determined by a relatively simple formula known as Ohm's Law. It was devised by a German physicist, George Simon Ohm, in 1848. He discovered that the current in any circuit is directly proportional to the voltage applied and is inversely proportional to the resistance.

$$I = \frac{E}{R}$$

or

$$R = \frac{E}{I}$$

or

$$E = I \times R$$

From Ohm's Law, the following can be derived:

1. If the resistance is kept the same and the voltage is increased, the current will increase.
2. If the voltage is kept the same and the resistance is increased, the current will decrease.

In honor of discovery of the law, the term for resistance measurement is the ohm; it is indicated by the Greek letter omega Ω. The symbol for resistance is the letter R. The symbol for voltage is the letter E (from EMF). The symbol for current is the letter I (from current Intensity).

By use of any one of the three forms of Ohm's Law, when two factors are known in a circuit, the third factor can always be found.

**EXAMPLE 3:** (I is the unknown)

A 12-ohm resistor has 24 volts across it. How much current is flowing through it (Figure 3–7)?

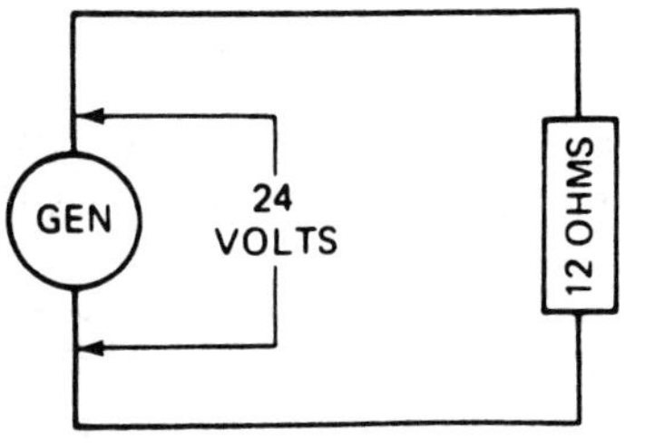


**Figure 3–7**

***Solution***

$$I = \frac{E}{R}$$

$$I = \frac{24}{12}$$

$$I = 2 \text{ amps}$$

Mathematically, the answer is 2 amperes. If an ammeter were connected in series with the 12-ohm resistor, as in Figure 3–8, the meter would indicate 2 amperes.

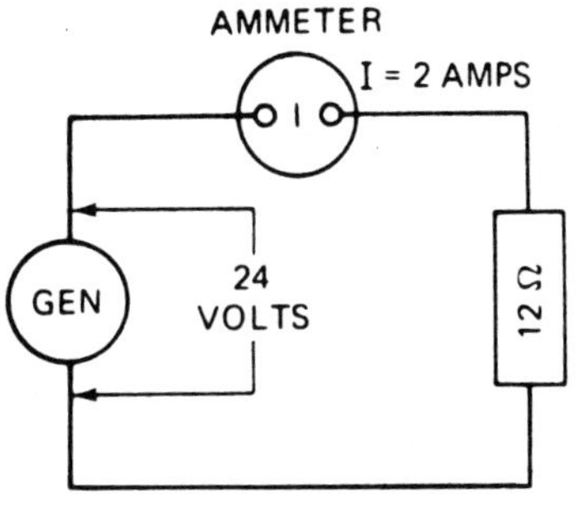


**Figure 3–8**

**EXAMPLE 4:** (E is the unknown)

A resistor of 60 ohms is connected across an electrical power source. An ammeter connected in series with the resistor indicates 2 amperes (Figure 3–9). What is the voltage of the power source?

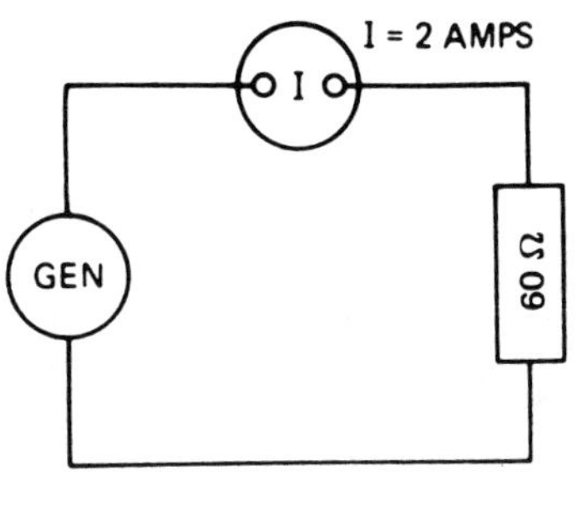


**Figure 3–9**

***Solution***

$$E = I \times R$$
$$E = 2 \times 60$$
$$E = 120 \text{ volts}$$

Mathematically, the answer is 120 volts. If a voltmeter were connected across the power source, it would indicate 120 volts (Figure 3–10).

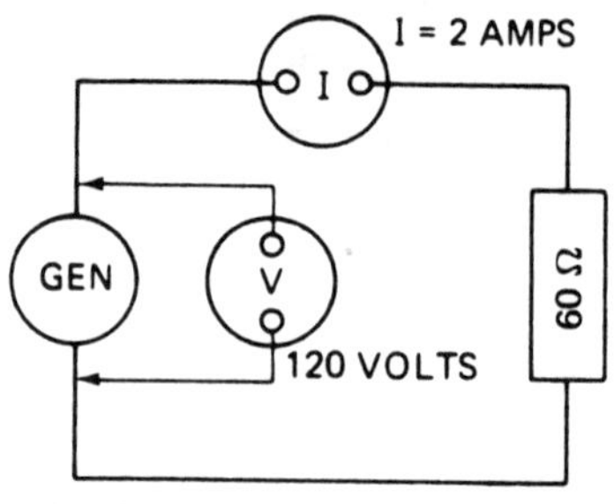


**Figure 3–10**

**EXAMPLE 5:** (R is the unknown)

A resistor is connected across a 220-volt power source. An ammeter connected in series with the resistor indicates 10 amperes (Figure 3–11). What is the value of the resistor?

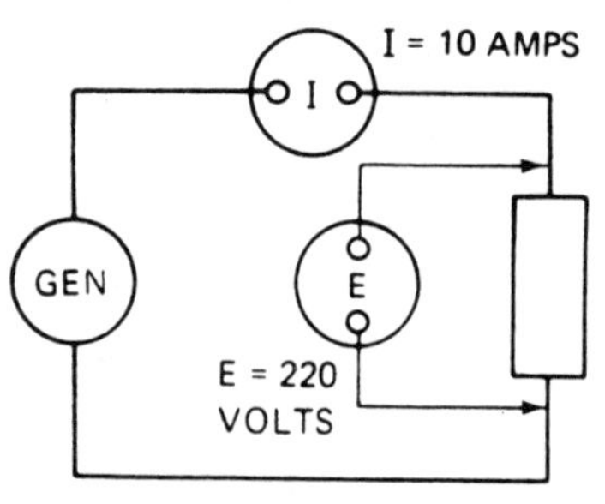


**Figure 3–11**

***Solution***

$$R = \frac{E}{I}$$
$$R = \frac{220}{10}$$
$$R = 22 \text{ ohms}$$

Mathematically, the answer is 22 Ω. If an ohmmeter were connected to the resistive device, it would indicate 22 Ω (Figure 3–12).

In normal everyday operations, the technician would not be required to solve Ohm's Law problems. An understanding of Ohm's Law does, however, provide the means to a better understanding of electricity. A good understanding of electricity is important to the technician when troubleshooting equipment.

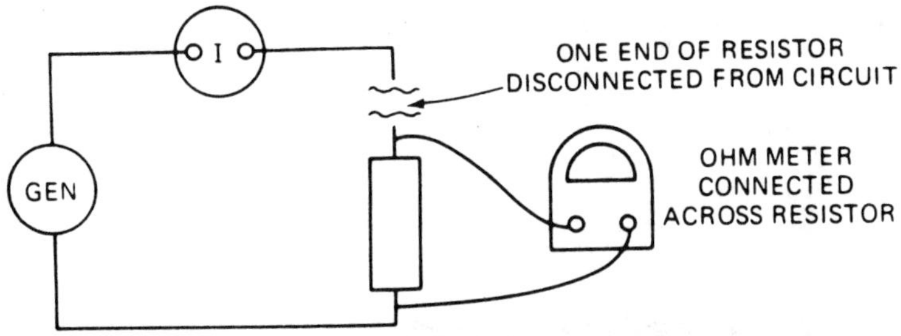


**Figure 3–12**

## REVIEW QUESTIONS

1. The resistance of a piece of wire is dependent on four things. What are they?
2. In an electric circuit, if the resistance is kept the same and the voltage is increased, the current will (increase, decrease) ______.
3. In an electric circuit, if the voltage is kept the same and the resistance is increased, the current will (increase, decrease) ______.
4. If a 6-ohm resistor has 20 amperes flowing through it, the voltage across the resistor will be ______ volts.
5. If 24 volts were measured across an 8-ohm resistor, how much current must be flowing through it?
6. An ammeter connected in series with a resistor indicates 11 amperes. A voltmeter connected across the resistor indicates 220 volts. What is the value of the resistor?
7. What voltage is needed to cause 4 amperes to flow through a 110-ohm resistor?
8. If the voltage across a 15-ohm resistor is measured at 90 volts, how much current is flowing through the resistor?
9. If the voltage applied to a resistor is doubled, will the current double or decrease to $\frac{1}{2}$ the original value?
10. What size resistor will cause a current of 3 amperes to flow when 120 volts are applied to it?

# UNIT FOUR

# *SERIES CIRCUITS*

In Unit 3, Ohm's Law and the application of Ohm's Law to simple electric circuits were discussed. In this unit, the discussion will be expanded to cover Ohm's Law as it relates to series circuits.

## *COMPLETE ELECTRIC CIRCUITS*

In order for current to flow in an electric circuit, the circuit must be complete. Another way of stating this is that current must be able to flow from the source through the load, back to the source, through the source back to the load, etc.

In Figure 4–1, an electric generator is the energy source, and a lamp is the load. As long as the circuit is complete, current will flow from the generator (source) through the lamp (load) and back through the generator. There is a continuous path for current to flow. It is a complete electric circuit.

If a switch is included in the circuit, the same current that flows through the lamp flows through the switch. Two conditions could exist. When the switch is closed, the circuit is complete and current can flow (Figure 4–2). When the switch is in the open position, the circuit is not complete (Figure 4–3). Current cannot flow through the open switch; therefore, there cannot be current flowing through the lamp.

In Figure 4–4, two circuits are shown. Note the position of the switches. In both circuits, the switches and the lamp are connected in series. In the circuit in Figure 4–4a, if either switch A-1 or A-2 is open, the circuit is not complete: The lamp will not light. Similarly, in the circuit in Figure 4–4b, if either switch B-1 or B-2 is open, the circuit is not complete. The lamp will not light.

Electrically, both circuits are the same. They are both series circuits. The same current must flow through each component of the circuit.

Shown in Figure 4–5 is a simple control circuit for an air conditioning compressor contactor. It is a series circuit. The contactor will be energized as long as *24 volts* are available from the source and both the *high-pressure cutout* and the *thermostat*

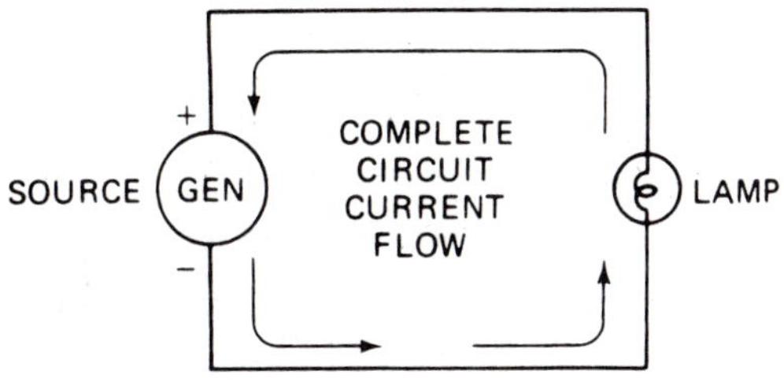


Figure 4–1

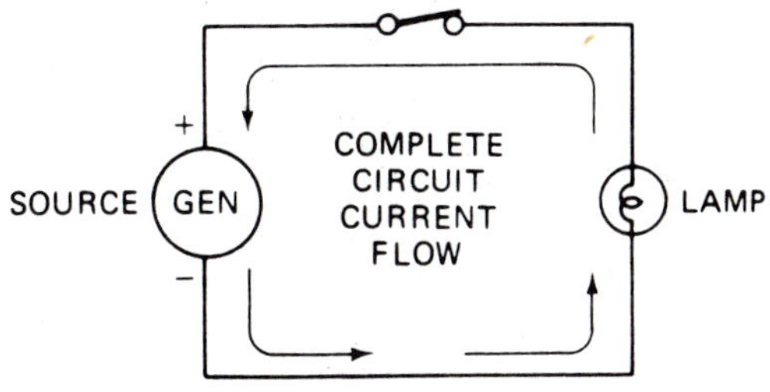


Figure 4–2

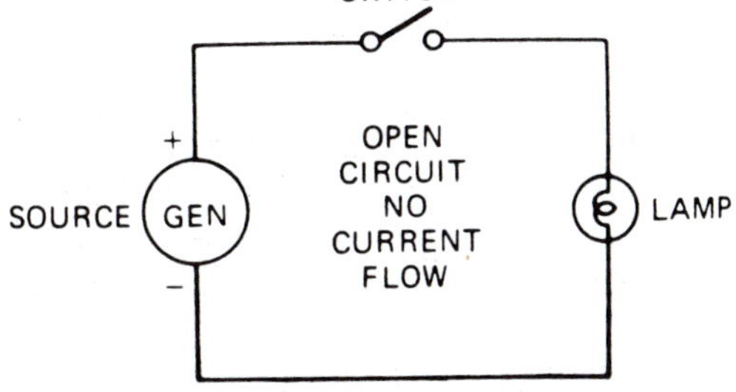


Figure 4–3

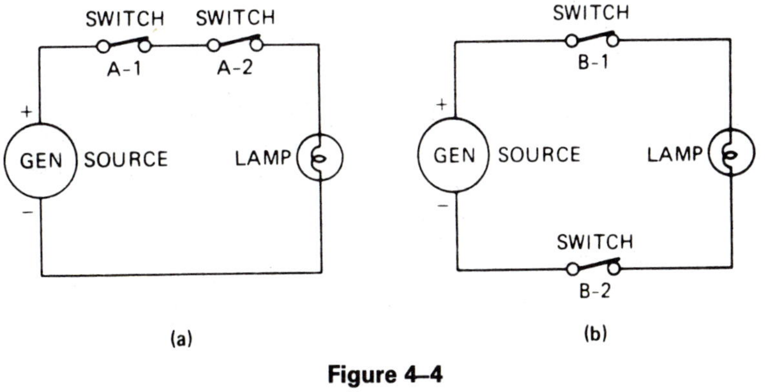


Figure 4–4

select switches are closed. If either switch should open, the circuit to the contactor coil would be open, no current could flow in the circuit, and the contactor would deenergize and remove power from the compressor motor.

One important law of series circuits has just been covered:

*The same current flows through each component in a series circuit.*

A complete circuit may be made up of the generator, the load, and the connecting wires.

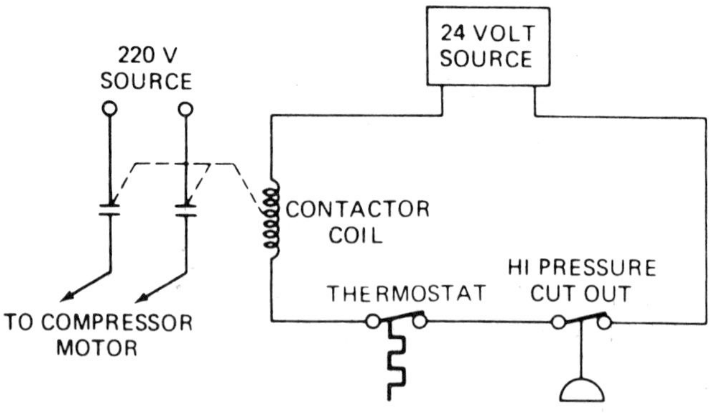


**Figure 4–5**

The connecting wires in a circuit are usually considered not to have resistance. Actually, wires do have resistance; but, for practical application, the resistance of wires (copper) is so small that it need not be considered in circuit calculation.

## *WIRE RESISTORS*

A common material used in the construction of wire-wound resistors is *nichrome*. Nichrome wire has a relatively high resistance for short lengths of wire. For example, 100 feet of nichrome wire could have a resistance of about 100 ohms. In order to make a practical, usable resistor out of the nichrome wire, it is wound on a core of ceramic with terminals at either end (Figure 4–6).

The circuit in Figure 4–7 consists of a source (a generator), connecting wires, and a load (a 100-ohm wire-wound resistor). According to Ohm's Law, if the generator

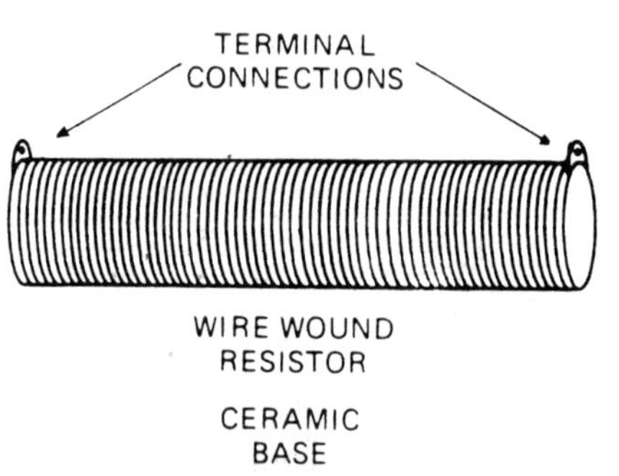


**Figure 4–6**

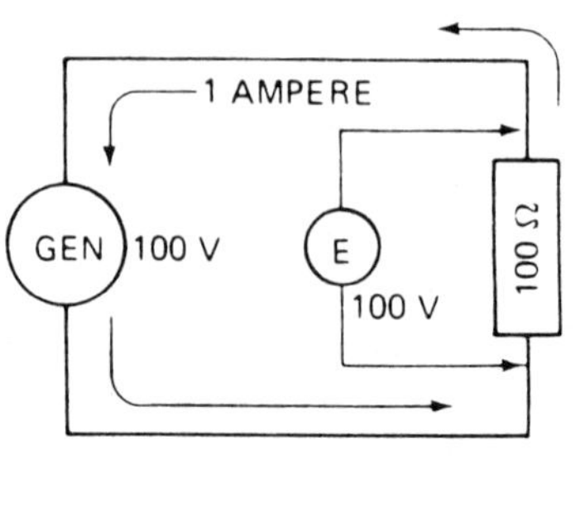


**Figure 4–7**

were producing 100 volts, then one ampere of current would flow in the resistor. A voltmeter connected across the resistor would read 100 volts.

$$I = \frac{E}{R}$$

$$I = \frac{100}{100}$$

$$I = 1 \text{ ampere}$$

If the resistor were broken at the exact center, there would be two resistors of 50 ohms each (two 50-feet lengths of the nichrome wire) (Figure 4–8). The two broken ends of wire could be clamped together to remake the 100-ohm resistor (Figure 4–9). The two 50-ohm resistors, tied in series, make the 100-ohm resistor: 50 + 50 = 100.

The two reconnected 50-ohm resistor sections may be connected into the original circuit (Figure 4–10). This circuit is a series circuit. The same current, 1 ampere,

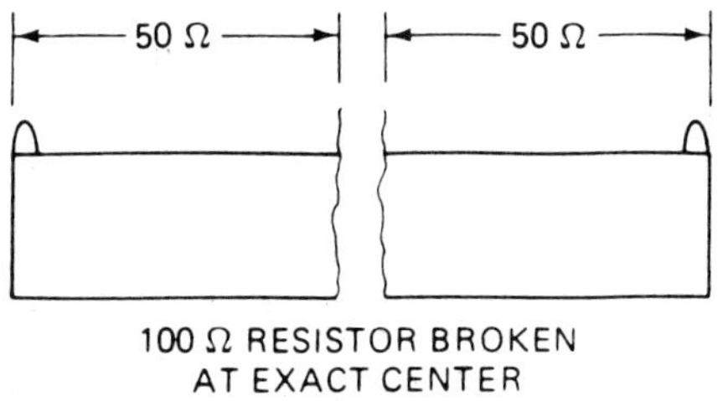


**Figure 4–8**

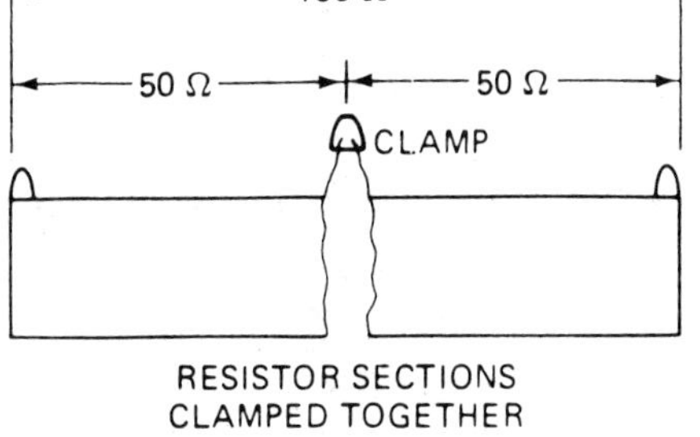


**Figure 4–9**

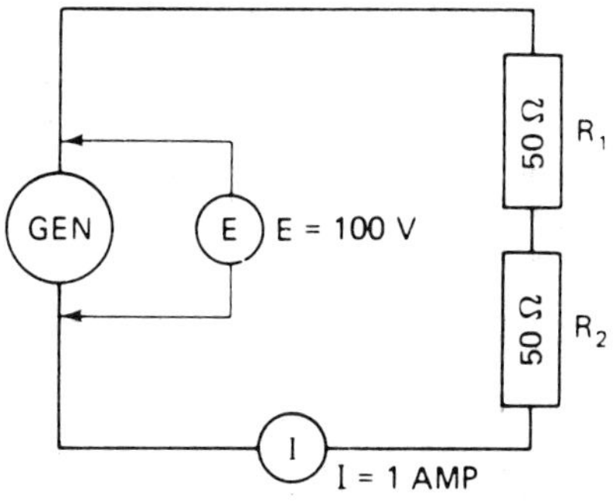


**Figure 4–10**

flows through each of the 50-ohm resistor sections. Ohm's Law may be used to calculate the voltage across $R_1$, the first resistor section.

$$E = I \times R$$
$$E = 1 \times 50$$
$$E = 50 \text{ volts}$$

Ohm's Law is also used to calculate the voltage across $R_2$, the second resistor section.

$$E = I \times R$$
$$E = 1 \times 50$$
$$E = 50 \text{ volts}$$

Note that the sum of the voltages across $R_1$ and $R_2$ equals the supplied voltage, 50 V + 50 V = 100 V.

$$E_T = E_{r1} + E_{r2} + \cdots$$

Also note that the total resistance, 100 ohms, is the sum of the individual resistors: 50 Ω + 50 Ω = 100 Ω.

$$R_t = R_1 + R_2 + \cdots$$

We can now state the laws of series circuits.

1. The same current flows through each component of a series circuit.
2. The total voltage in a series circuit is equal to the sum of the voltages across the individual components.
3. The total resistance of a series circuit is equal to the sum of the resistances of the individual components.

---

**EXAMPLE 1**

A circuit consists of a 25-ohm resistor and a 50-ohm resistor connected in series across a 150-volt generator (Figure 4–11). Determine the current flow in the circuit and the voltage across each resistor.

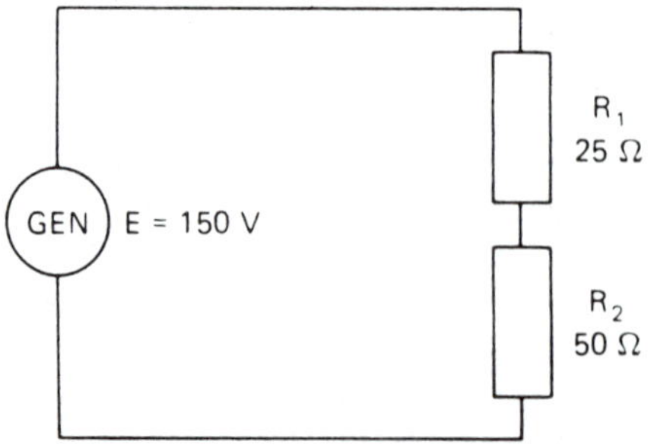


**Figure 4–11**

***Solution:*** The total resistance is equal to the sum of the individual resistances.

$$R_t = R_1 + R_2$$
$$R_t = 25 + 50$$
$$R_t = 75 \text{ ohms}$$

Use Ohm's Law on the whole circuit:

$$I = \frac{E}{R}$$

$$I = \frac{150\ V}{75}$$

$$I = 2 \text{ amperes}$$

Because the same current flows through each component, the voltage across each component may be determined by using Ohm's Law:

$$E_{R1} = I \times R \qquad E_{R2} = I \times R$$

$$E_{R1} = 2 \times 25 \qquad E_{R2} = 2 \times 50$$

$$E_{R1} = 50 \text{ volts} \qquad E_{R2} = 100 \text{ volts}$$

The supply voltage is equal to the sum of the voltages across the individual components.

$$E_T = E_{R1} + E_{R2}$$

$$E_T = 50 + 100$$

$$E_T = 150$$

---

## EXAMPLE 2

Determine the value of each resistor, the voltage across each resistor, and the total resistance of the circuit shown in Figure 4–12. The generator supplies 240 volts; the ammeter indicates 4 amperes.

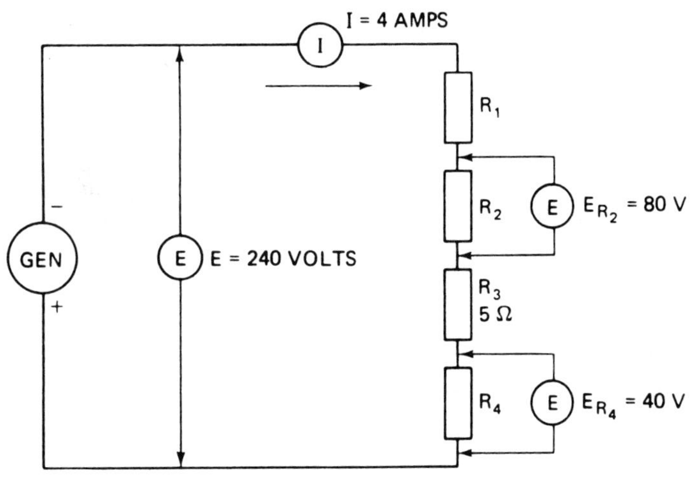


**Figure 4–12**

***Solution:*** Solve for $R_T$, using Ohm's Law for the whole circuit:

$$R_T = \frac{E_T}{I}$$

$$R_T = \frac{240}{4}$$

$$R_T = 60 \text{ ohms}$$

Solve for $R_2$:

$$R_2 = \frac{E_{R2}}{I}$$

$$R_2 = \frac{80}{4}$$

$$R_2 = 20 \text{ ohms}$$

Solve for $E_{R3}$:

$$E_{R3} = I \times R_3$$

$$E_{R3} = 4 \times 5$$

$$E_{R3} = 20 \text{ volts}$$

Solve for $R_4$:

$$R_4 = \frac{E_{R4}}{I}$$

$$R_4 = \frac{40}{4}$$

$$R_4 = 10 \text{ ohms}$$

Solve for $R_1$, using the total resistance law:

$$R_T = R_1 + R_2 + R_3 + R_4$$

$$60 = R_1 + 20 + 5 + 10$$

$$R_1 = 60 - 35$$

$$R_1 = 25 \text{ ohms}$$

Check the solution, using total voltage:

$$R_1 = 25 \text{ ohms}$$

$$E_{R1} = I \times R$$

$$E_{R1} = 4 \times 25$$

$$E_{R1} = 100 \text{ V}$$

$$E_T = E_{R1} + E_{R2} + E_{R3} + E_{R4}$$

$$240 = 100 + 80 + 20 + 40$$

$$240 = 240$$

The solution checks.

## REVIEW QUESTIONS

(Draw a sketch of each circuit before attempting solutions.)

1. A circuit consists of four 20-ohm resistors connected in series. The total resistance of the circuit is ______ ohms.

2. A circuit consists of an 8-ohm resistor and a 4-ohm resistor connected in series. The supply voltage to the circuit is 24 volts. How much current will flow in the circuit? How much voltage would be measured across each resistor?
3. A circuit consists of a 10-ohm resistor and a 2-ohm resistor connected to a 24-volt power source. The 10-ohm resistor has 20 volts across it. How much current is flowing in the 2-ohm resistor?
4. A circuit consists of three resistors connected in series to a power source. Each resistor has 6 volts across it. What is the voltage of the power source?
5. A series circuit consists of two resistors connected to a power source of 24 volts. The first resistor has 16 volts across it; the second resistor is a 4-ohm resistor. What is the total resistance of the circuit?
6. The (same, different) ______ current flows through each resistor in a series circuit.
7. Two 10-ohm resistors are connected in a series circuit. The voltage across the circuit is 40 volts. The current through each resistor is ______ amps.
8. Four 8-ohm resistors are connected in a series circuit. The current through one of the resistors is 2 amperes. The voltage supplied to the circuit is ______ volts.
9. A current flow of 3 amperes is flowing in a two-resistor series circuit. The supply voltage is 24 volts. One of the resistors has a value of 6 ohms. The other resistor is ______ ohms.
10. A three-resistor series circuit is made up of 6-ohm resistors. A voltmeter placed across one of the resistors reads 9 volts. The source voltage is ______ volts.
11. An 8-kilohm resistor and a 12-kilohm resistor are connected in parallel with a third resistor. The supply voltage is 24 volts. The line current is 6 ma. What is the value of the third resistor?
12. Twelve milliamperes are flowing through a circuit made up of an 8-kilohm and a 6-kilohm resistor in series. What is the applied voltage?
13. A three-resistor series circuit is made up of two 14-kilohm resistors and one 8-kilohm resistor. The voltage across the 8-kilohm resistor is 19 volts. What is the current in the circuit?
14. What is the supply voltage of the circuit in question 13?
15. A three-resistor circuit has 36 volts across it. The current in the circuit is $\frac{1}{2}$ ma. One of the resistors is 8 kilohms and another is 3 kilohms in value. What is the value of the third resistor?

# UNIT FIVE

# PARALLEL AND SERIES PARALLEL CIRCUITS

## PARALLEL CIRCUITS

Parallel circuits are found when the same source voltage is required across two or more components. A typical parallel circuit is shown in Figure 5–1. If the generator were producing 120 volts at its output terminals, there would be 120 volts across each lamp. Current would flow through each lamp. If each lamp had a resistance of 60 ohms, the current would be 2 amperes in each, according to Ohm's Law:

$$\frac{120 \text{ volts}}{60 \text{ ohms}} = 2 \text{ amperes}$$

The current for both lamps would come through the generator, the power source. There would be 4 amperes flowing through the generator (2 + 2 = 4).

Current flow in a parallel circuit is quite similar to water flow in a pipe system. In the water-pipe junction shown in Figure 5–2, 2 gallons of water per minute are flowing in pipe A, entering from the left. In pipe B, 5 gallons of water per minute are flowing, entering from the left. There must be 7 gallons of water per minute flowing out of pipe C. The water cannot disappear. The sum of the water flows into the junction must equal the water flow out of the junction.

In an electric circuit, the sum of the currents entering a junction must equal the current leaving the junction. In Figure 5–3, the lower junction of Figure 5–1 is shown.

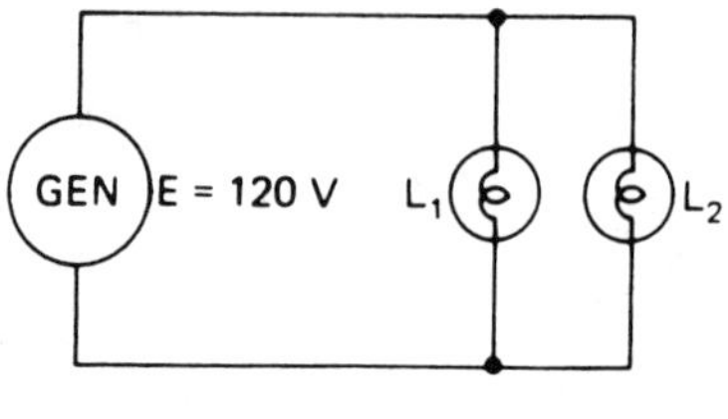


**Figure 5–1**

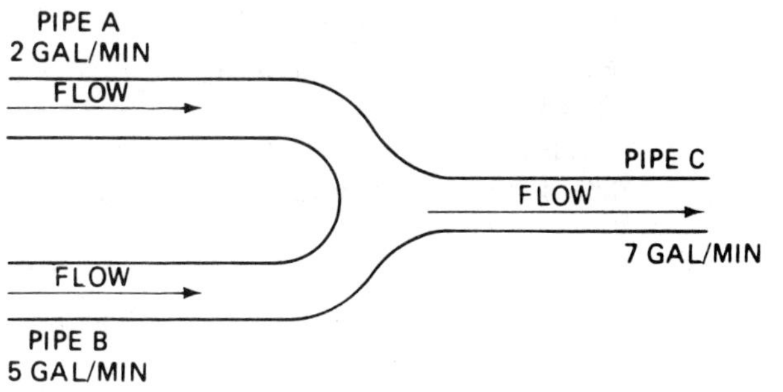


**Figure 5–2**

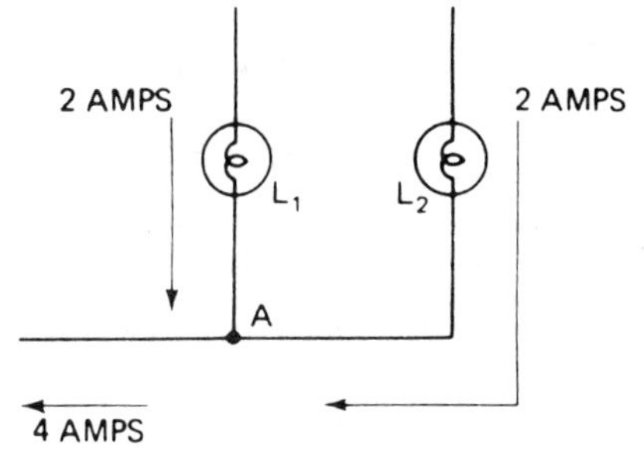


**Figure 5–3**

If 2 amperes are flowing to junction A from lamp $L_1$, and 2 amperes are flowing to it from lamp $L_2$, there must be 4 amperes leaving junction A.

Sometimes it is necessary to use this law to find a branch current value. (A branch is one of the paths for current flow in a parallel circuit.) The circuit in Figure 5–4 is a portion of a total circuit. Ammeter 1 indicates 3 amperes, and ammeter 2, in the lamp circuit, indicates 2 amperes. Since there are 3 amperes leaving junction B and only 2 amperes coming to the junction B through the lamp, there must be 1 ampere flowing through resistor $R_1$ to the junction B.

The total circuit of Figure 5–4 could contain another branch, as in Figure 5–5.

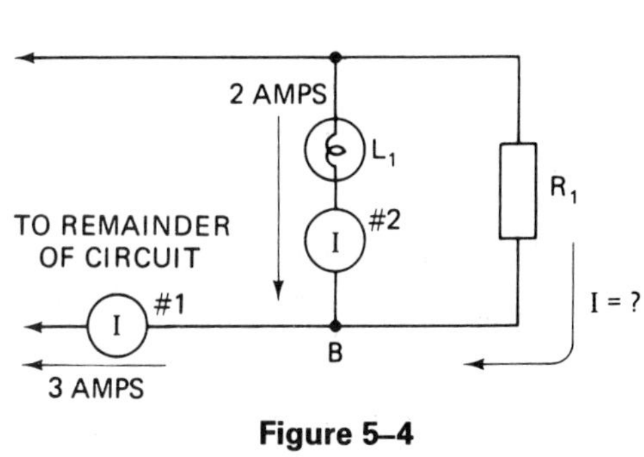


**Figure 5–4**

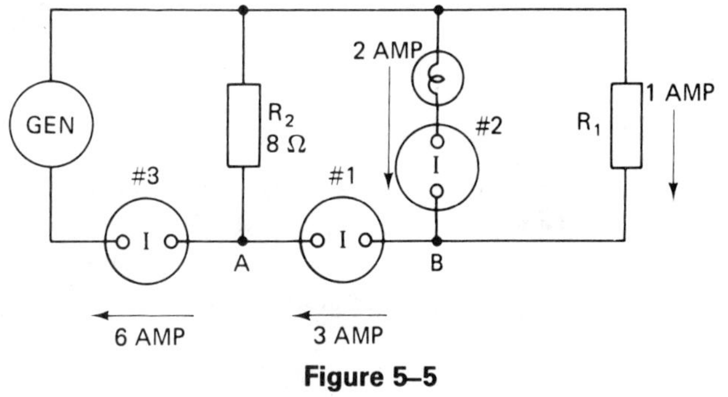


**Figure 5–5**

The current entering junction A, coming from junction B, is 3 amperes. The current leaving junction A, going to the generator, is 6 amperes. Where did the other 3 amperes come from? There must be 3 amperes flowing into junction A through $R_2$. Given that $R_2$ is an 8-ohm resistor, other unknown values of the circuit may be determined. To determine the voltage across $R_2$, use Ohm's Law for E:

$$E = I \times R$$

$$E = 3 \times 8$$

$$E = 24\ V$$

The voltage across each component in parallel is the same.

The voltage across $R_2$ is 24 volts.

The voltage across $L_1$ is 24 volts.

The voltage across $R_1$ is 24 volts.

To determine the resistance of Lamp $L_1$, use Ohm's Law for R:

$$R_{L1} = \frac{E}{I}$$

$$R_{L1} = \frac{24}{2}$$

$$R_{L1} = 12 \text{ ohms}$$

To determine the resistance of $R_1$:

$$R_1 = \frac{E}{I}$$

$$R_1 = \frac{24}{1}$$

$$R_1 = 24 \text{ ohms}$$

To determine the resistance of the circuit, the total current and voltage must be used. The total current as indicated by ammeter 2 (Figure 5–5) is 6 amperes. The total voltage is 24 volts.

$$R_T = \frac{E}{I}$$

$$R_T = \frac{24}{6}$$

$$R_T = 4 \text{ ohms}$$

The total resistance of the circuit is 4 ohms, which is less than the resistance of any of the three components: $R_1$ (24 ohms), $L_1$ (12 ohms), or $R_2$ (8 ohms).

*The total resistance of a parallel circuit is less than the smallest branch resistance.*

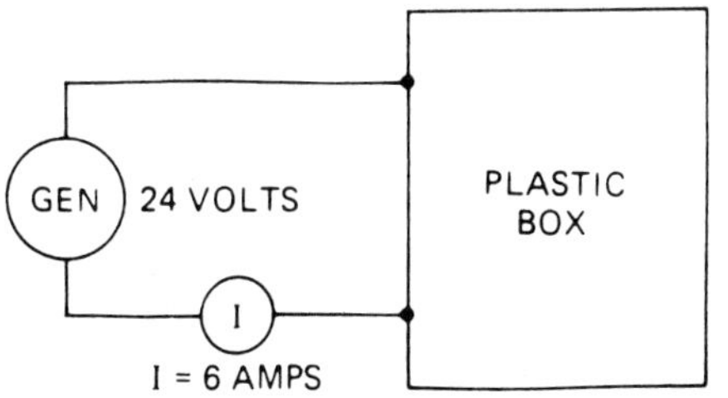


**Figure 5–6**

There are other interesting relationships in parallel circuits. The circuit in Figure 5–6 contains a generator producing 24 volts, wires from the generator to a plastic box, and an ammeter indicating 6 amperes. There is no indication as to what is in the box. Solve for $R_T$ of the circuit in the box:

$$R_T = \frac{E}{I}$$

$$R_T = \frac{24}{6}$$

$$R_T = 4 \text{ ohms}$$

When the box is opened, a 12-ohm resistor is found connected in parallel with a 6-ohm resistor (Figure 5–7). The resistance of the parallel combination could also be found using the product-over-the-sum method as in the formula:

$$R_T = \frac{R_1 \times R_2}{R_1 + R_2} \frac{\text{The product of } R_1, R_2}{\text{The sum of } R_1, R_2}$$

For the parallel combination in Figure 5–7:

$$R_T = \frac{12 \times 6}{12 + 6}$$

$$R_T = \frac{72}{18}$$

$$R_T = 4 \text{ ohms}$$

This is the same total resistance value that was calculated using Ohm's Law.

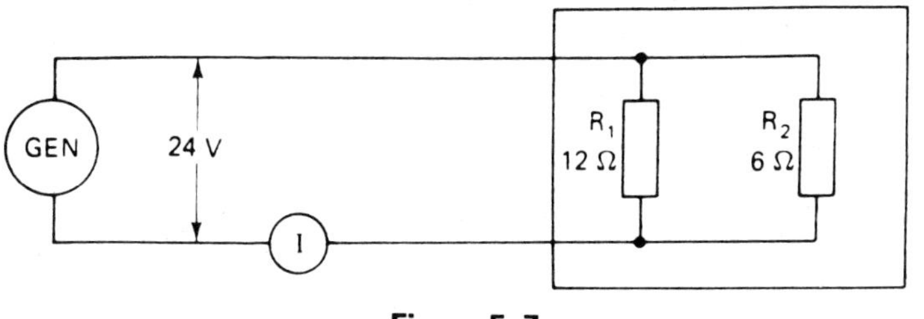


**Figure 5–7**

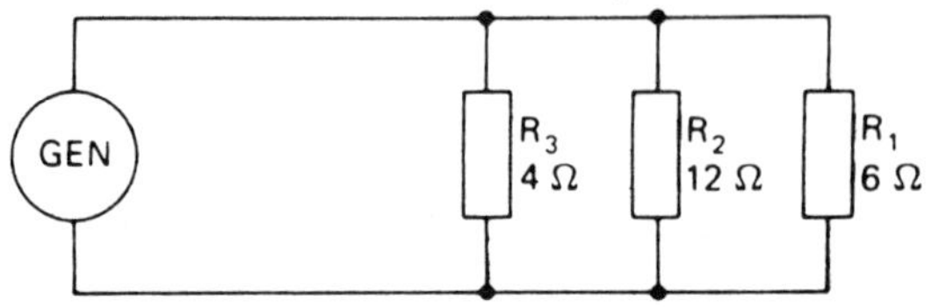


**Figure 5–8**

## Product Over Sum Method

If more than two resistors are connected in parallel, the *product-over-the-sum method* may be used on two resistors at a time. An example of this procedure is given using Figure 5–8. What is the total resistance of the circuit? It has been shown that the combination of $R_1$ and $R_2$ is equal to a 4-ohm resistor. Figure 5–9 shows the equivalent circuit after resistors $R_1$ and $R_2$ are combined. The product-over-the-sum method may now be used to determine $R_T$ of the total circuit.

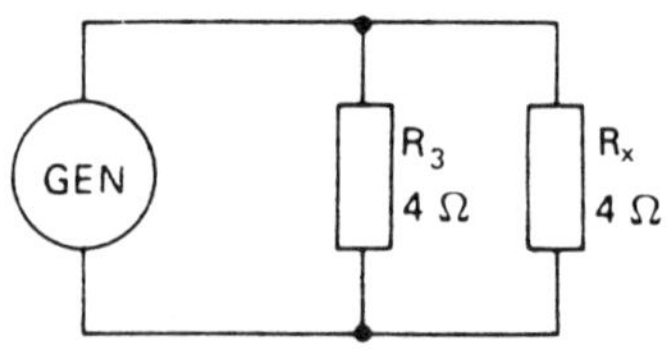


**Figure 5–9**

$$R_T = \frac{R_x \times R_3}{R_x + R_3}$$

$$R_T = \frac{4 \times 4}{4 + 4}$$

$$R_T = \frac{16}{8}$$

$$R_T = 2 \text{ ohms}$$

Thus, the $R_T$ of two resistors of equal value connected in parallel is equal to the value of one resistor divided by the number of equal resistors. If three 9-ohm resistors were connected in parallel, the $R_T$ would be equal to nine divided by three (Figure 5–10).

Another method that may be used to solve for total resistance of a parallel circuit is to assume a voltage across the circuit. For example, in Figure 5–8, 4-ohm, 12-ohm, and 6-ohm resistors are connected in parallel. If an assumed supply voltage of 12 volts is considered, the current through each resistor by Ohm's Law would be:

$$I_{R1} = \frac{12V}{R_1} \qquad I_{R2} = \frac{12V}{R_2} \qquad I_{R3} = \frac{12V}{R_3}$$

$$I_{R1} = \frac{12V}{6\Omega} \qquad I_{R2} = \frac{12V}{12\Omega} \qquad I_{R3} = \frac{12V}{4\Omega}$$

$$I_{R1} = 2 \text{ Amp} \qquad I_{R2} = 1 \text{ Amp} \qquad I_{R3} = 3 \text{ Amp}$$

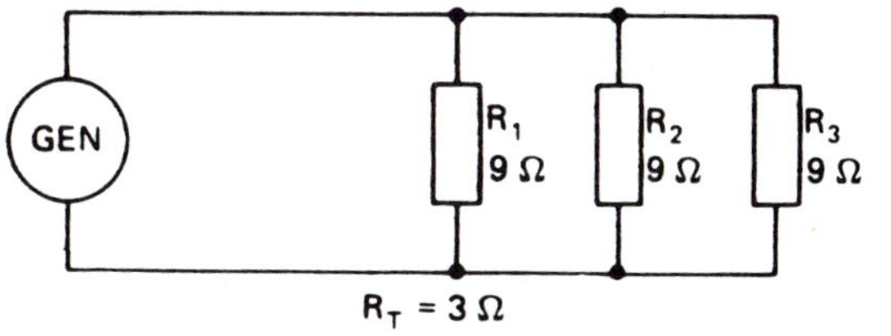


**Figure 5–10**

The total current is the sum of the currents in the branches.

$$I_T = I_1 + I_2 + I_3$$

$$I_T = 2A + 1A + 3A$$

$$I_T = 6 \text{ Amp}$$

To solve Ohm's Law for the complete circuit:

$$R_T = \frac{E_T}{I_T}$$

$$R_T = \frac{12V}{6A}$$

$$R_T = 2\Omega$$

Because the applied voltage is an assumed value, the currents in the branches are also assumed values. The calculated total resistance is the correct value. Supply voltage does not influence the resistance of resistors or the total circuit resistance.

The use of assumed voltage values leads to another important formula for parallel circuits. The formula is:

$$R_T = \frac{1}{\frac{1}{R_1} + \frac{1}{R_2} + \frac{1}{R_3} \cdots \frac{1}{R_N}}$$

### Reciprocal Method

Consider again the circuit of Figure 5–8, in which 4-ohm, 12-ohm, and 6-ohm resistors are connected in parallel. If 1 volt is assumed across the circuit, the current in the individual resistors becomes:

$$I_{R1} = \frac{1}{R_1}$$

$$I_{R2} = \frac{1}{R_2}$$

$$I_{R3} = \frac{1}{R_3}$$

The total current is the sum of the currents in the branches.

$$I_T = \frac{1}{R_1} + \frac{1}{R_2} + \frac{1}{R_3}$$

The total resistance is the supply voltage (1V) divided by the total current.

$$R_T = \frac{1}{\frac{1}{R_1} + \frac{1}{R_2} + \frac{1}{R_3} + \cdots \frac{1}{R_N}}$$

Because 1 divided by a number is called the *reciprocal of the number,* the formula for $R_T$ is called the *reciprocal of the sum of the reciprocals*. This formula provides the most important method for the solving of parallel circuits particularly since scientific calculators have come into general use.

### Scientific-Calculator Solution

A *scientific-calculator solution* of a parallel circuit problem follows:

PROBLEM: Determine the equivalent resistance of 8Ω, 12Ω, 9Ω, and 17Ω in parallel using the reciprocal of the sum of the reciprocals method.

| STEP | ENTER | PRESS | DISPLAY |
|---|---|---|---|
| 1 | 8 | 1/x | 0.125 |
| 2 | | + | 0.125 |
| 3 | 12 | 1/x | 0.0833 |
| 4 | | + | 0.2083 |
| 5 | 9 | 1/x | 0.1111 |
| 6 | | + | 0.3194 |
| 7 | 17 | 1/x | 0.5882 |
| 8 | | = | 0.3782 |
| 9 | | 1/x | 2.6463 |

The resistance of the circuit is 2.6463 ohms.

Here is a scientific calculator solution of a parallel circuit problem containing resistors in the kilohm range. This requires the use of the exponent key (EXP) for proper entry.

PROBLEM: What is the equivalent resistance of a parallel circuit containing a 2kΩ, 6kΩ, 11kΩ, and a 7kΩ resistor?

$$R_{eq} = \frac{1}{\frac{1}{R_1} + \frac{1}{R_2} + \frac{1}{R_3} + \frac{1}{R_4}} = R_{eq} = \frac{1}{\frac{1}{2k} + \frac{1}{6k} + \frac{1}{11k} + \frac{1}{7k}}$$

| STEP | ENTER | PRESS | DISPLAY |
|---|---|---|---|
| 1 | 2 | EXP | 2.00 |
| 2 | 3 | 1/x | 0.0005 |
| 3 | | + | 0.0005 |
| 4 | 6 | EXP | 6.00 |
| 5 | 3 | 1/x | 0.0001666 |
| 6 | | + | 0.0001666 |
| 7 | 11 | EXP | 11.00 |
| 8 | 3 | 1/x | 0.0000909 |
| 9 | | + | 0.0007575 |
| 10 | 7 | EXP | 7.00 |
| 11 | 3 | 1/x | 0.0001428 |
| 12 | | = | 0.0009004 |
| 13 | | 1/x | 1110.58 |

The effective resistance of the circuit is 1110.58 ohms.

This problem could have been solved leaving out the exponent since all of the resistors were given in the kilohm range. If the numbers 2, 6, 11, and 7 were simply entered in the calculator, the final answer would have been 1.110. The k factor could then have been reinserted providing 1.110 k as the answer. In many problems, the resistors will be mixed, ohms with kilohms, or kilohms with megohms. It is good practice to enter the actual resistor value into the calculator using the EXP key when needed. With practice, you will develop individual techniques.

At this time you should understand:

1. The same voltage appears across each component of a parallel circuit.
2. The total current in a parallel circuit is equal to the sum of the currents in the branches.
3. The total resistance ($R_T$) of a parallel circuit is always smaller than the smallest resistor connected in the parallel combination.

## *SERIES PARALLEL CIRCUITS*

A series parallel circuit is made up of a combination of components; some are connected in series, and others are connected in parallel. When working with series parallel circuits, it is necessary to use the laws of series circuits on the series elements and the laws of parallel circuits on the parallel elements.

The circuit in Figure 5–11 is a series parallel circuit. The 10-ohm resistor $R_3$ is connected in series with the parallel combination $R_1$ and $R_2$. In order to solve for

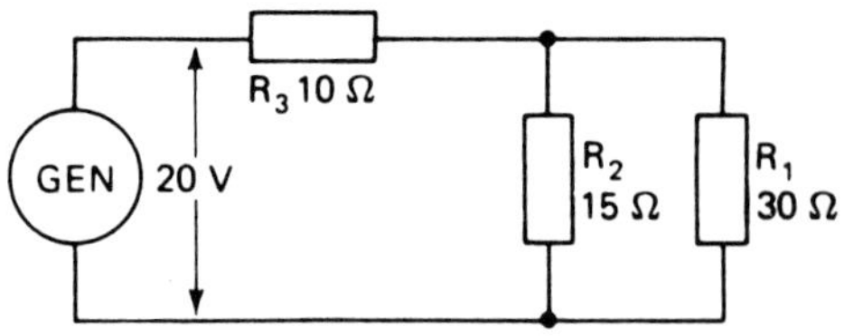


**Figure 5–11**

the current flow in the circuit, the resistors must be combined. Resistors $R_1$ and $R_2$ are in parallel.

$$R_T = \frac{R_1 \times R_2}{R_1 + R_2}$$

$$R_T = \frac{15 \times 30}{15 + 30}$$

$$R_T = \frac{450}{45}$$

$$R_T = 10 \text{ ohms}$$

The equivalent circuit is a series circuit (Figure 5–12).

$$R_T = R_1 + R_2$$

$$R_T = 10 + 10$$

$$R_T = 20 \text{ ohms}$$

The generator is producing 20 volts. According to Ohm's Law, the current through the generator is:

$$I = \frac{E}{R}$$

$$I = \frac{20}{20}$$

$$I = 1 \text{ ampere}$$

Going back to the original circuit (Figure 5–11), the current through the series 10-ohm resistor $R_3$ is 1 ampere, and there are 10 volts across it. With 10 volts across $R_3$, there are 10 volts left (from the 20 volts supplied) to appear across $R_1$ and $R_2$.

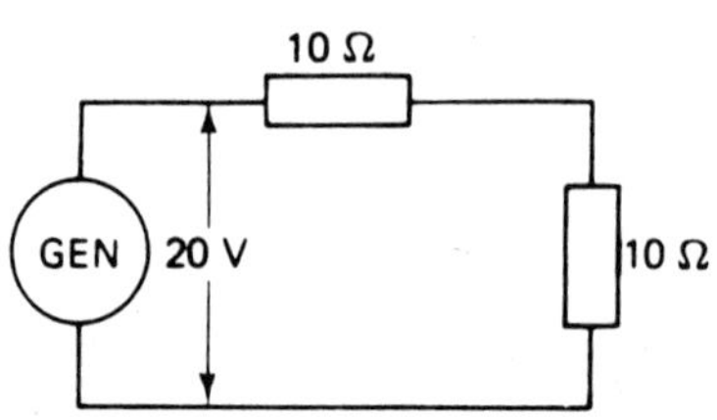


**Figure 5–12**

Since the *same* voltage appears across resistors in parallel, the current flow through $R_1$ is:

$$I = \frac{E}{R}$$

$$I = \frac{10}{30}$$

$$I = \tfrac{1}{3} \text{ or } 0.333 + \text{ ampere}$$

The current flow through $R_2$ is:

$$I = \frac{E}{R}$$

$$I = \frac{10}{15}$$

$$I = \tfrac{2}{3} \text{ or } 0.666 + \text{ ampere}$$

The total current entering the junction (1 ampere) is equal to the total current in the branches.

$$\tfrac{1}{3} \text{ ampere} + \tfrac{2}{3} \text{ ampere} = 1 \text{ ampere or}$$

$$0.333 + \text{ ampere plus } 0.666 + \text{ ampere} = 0.999 + \text{ or } 1 \text{ ampere}$$

There are times when the information known about a circuit is other than the resistance values. For example, the circuit in Figure 5–11 might be encountered, but the known information could be that in Figure 5–13: The generator is producing 20 volts; the value of resistor $R_1$ is 30 ohms; the value of resistor $R_2$ is 20 ohms; and a voltmeter across $R_1$ indicates 10 volts. The unknowns may be found by using the laws of series circuits, parallel circuits, and Ohm's Law.

The current flow through $R_1$ could be found using Ohm's Law. It has already been shown that with 10 volts across 30 ohms, the current is $\frac{1}{3}$ ampere. According to the laws of parallel circuits, the voltage across $R_2$ is the same as across $R_1$, 10 volts. The current through $R_2$ can be determined: 10 volts across 15 ohms provides $\frac{2}{3}$ ampere. The current entering junction point A is equal to the sum of the currents leaving it. Therefore, 1 ampere ($\frac{2}{3}$ ampere plus $\frac{1}{3}$ ampere) enters junction point A.

This 1 ampere is the current flowing through the unknown series resistor $R_3$. Since the supply voltage is 20 volts and there are 10 volts across the parallel resistors $R_1$ and $R_2$, there must be 10 volts across $R_3$ (the supply of 20 volts minus the 10 volts across the parallel branch leaves 10 volts).

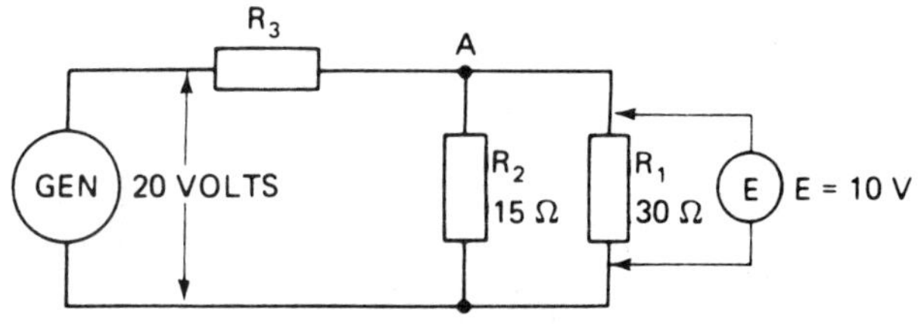


**Figure 5–13**

According to Ohm's Law:

$$R_3 = \frac{E}{I}$$

$$R_3 = \frac{10}{1}$$

$$R_3 = 10 \text{ ohms}$$

## *SWITCHES, SERIES PARALLEL CIRCUITS*

In electrical systems, switches are used to control the operation of devices such as compressor motors, fan motors, heaters, relays, and indicator lights. These switches are connected in series, parallel, or series parallel (as required) in order to accomplish the desired circuit action. Some examples of switch circuit combinations are shown in Figure 5–14.

The lamps shown in these circuit diagrams could easily have been compressor motors or fan motors. The switches could have been circuit breakers, temperature

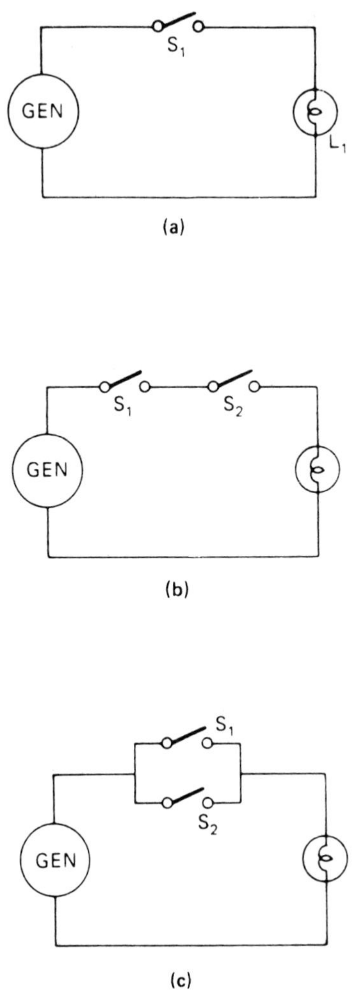


**Figure 5–14** (a) Switch $S_1$ must be closed if lamp $L_1$ is to light. (b) Switches $S_1$ and $S_2$ must be closed if lamp $L_1$ is to light. (c) Either switch $S_1$ or $S_2$ must be closed if lamp $L_1$ is to light.

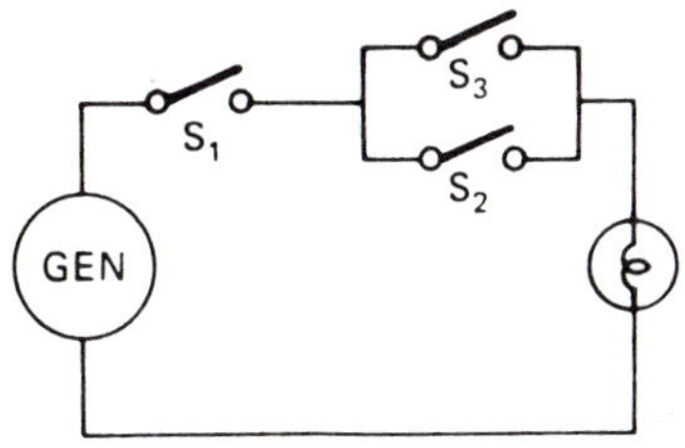


(d)

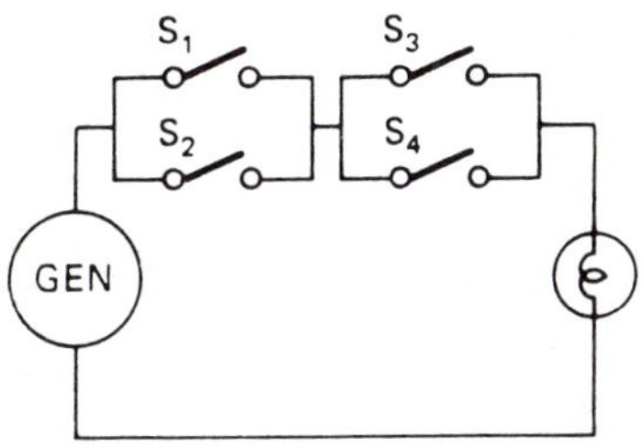


(e)

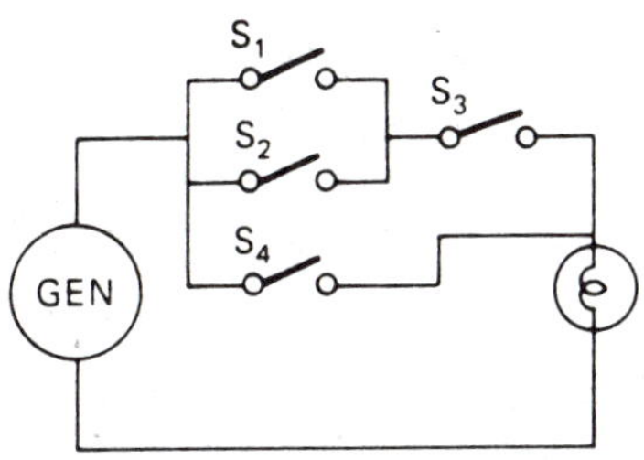


(f)

**Figure 5–14** (continued) (d) Switches $S_1$ and $S_2$ or $S_3$ must be closed if lamp $L_1$ is to light. (e) Switches $S_1$ or $S_2$ and $S_3$ or switch $S_4$ must be closed if lamp $L_1$ is to light. (f) Switches $S_1$ or $S_2$ and $S_3$ or $S_4$ must be closed if lamp $L_1$ is to light.

controls, or pressure controls. The important point to remember is how switches control various electric ac circuits.

1. In parallel restrictive circuits, the total current is the sum of the currents in the branches.
2. The same voltage appears across each component in parallel.
3. The total resistance may be found by using the product-over-the-sum method.
4. In series parallel circuits, the laws of series circuits apply to the portions of the circuit connected in series.
5. In series parallel circuits, the laws of parallel circuits apply to the portion of the circuit connected in parallel.
6. Ohm's Law may be applied to the whole circuit or any part of the circuit.

## *SNEAK CIRCUITS*

Whenever switches are connected in series parallel, the possibility of *sneak circuits* exists. A sneak circuit is one that provides for the operation of a component at a time when operation is not wanted. Sneak circuits are not always obvious at the time a circuit is designed. Some sneak circuits only cause operation under odd conditions and are therefore difficult to discover.

Consider the circuit of Figure 5–15. The path through the switch is normally

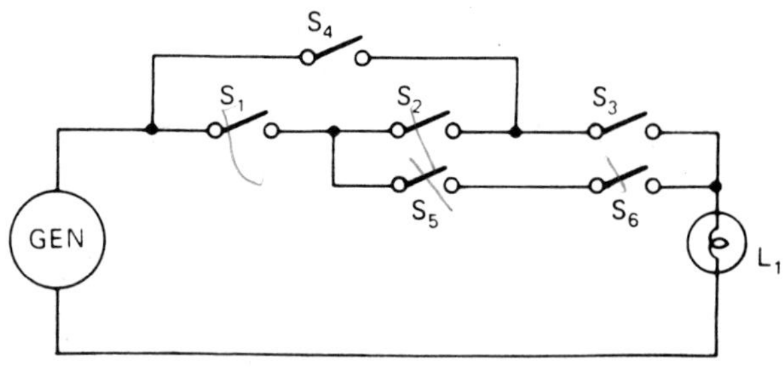


**Figure 5–15**

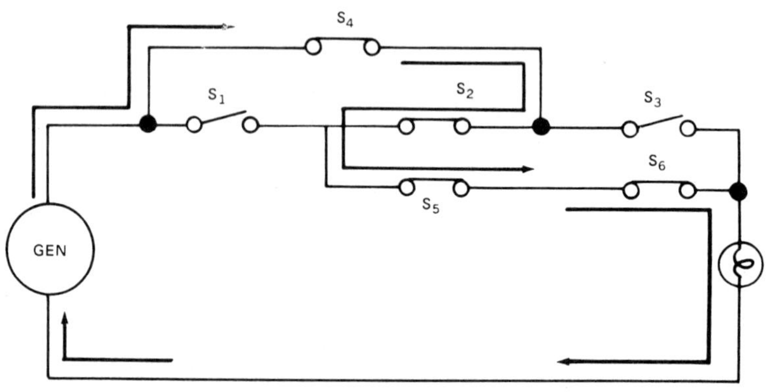


**Figure 5–16** Sneak circuit.

considered to start from the left top and proceed through the switches to the lamp. Different paths and switches might be considered, but most people would follow a path from left to right. The circuit designer no doubt drew the circuit while considering paths from left to right.

A sneak circuit might exist in a path through $S_2$ from right to left. If switches $S_4$, $S_2$, $S_5$, and $S_6$ are closed, there is a complete path for current flow, as shown in Figure 5–16. The circuit path may be an unwanted path causing lamp $L_1$ to be on during a period when it should not be on. When the sneak circuit is discovered, a redesign is required.

Most sneak circuits would not be as obvious as the one shown in this example. In some cases, a considerable amount of circuit investigation is necessary before the problem can be discovered. Keep in mind that, when a sneak circuit is causing a problem, the technician investigating the problem normally assumes that something is malfunctioning, when, in fact, every component is operating correctly.

Sneak circuits are normally in-plant manufacturing problems, not field problems. However, they do sometimes occur in the field and technicians must be aware of them.

## REVIEW QUESTIONS

(Draw a sketch of each circuit before attempting solutions.)

1. Four 40-ohm resistors are connected in parallel. What is the total resistance of the combination?
2. A 40-ohm resistor and a 20-ohm resistor are connected in parallel. The voltage across the 40-ohm resistor is 80 volts. What is the total current in the circuit?

3. A 10-ohm resistance is needed in a control circuit. Three 30-ohm resistors are available. How could these resistors be connected in order to obtain the required resistance?
4. A resistance is needed in a control circuit. The resistance must be more than 2 ohms but less than 4 ohms. Three 2-ohm resistors are available. How would you connect the 2-ohm resistors in order to obtain the required resistance?
5. Figure 5–15 is a lamp and switch circuit. Which switches must be closed in order to light the lamp, using the least number of switches?
6. In Figure 5–17, if the voltage across $R_1$ is 8 volts and the voltage across $R_3$ is 6 volts, the voltage across $R_2$ must be ______ volts.

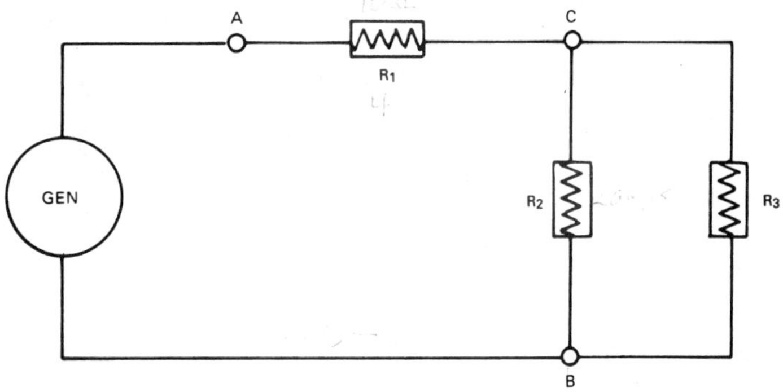


**Figure 5–17**

7. Using the same voltages given in question 6, the generator voltage must be ______ volts.
8. In Figure 5–17, if the voltage from point A to point C is 18 volts and the voltage from point A to point B is 24 volts, then the voltage from point C to point B must be ______ volts.
9. In Figure 5–17, if the current through $R_1$ is 4 amperes and the current through $R_2$ is 3 amperes, then the current through $R_3$ must be ______ amperes.
10. In Figure 5–17, if the current through $R_2$ were 2 amperes and the current through $R_3$ were 3 amperes and $R_1$ were a 10-ohm resistor, then the voltage across $R_1$ would be ______ volts.
11. In Figure 5–15, will lamp $L_1$ light if switches $S_1$, $S_5$, and $S_3$ are closed?
12. In Figure 5–17, the supply voltage is 18V. If the value of $R_1$ is 3 kilohms and the current through $R_1$ is 4 ma, what is the voltage across $R_2$?
13. In question 12, if $R_2$ is a 2-kilohm resistor, what is the value of $R_3$?
14. If $R_1$, $R_2$, and $R_3$ of Figure 5–17 are 8-kilohm resistors, what is the total resistance of the circuit?
15. The generator of Figure 5–17 is supplying 18 volts. Resistor $R_1$ is a 22-kilohm resistor. Resistors $R_2$ and $R_3$ are both 14-kilohm resistors. The desired generator current is 1 ma. What value of resistance could be connected in parallel with $R_1$ that would provide for the required 1 ma?

# UNIT SIX

# *ELECTRIC METERS*

An electric meter is used to determine the characteristics of an electric circuit or component. The common types of meters used by the technician include the ammeter, voltmeter, and ohmmeter. Occasionally, the technician will have the opportunity to use a wattmeter. The meters used in the trade today include the analog and digital types.

## *ANALOG METER*

An analog electric meter uses a meter movement that operates on the electric motor principle. In order to obtain meter pointer deflection, a current must pass through the meter-movement coil. This current sets up a magnetic field in the meter-movement coil which is repelled by the fixed magnetic field provided by the permanent magnets. The stronger the current flow through the meter movement coil, the greater the deflection of the coil and pointer as shown in Figure 6–1.

In the meter movement shown in Figure 6–1, the permanent magnets are indicated by the capital N for north pole and capital S for south pole. The movement deflection coil is pivoted at the center and has a pointer attached to it. A coil spring is attached to the meter movement holding the movement and pointer at zero (0) deflection when there is no current flow through the movement coil.

Physical stops in the form of pins are located just beyond the zero and full-scale deflection points in order to limit the movement of the pointer. When current flows through the meter movement deflection coil, a magnetic field is set up around this coil. The magnetic field of the deflection coil will be repelled by the field of the permanent magnets. The meter movement will rotate around the pivot point causing the pointer to move up scale.

As the meter movement rotates up scale, pressure will increase on the coil spring attached to the movement. The rotation will stop when the pressure of the repelling magnetic fields is balanced by the reverse pressure offered by the coil spring.

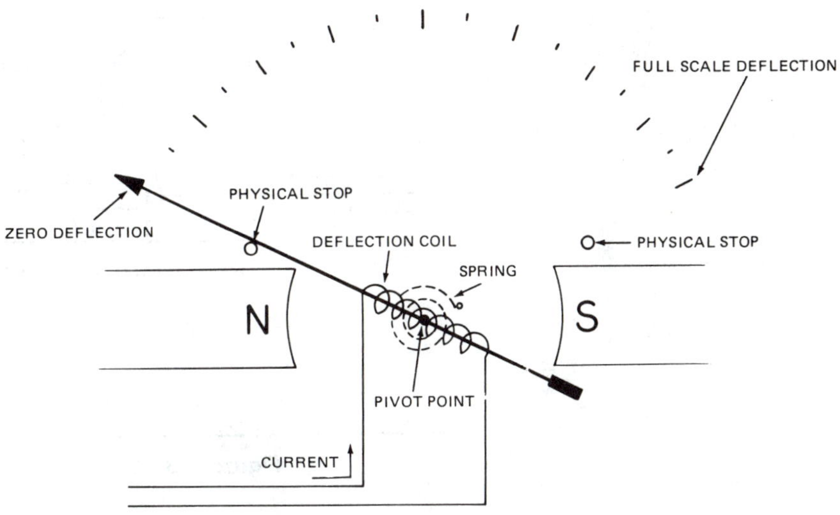


**Figure 6–1** Meter movement.

A different amount of current flow through the meter movement coil will cause a different amount of meter movement and pointer deflection.

## *AMMETER*

An ammeter is a meter used to measure current flow. Current flow is the movement of electrons through the circuit. In order to measure current flow, the circuit must be opened and the ammeter inserted so that the current to be measured flows through the ammeter. The circuit connections are shown in Figure 6–2.

Normally, the current to be measured is much higher than the amount needed to deflect the meter movement to full scale. In order to make meters useful in line-current measuring systems, a *shunt* is installed within the meter. The shunt is a low-resistance current path that allows a proportional amount of current to flow through the meter movement. The remaining current bypasses the meter movement by going through the shunt. This is shown in Figure 6–3.

Suppose that a 1-ampere full-scale meter were to be designed using a meter movement that required 0.001 ampere (1 ma) full-scale deflection. The meter movement resistance is 100 ohms.

According to Ohm's Law, there would be $0.001 \text{ A} \times 100 \ \Omega = 0.1 \text{ V}$ across the meter movement at full-scale deflection.

The scale of the meter would be changed to indicate 1 ampere instead of 1 ma, but it would still take only 1 ma through the meter movement to cause full-scale deflection. This arrangement is shown in Figure 6–4.

One ampere is coming down the line. Only 1 ma goes through the meter

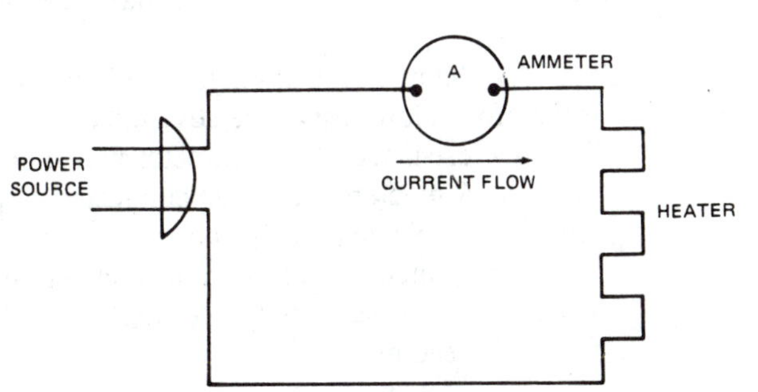


**Figure 6–2** Ammeter connection.

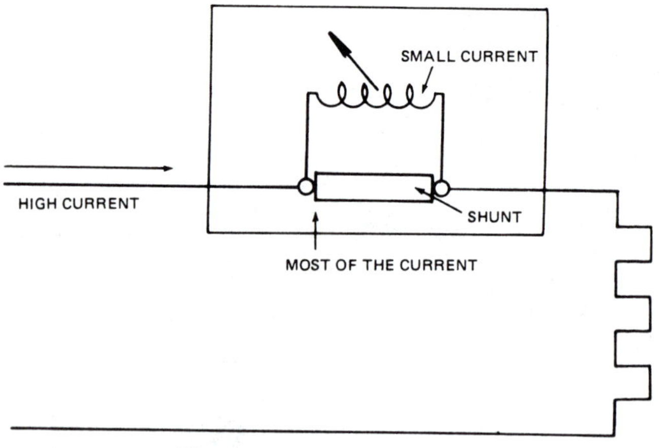


**Figure 6–3** Shunt circuit.

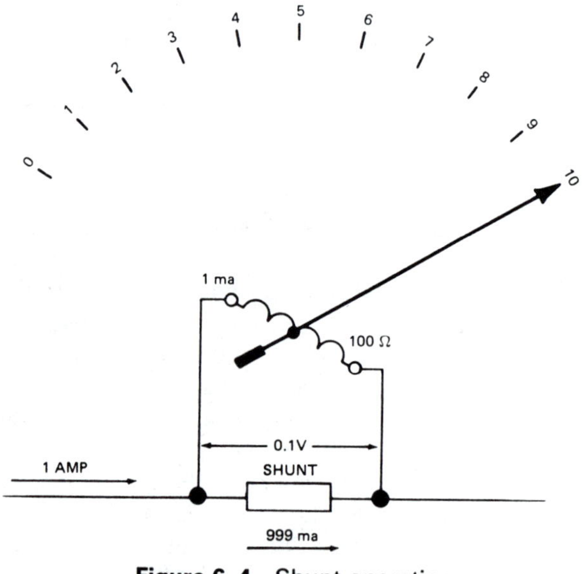


**Figure 6–4** Shunt operation.

movement while 999 ma go through the shunt. The shunt would be in parallel with the meter movement. The resistance of the shunt would be

$$\frac{0.1\text{ V}}{999\text{ ma}} = \frac{0.1\text{ V}}{.99\text{ A}} = .1001\ \Omega$$

The shunt in the ammeter is a low-resistance path for current flow. An ammeter is therefore a low-resistance device. An ammeter is *never* connected across the circuit. The low resistance of the ammeter will short the circuit, usually causing the line fuse to open. Ammeters are *never* connected in parallel with any component. Ammeters are *always* connected in *series*.

Many meters will contain a switch system that provides more than one full-scale deflection sensitivity. A single meter could have a 1-, 10-, and 100-ampere full-scale reading.

## CLAMP-ON AMMETER

One of the problems in using a standard ammeter is that the circuit must be opened in order to insert the ammeter. In large high-current situations, opening the circuit can be a time-consuming project. This is especially true if current is to be measured in a number of different current-carrying lines. The clamp-on ammeter was developed in order to simplify the current measurement procedure.

It was shown in Unit 2 that any wire carrying current has a magnetic field about it. If the current is an alternating current (ac), the current will be varying in the form of a sine wave. The varying current will produce a varying magnetic field about the wire. The strength of the magnetic field is directly proportional to the strength or magnitude of the current flow.

The clamp-on ammeter uses a voltage step-up, current step-down transformer in the current measuring process.

An iron-clamping mechanism is placed around the wire carrying ac current as shown in Figure 6–5.

Inside the clamp-on ammeter, the iron clamp completes a magnetic circuit through a current transformer. The transformer produces an output that is connected to a standard meter movement. The meter movement has a current scale that indicates the current flow in the ac current-carrying wire.

After the current is measured, the clamp-opening lever may be depressed, which will open the clamp, allowing the meter to be removed from the current-carrying wire.

A number of manufacturers provide clamp-on ammeters. The meters have different specifications, particularly with regard to current ranges. Always read equipment specifications before using the equipment.

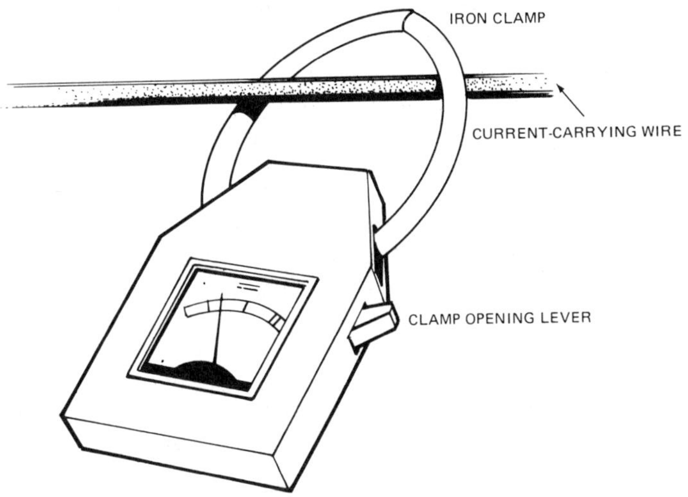


**Figure 6–5** Clamp-on ammeter.

## VOLTMETER

A voltmeter can be developed with the same meter movement used in the ammeter. That meter movement required 1 ma for full-scale deflection and had 100 ohms of resistance.

In Figure 6–6, a 300-volt meter is shown. The scale has been drawn using 300 V as the full-scale deflection. Because voltage is measured across a circuit, the

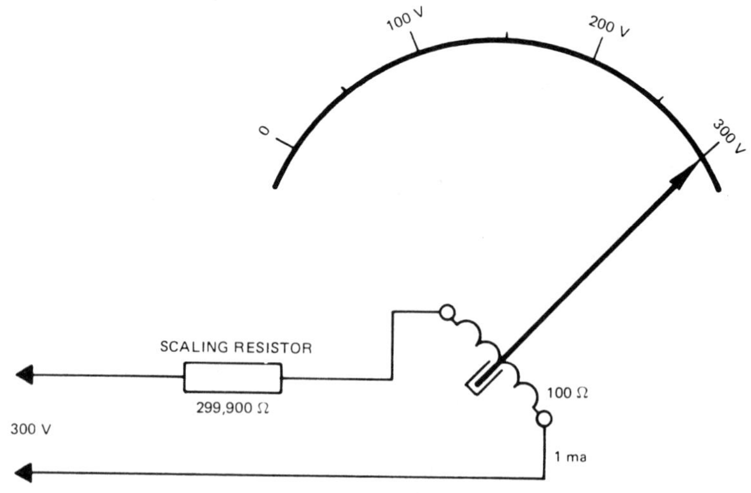


**Figure 6–6** Voltmeter.

voltmeter will have to be placed across the circuit in order to measure voltage. There will be 300 volts at the terminals of the voltmeter. According to Ohm's Law, the total resistance will equal the voltage divided by the current:

$$R = \frac{300}{.001} \text{ or } 300{,}000 \text{ ohms}$$

The scaling resistor would be 299,900 ohms with the meter-movement resistance of 100 ohms added to it for the 300,000-ohm total. The procedure used in making a voltmeter is to determine the required resistance by dividing the full-scale voltage wanted by the meter movement full-scale current. The scaling resistor is then the total resistance minus the resistance of the meter movement.

## *VOLTMETER LOADING*

A very common problem occurs in using a voltmeter to measure voltages across resistances that are of the same order as the resistance of the meter. Remember that when a voltmeter is connected to a circuit, the internal resistance of the meter is connected across (in parallel) with that portion of the circuit.

Consider two circuits where voltage is to be measured. In Figure 6–7a, the voltmeter of Figure 6–6 will indicate the correct voltage.

The meter in Figure 6–7b will only indicate 10 volts, which is the voltage across $R_2$ with the meter connected. Remember the internal resistance of the meter is 300 kilohms. When it is connected to the circuit, the circuit changes. See Figure 6–8.

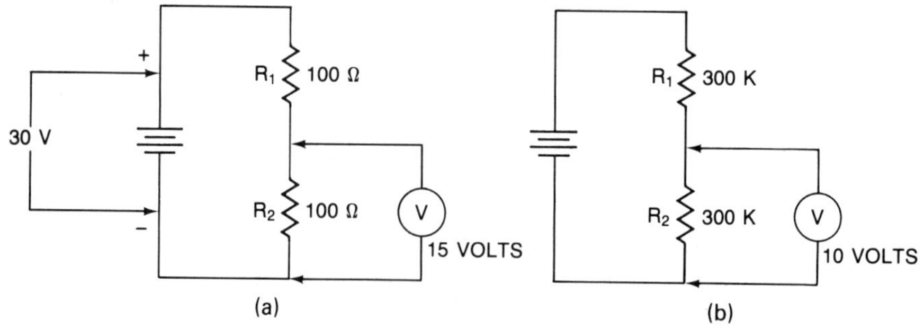


**Figure 6–7**

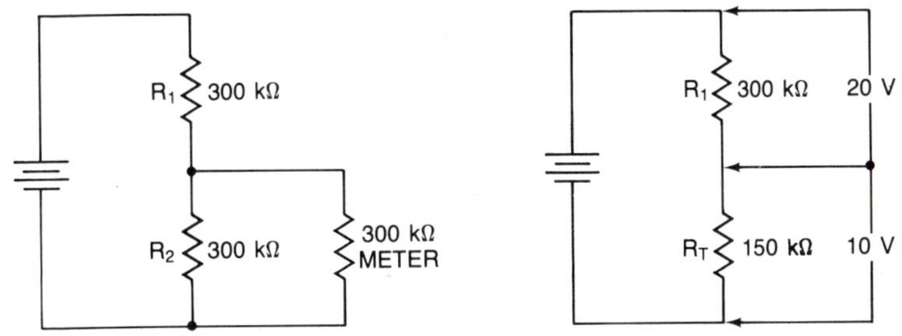


**Figure 6–8**

The meter indicates the voltage that is present after the meter is connected. Whenever a voltmeter is used to make measurements in a circuit, it is good practice to use a meter that has an internal resistance 10 times higher than the resistance of the components in the circuit.

Many meters now include amplifying components. The input resistance in some of these meters is 10 megohms or higher. This level of input resistance makes the meter very useful in high-resistance circuits.

## *OHMMETER*

The ohmmeter will be developed using the same 1 ma full-scale meter movement. An ohmmeter requires the use of an internal power source. The standard source is a 1.5-volt cell. Observe the circuit shown in Figure 6–9. If the terminals of the meters were shorted together, a complete series circuit would exist. The power source is 1.5 volts. The series resistances are 100 ohms meter resistance, the 1,300-ohm resistor, and the 300-ohm potentiometer. The potentiometer would be adjusted to provide a total resistance of 1,500 ohms for the series circuit. Using Ohm's Law, we would then find

$$I = \frac{E}{R} = \frac{1.50}{1,500}$$

The current flow through the circuit would be 1 ma, the full-scale current of the meter movement. The meter movement would rotate to full scale where the scale is marked

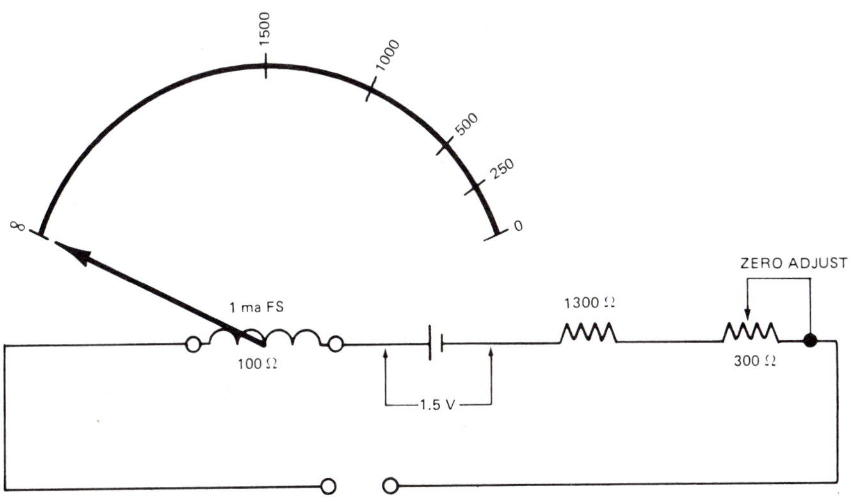


**Figure 6–9** Ohmmeter.

zero (0) ohms. The meter leads are shorted together. The resistance is zero (0) ohms. The meter indicates 0 ohms.

If the meter leads were connected across a 1,500-ohm resistor, the series circuit would contain the 1,500-ohm resistor, the 100-ohm meter movement, the 1,300-ohm resistor, and the potentiometer set at 100 ohms. The total resistance would be 3,000 ohms. According to Ohm's Law,

$$I = \frac{E}{R} = 0.0005 = \frac{1.5}{3{,}000}$$

there would be 0.5 ma flowing through the circuit. The meter movement would move to approximately half scale where the marking is 1,500 ohms.

It is obvious that the value of the resistor connected between the terminals of the meter will determine the total current flow through the series circuit and therefore the amount of current through the meter movement. The current determines how far up scale the meter will rotate. A high resistance will keep the current low and provide little meter movement. A low resistance will provide a higher current and the meter may rotate close to zero (0) ohms. Many ohmmeters provide three or four multipliers for resistance measurement. Common multipliers include $R \times 1$, $R \times 10$, $R \times 100$, $R \times 1{,}000$, and $R \times 10{,}000$.

## *DIGITAL METERS*

Many new meters are becoming available as new digital electronic circuitry is developed. Digital meters are easier to read than the standard meters because the meter reader has only to read the digits. With the older analog meters, the meter reader must interpolate, that is, decide what the meter is indicating. This process takes practice if correct answers are to be obtained.

## REVIEW QUESTIONS

The meter shown in Figure 6–10 is a volt-ohm-ammeter. The switch in the meter provides for the selection of *volts*, *current*, or *resistance*. The position of the switch selects the range of the meter.

The meter is shown with a pointer placed at different positions. Read the statement of the question and indicate what value the meter is indicating.

1. The selector switch is in the 30V position. Pointer B position indicates _______ volts.
2. The selector switch is in the 3A position. Pointer A indicates _______ amperes.
3. The selector switch is in the 3V position. Pointer C indicates _______ volts.
4. The selector switch is in the 300V position. Pointer A indicates _______ volts.
5. The selector switch is in the $R \times 10$ position. Pointer A indicates _______ ohms.
6. The selector switch is in the $R \times 1{,}000$ position. Pointer B indicates _______ ohms.
7. The selector switch is in the 30A position. Pointer B indicates _______ amperes.
8. The selector switch is in the $R \times 100$ position. Pointer C indicates _______ ohms.
9. The selector switch is in the $R \times 1$ position. Pointer B indicates _______ ohms.
10. The selector switch is in the 300 ma position. Pointer A indicates _______ ma.

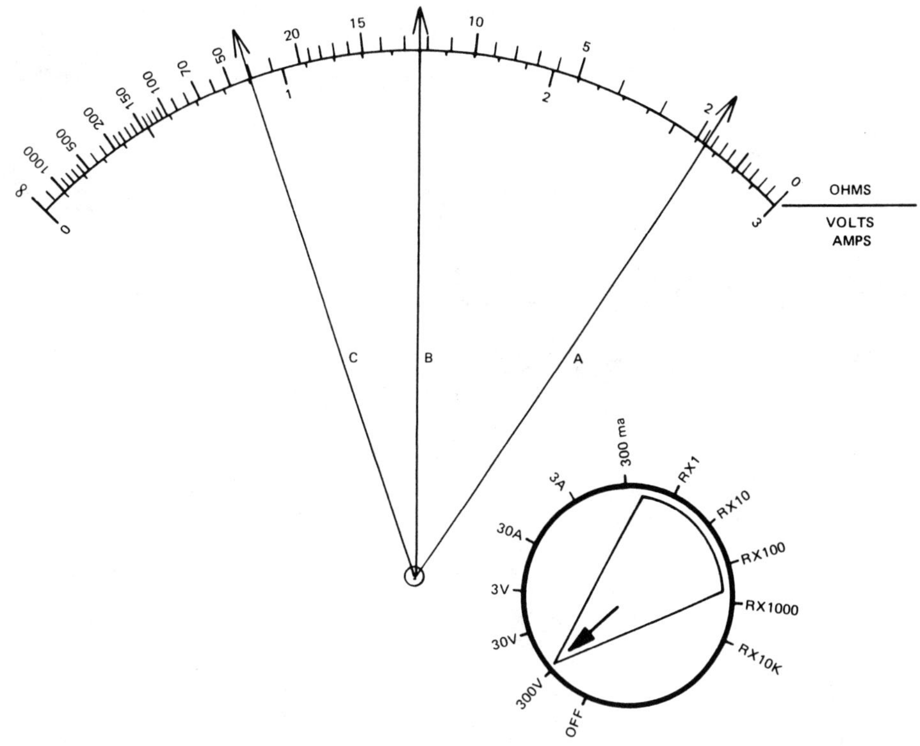


Figure 6–10

11. A 0.1 ma full-scale meter movement is to be used as a 200-volt voltmeter. The resistance of the meter movement is 1,000 ohms. What is the value of the scaling resistor needed in series with the meter movement?
12. The 0.1 ma full-scale meter movement is to be used as an ammeter of 10 ma full scale. The resistance of the meter movement is 1,000 ohms. What is the resistance value of the shunt required?
13. A 100-volt voltmeter with an internal resistance of 100 kilohms is connected as shown in Figure 6–11. What will the meter indicate?
14. If the meter in Figure 6–11 were changed to a 100-volt voltmeter having an input resistance of 10 megohms, what would the meter indicate?

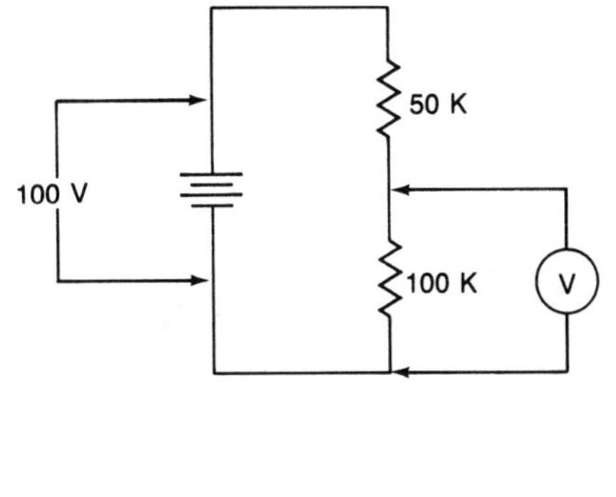


Figure 6–11

# UNIT SEVEN

# *BATTERIES AND ELECTROMOTIVE FORCE*

## *BATTERIES*

The purpose of this unit is not to explain how chemical energy is stored, converted, and then released as electrical energy when needed, but to cover the connection of cells, internal resistance, and the terminal voltage of batteries.

An Italian chemist-physicist, Count Alessandro Volta (1745–1827) invented the battery in 1780. His battery, the "voltaic pile," was the first source of "constant current electricity." For his discovery, the *volt*, a unit of electromotive force, was named in his honor in 1881.

Whenever two different metals are placed in an acid or salts solution in an insulated case, a voltage is produced between the metals by chemical action. This combination, the two metals and the solution, is known as a cell (Figure 7–1). One of the plates will be positive, whereas the other is negative. All cells and batteries provide a source of direct current.

The term *cell* refers to a single unit, the two metals in a solution. The term *battery* refers to cell combinations in series or in parallel that provide for higher current or voltage capabilities. A battery is used for storing and converting chemical

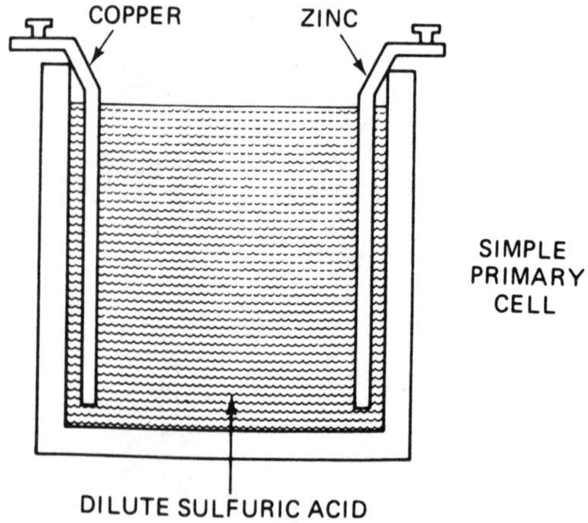


**Figure 7–1**

**Figure 7–2** Primary cells.

energy into electrical energy. Commonly, the term *battery* is now used instead of *cell*; for example, a flashlight *battery* is actually a flashlight *cell*.

There are two main types of cells—primary and secondary. Primary cells are temporary (nonrechargeable) and use up the materials of which they are composed while providing electrical energy. Secondary cells change in chemical composition while providing electrical energy but may be recharged (Figure 7–2). Recharging converts the cell back to its original condition by passing current through it in the opposite direction. A flashlight battery is a good example of a primary cell, and a car battery is a good example of a secondary cell.

Chargers have become available that will recharge primary cells, including some flashlight batteries. Most of the batteries are not returned to "original" condition, however. By design, some batteries are intended to be recharged, whereas others are intended to be discarded when discharged.

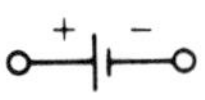


**Figure 7–3** Symbol for cell/battery.

The symbol used for a cell is shown in Figure 7–3. The longer line is the positive terminal of the cell; the shorter line is the negative terminal.

The voltage output of a cell is determined by the materials used to make up the cell. The standard carbon-zinc (flashlight) cell comes in many sizes. The small pen-light cell has an output of 1.5 volts. The common D size flashlight cell also has an output of 1.5 volts. The D size cell can, however, deliver higher current for a longer time than the pen-light cell.

## *INTERNAL RESISTANCE*

Everything through which current flows has resistance. Batteries have an internal resistance. If the open circuit voltage of a cell is measured and current is then supplied (by that cell) to an external circuit, its voltage will decrease. This decrease in voltage is equal to the voltage developed across the internal resistance of the cell. Figure 7–4 illustrates a battery with the internal resistance shown before the output terminals.

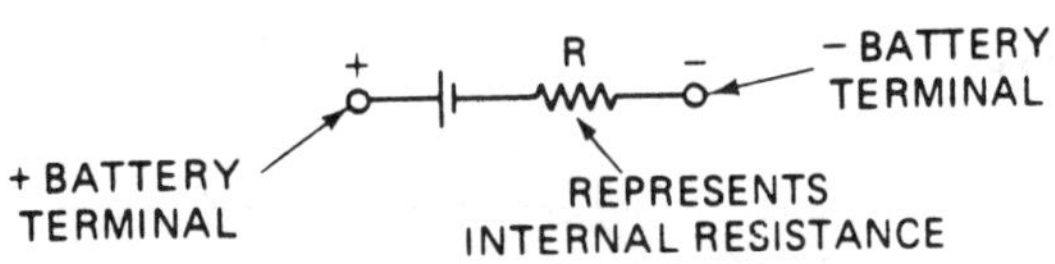


BATTERY OR CELL WITH INTERNAL RESISTANCE

**Figure 7–4**

A cell with a voltmeter in parallel, an ammeter in series, and a load resistor of 10 ohms in series controlled by a switch is shown in Figure 7–5. When the switch is open, the cell voltage is 1.5 volts. When the switch is closed, the ammeter indicates 0.14 ampere of current. The voltmeter indicates 1.4 volts. This means that 0.1 volt

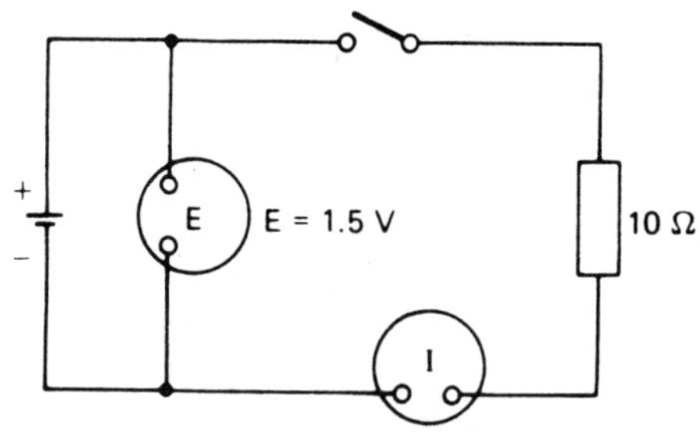


Figure 7–5

is being developed across the internal resistance of the battery (1.5 − 1.4 = 0.1). The internal resistance of the cell may be determined by using Ohm's Law:

$$R = \frac{E}{I}$$

$$R = \frac{0.1}{0.14}$$

$$R = 0.714 \text{ ohm}$$

All batteries do not have the same internal resistance. As a cell becomes discharged, its internal resistance increases. As the internal resistance of the cell increases, its ability to supply current decreases.

### *BATTERIES IN SERIES*

Cells are connected in series whenever a voltage higher than that which can be supplied by a single cell is needed. Two flashlight cells are shown in series in Figure 7–6. The total voltage available is 1.5 plus 1.5, or 3 volts. This is standard for a two-cell flashlight. If three cells were connected in series, as in Figure 7–7, the available voltage would be 4.5 volts (1.5 + 1.5 + 1.5 = 4.5).

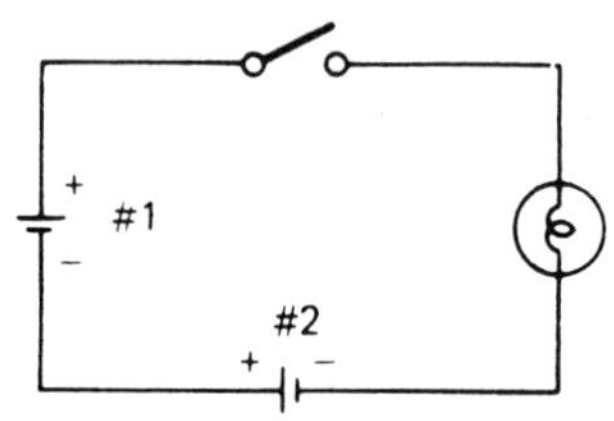


Figure 7–6

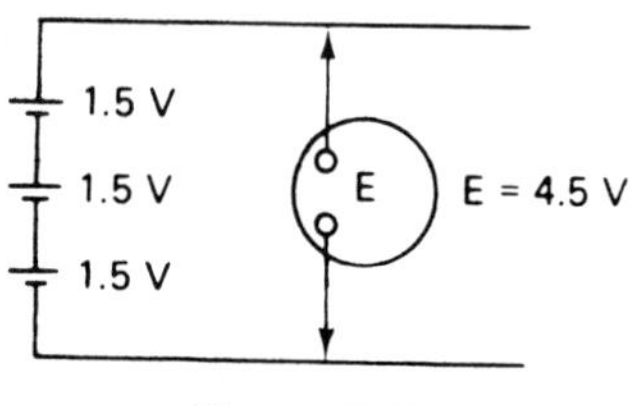


Figure 7–7

### *CELLS IN PARALLEL*

Whenever higher current capability is needed, cells are connected in parallel. Certain precautions must be taken when connecting cells in parallel. Most importantly, the cells must have the same terminal voltage. For example, a 1.2-volt cell should *not* be connected in parallel with a 1.5-volt cell.

A load device drawing a constant current of 3 amperes requires a voltage of at least 1.3 volts (Figure 7–8). Cells of 1.5 volts are available. Each cell has an internal resistance of 0.1 ohm. With one cell connected to the load device, there is 0.3 volt developed across the internal resistance of the cell, leaving 1.2 volts to appear across the load. If two cells are connected in parallel, the situation is improved.

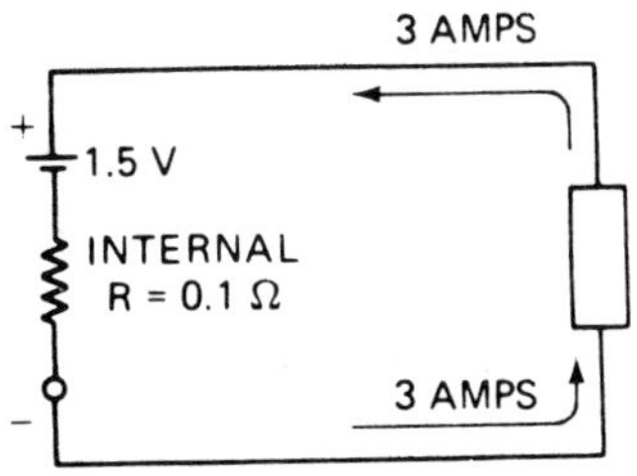


Figure 7–8

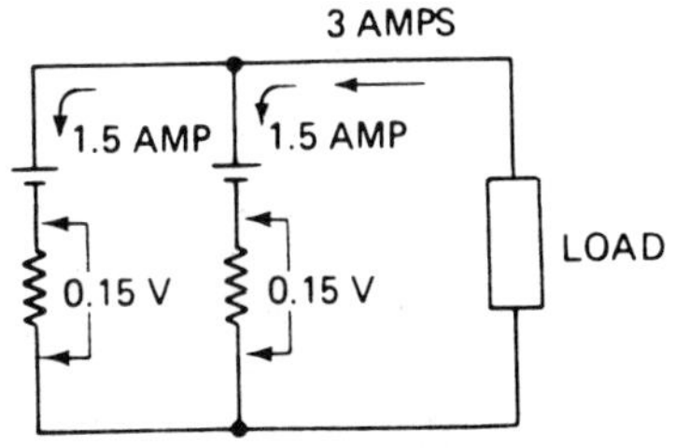


BATTERIES IN PARALLEL CURRENT SPLITS

Figure 7–9

In Figure 7–9, the load device is still drawing 3 amperes. Each cell, however, is supplying only 1.5 amperes. The voltage developed across the internal resistance is:

$$E = I \times R$$
$$E = 1.5 \times 0.1$$
$$E = 0.15 \text{ volt}$$

The voltage at the terminals of the cells and across the load is:

$$E_{out} = 1.5 - 0.15$$
$$E_{out} = 1.35 \text{ volts}$$

This voltage meets the original requirement of at least 1.3 volts across the load.

Whether or not more cells should be connected in parallel depends on how long current is to be drawn. If three cells are connected in parallel, each has to supply only 1 ampere. The voltage developed across the internal resistance is only 0.1 volt. The terminal voltage of each cell and the voltage to the load is 1.4 volts (Figure 7–10). The three-cell combination will last longer than the two-cell combination.

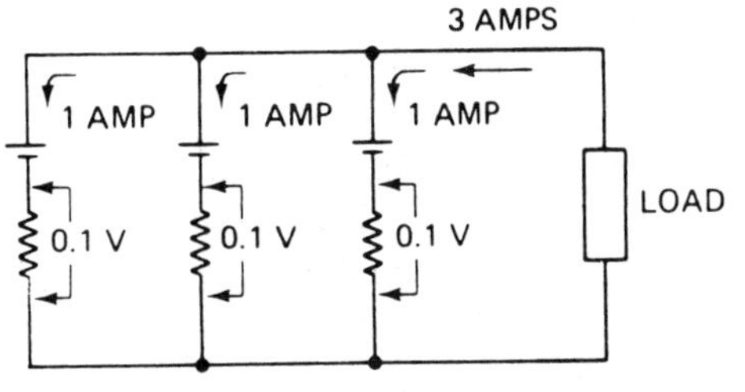


LINE CURRENT SPLITS IN 3 EQUAL PARTS

Figure 7–10

*Internal resistance is present in every device that produces electricity, whether it is a battery, generator, bimetal strip, or a solar cell. The effect of internal resistance is the same on these other devices as on batteries.*

## MAXIMUM POWER TRANSFER

In a later unit, the importance of matching source resistance to load resistance will be emphasized. The facts regarding the matching requirement can be shown using a simple battery with its internal resistance and a number of load resistors.

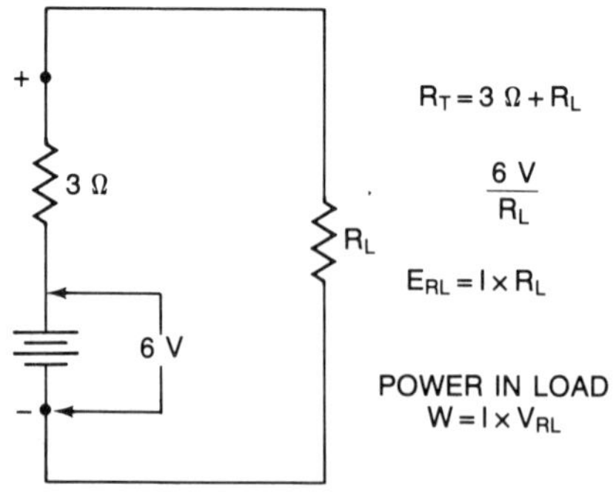


**Figure 7–11**

| Ohms RL | Ohms RT | Amps 6V RT | Volts R Int. | Volts RL I x RL | Power Load I x ERL |
|---|---|---|---|---|---|
| 20 | 23 | 0.26 | 0.78 | 5.22 | 1.36 |
| 15 | 18 | 0.33 | 1.0 | 5.0 | 1.66 |
| 10 | 13 | 0.46 | 1.4 | 4.6 | 2.12 |
| 8 | 11 | 0.55 | 1.6 | 4.4 | 2.4 |
| 6 | 9 | 0.66 | 2.0 | 4.0 | 2.64 |
| 5 | 8 | 0.75 | 2.25 | 3.75 | 2.81 |
| 4 | 7 | 0.86 | 2.58 | 3.42 | 2.95 |
| 3 | 6 | 1.0 | 3.0 | 3.0 | 3.0*** |
| 2.99 | 5.99 | 1.0+ | 3.01 | 2.99 | 2.99 |
| 2 | 5 | 1.2 | 3.6 | 2.4 | 2.88 |
| 1 | 4 | 1.5 | 4.5 | 1.5 | 2.25 |

**Figure 7–12**

In this example (Figure 7–11), a 6-volt battery will be considered. For the sake of simplicity, the battery will be considered to have an internal resistance of 3 ohms. Resistors of from 20 to 1 ohm will be considered as the load on the battery. A chart, Figure 7–12, shows the relation of resistor value, current, load voltage, and power in the load resistor.

When the 20-ohm resistor is used as a load, the power in the load is 1.36 watts. The power in the load increases as the load resistor decreases in value. The maximum is 3 watts (W) with the 3-ohm load. The power in the load immediately starts to decrease as the load resistor is reduced in value below 3 ohms.

The maximum power is developed in the load when the resistance of the load matches the resistance of the source; in this case, it is the 3-ohm internal battery resistance.

## *UNBALANCED AND BALANCED CIRCUITS*

It has been shown that when two batteries are connected in series, the total voltage available is the sum of the two battery voltages. The individual battery voltages are also available (Figure 7–13). This is quite similar to regular household electric power where 240 volts ac may be available between the two hot lines, and 120 volts is available from either hot line to the neutral or ground wire (Figure 7–14).

### Unbalanced Circuits

An example of an unbalanced circuit is shown in Figure 7–15. One resistor of 6 ohms is connected across a 12-volt section of the power source. A current of 2 amperes

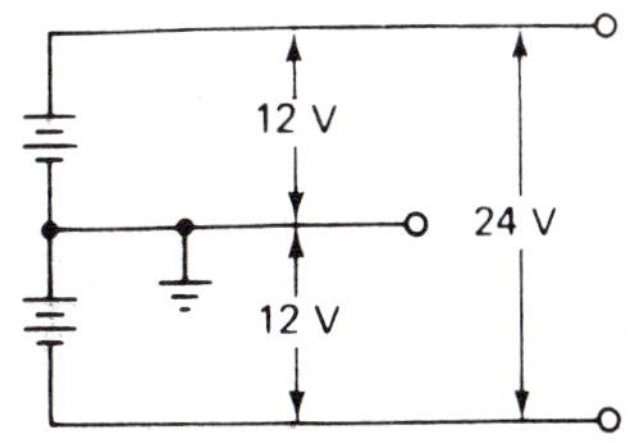


Figure 7–13

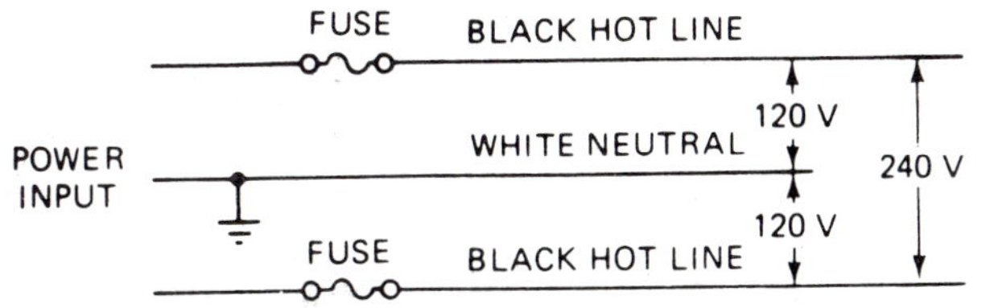


Figure 7–14

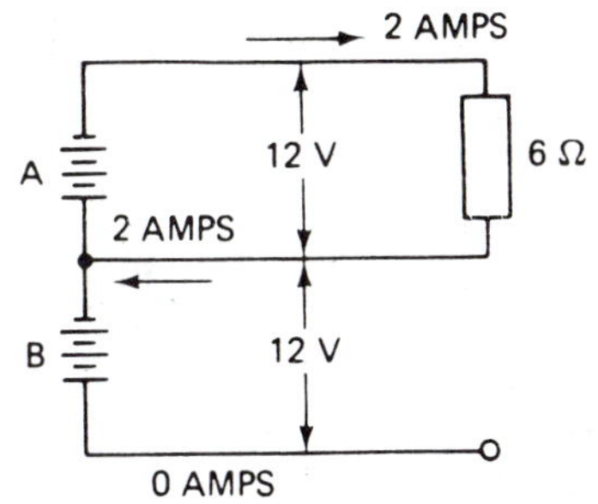


Figure 7–15

will flow in the circuit of battery A and in the 6-ohm resistor when there is no current flow in battery B.

If a 12-ohm resistor were added to the system, the circuit would still be unbalanced, as in Figure 7–16. Note that, at the junction below the 6-ohm resistor, the current splits: 1 ampere returns to battery A while the other 1 ampere flows down through the 12-ohm resistor and through battery B. This 1 ampere joins with the 1 ampere returning on the neutral wire to provide the 2 amperes flowing through battery A.

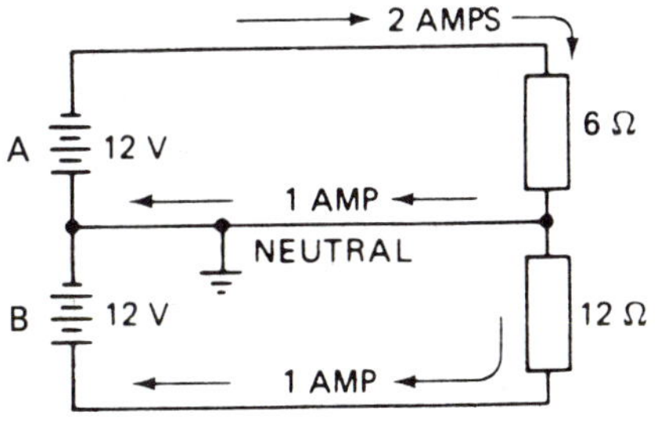


Figure 7–16

### Balanced Circuits

If both resistors are of the same value, there is no current flowing in the neutral wire (Figure 7–17). Although balanced circuits are possible under laboratory conditions, they are almost never obtained in actual practice in household situations. The unbalanced circuit is the most common.

In the circuit in Figure 7–18, the lower voltage as well as the higher voltage components are shown. In each junction, the current entering the junction is equal

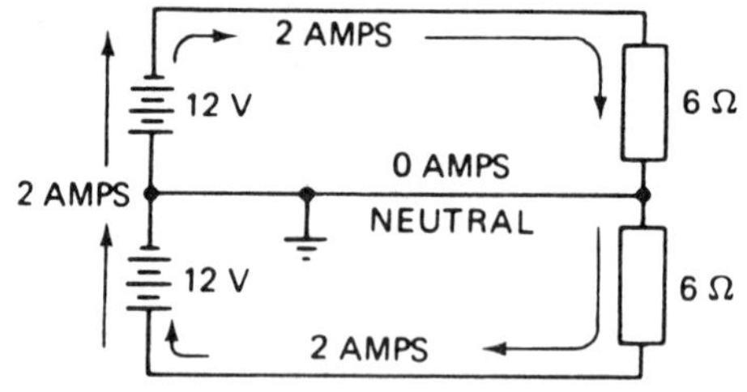


Figure 7–17

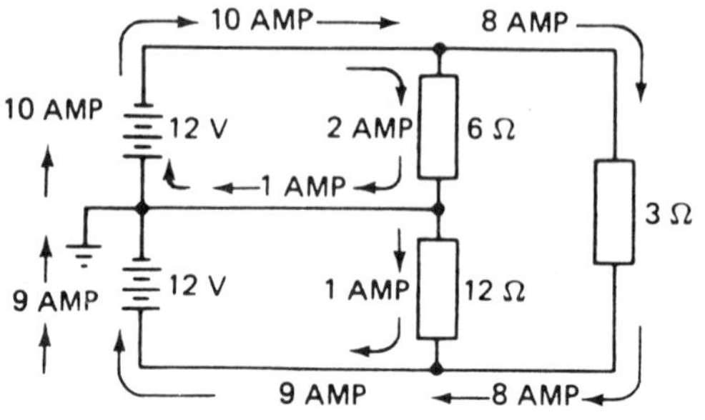


Figure 7–18

to the current leaving the junction. Ohm's Law is true for each component of the system.

The relationship shown in Figure 7–18 parallels that of the 240/120 systems used in homes. The voltage levels in homes are different from those shown with batteries, but circuit relationships are the same. Figure 7–19 shows a portion of a system of electrical distribution in a home.

The electrical system includes a clothes dryer (5,600 W) and a 2 horsepower (HP) water pump (2,486 W) across the 220-volt source. One 110-volt circuit includes a 100-watt lamp, a 75-watt lamp, and a 1,250-watt hair dryer. The other 110-volt circuit includes the toaster at 900 watts and a 60-watt lamp. Current for each component is indicated in Figure 7–19. A current of 49.74 amperes is indicated in the top hot line while a current of 45.53 amperes is shown in the bottom hot line. The unbalanced current of 4.21 amperes is shown in the neutral line.

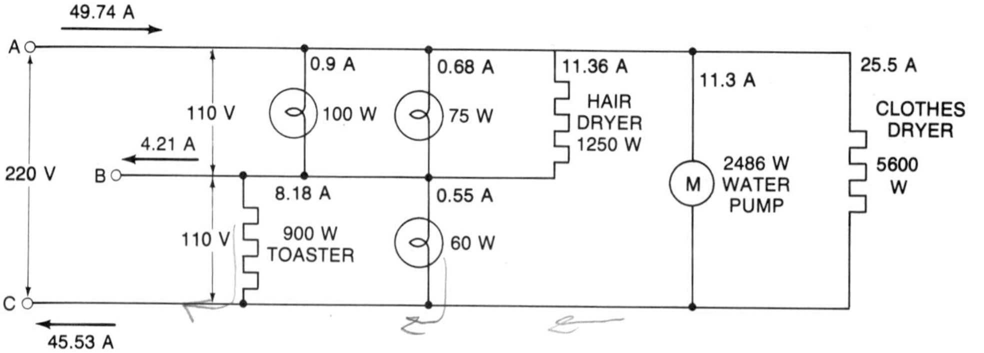


Figure 7–19

## REVIEW QUESTIONS

1. Whenever a higher current capability is needed, batteries are connected in ______ .

2. When a higher voltage is needed, batteries are connected in ______ .

3. Current flow through the internal resistance of a battery causes the output voltage to ______ .
4. A battery has a no-load terminal voltage of 13 volts. When 1 ampere of current is drawn from the battery, the terminal voltage is 12 volts. What is the internal resistance of the battery?
5. As a battery becomes discharged, its internal resistance ______ .
6. A 12-volt battery has an internal resistance of 0.5 ohm. How much current can be drawn from the battery before the terminal voltage decreases to 10 volts?
7. How many 1.5-volt batteries must be connected in series in order to obtain 13.5 volts?
8. In a balanced circuit, ______ current flows in the neutral wire.
9. Is the balanced or unbalanced circuit more common in actual practice?
10. Refer to Figure 7–18. The voltage across the 3-ohm resistor is ______ volts.
11. In the circuit in Figure 7–19, what would the current in the neutral line (B) become if another lamp of 100 watts were connected from line A to line B and a 60-watt lamp were added from line C to line B?
12. Would the current in the neutral line of the circuit of Figure 7–19 change if the clothes dryer were disconnected from the system?

# UNIT EIGHT

# *KIRCHHOFF'S LAWS AND NETWORK THEOREMS*

Before attempting a coverage of the material in this unit, some ground rules must be established.

1. The polarity of voltage across a resistor is taken with electrons entering the negative terminal of the resistor and exiting from the positive terminal.
2. The polarity of currents entering a junction will be considered as negative. The currents exiting the junction will be considered positive.

## *KIRCHHOFF'S VOLTAGE LAW*

Simply stated, Kirchhoff's Voltage Law is: *The sum of the voltages around any complete loop is equal to zero.*

In Figure 8–1a, follow the loop counterclockwise from point A and take as polarity the sign of the voltage first encountered:

$$+\ 12V - 8V - 4V = 0$$

## *KIRCHHOFF'S CURRENT LAW*

Kirchhoff's Current Law simply stated is: *The algebraic sum of the currents entering a junction is equal to zero.*

In Figure 8–1b, for junction A, $-\ 4A + 3A + 1A = 0$. For junction B, $-3A - 1A + 4A = 0$.

An obvious method of stating the formula is: *The sum of the currents entering a junction is equal to the currents leaving the junction*, as seen in Figure 8–2.

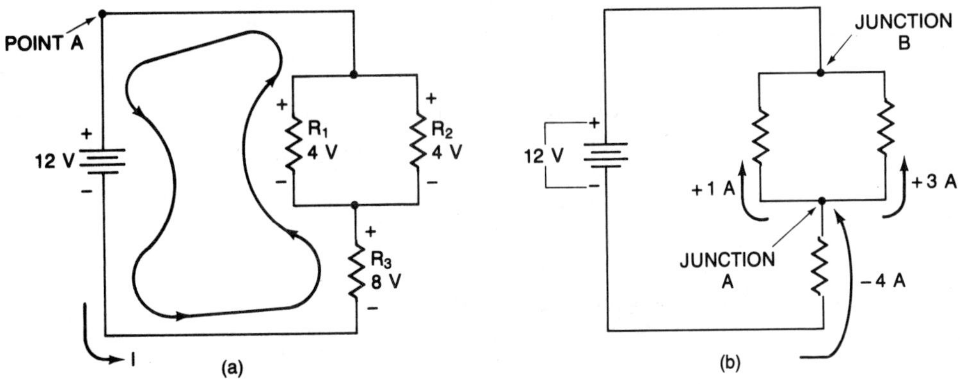


Figure 8–1

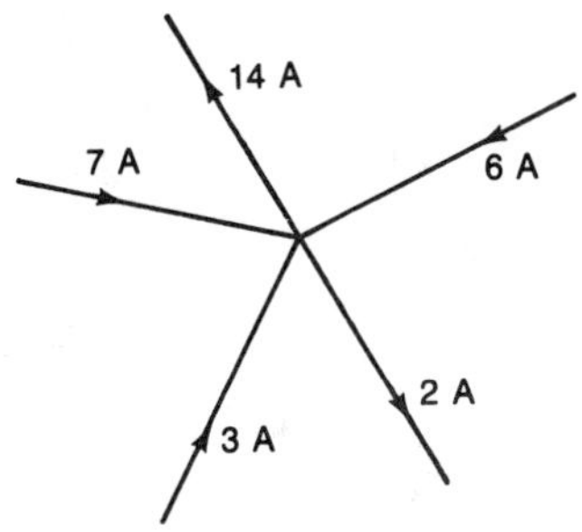


Figure 8–2

There are 7 amps, 3 amps, and 6 amps, for a total of 16 amps, entering the junction (Figure 8–2). There are 14 amps and 2 amps, for a total of 16 amps, leaving the junction.

Kirchhoff's Voltage and Current laws provide a means for solving problems that include more than one voltage source. As an example, consider Figure 8–3. Here is a circuit containing two voltage sources. Kirchhoff's laws provide for the development of formulas leading to the solution of the problem.

$V_{R1} = 6I_1$ Ohm's Law

$V_{R2} = 8I_2$ Ohm's Law

$V_{R3} = 4I_3$ Ohm's Law

Loop Equation A: $+19 - 6I_1 - 4I_3 = 0$ Sum of Voltages

Loop Equation B: $+24 - 8I_2 - 4I_3 = 0$ Sum of Voltages

Current $I_1 + I_2 = I_3$ Current Law

$I_2 = I_3 - I_1$

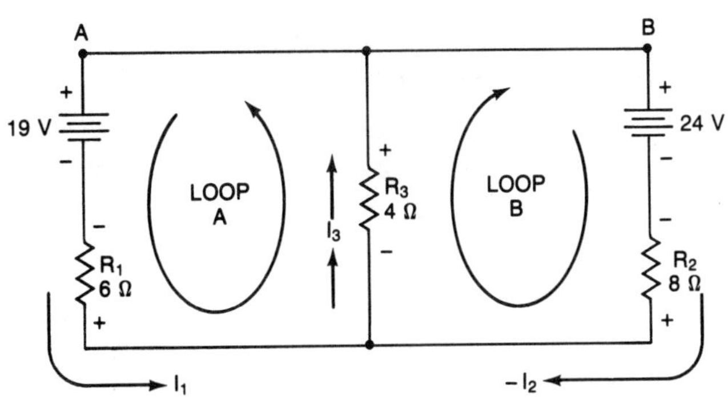


Figure 8–3

Equality for $8I_2$ in equation B

$$-8I_2 = -8I_3 + 8I_1$$

Substituting for $8I_2$ in equation B

$$24 - 8I_3 + 8I_1 - 4I_3 = 0$$

Combining and rearranging equation B

$$+8I_1 - 12I_3 = -24$$

Rearranging equation A

$$-6I_1 - 4I_3 = -19$$

Multiplying equation A by $-3$ and adding to equation B

$$\begin{array}{r} +8I_1 - 12I_3 = -24 \\ +18I_1 + 12I_3 = +57 \\ \hline +26I_1 \qquad\quad = +33 \end{array}$$

$$I_1 = 1.27 \text{ A}$$

| | | |
|---|---|---|
| $E_{R1} =$ | 7.62V | Ohm's Law |
| $E_{R2} =$ | 11.38V | Sum of V loop A |
| $E_{R3} =$ | 12.62V | Sum of V loop B |
| $I_{R3} =$ | 2.845A | Ohm's Law |
| $I_{R2} =$ | 1.58A | Ohm's Law |
| $I_1 + I_2 =$ | $I_3$ | Sum of currents for check |
| 1.27A + 1.58A = | 2.854A | Within limits |

A number of theorems have been developed that assist in the solving of multivoltage source problems. A knowledge of their use will be beneficial to all engaged in electrical work.

## SUPERPOSITION THEOREM

The *superposition theorem* provides a means of solving multivoltage-source problems.

Simply stated: *The voltage across any resistor in a multivoltage-source system is the algebraic sum of the individual voltages resulting from each source considered separately while all other sources are replaced by a short circuit.* (Actually each source should be replaced by the internal resistance of the source, but in most cases, this resistance value is insignificant.) A sample problem is shown as the circuit in Figure 8–4. First consider the 12-volt source with the other two sources shorted.

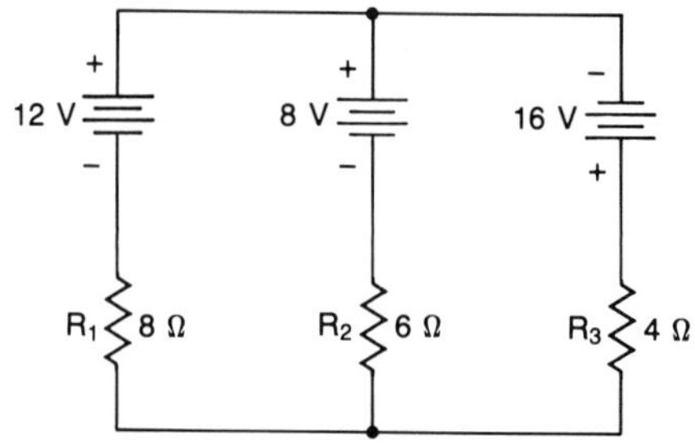


**Figure 8–4**

The resistors $R_2$ and $R_3$ are combined in parallel and then the problem is solved for the voltage across each resistor. The voltages are shown in Figure 8–5 and also in Figure 8–8 where the voltages from the other two voltage sources are also shown.

The final total resistance is 10.4 ohms (Figure 8–5c). The battery current is 12 volts/10.4, or 1.15 amps. The voltage across $R_1$ is 9.23 volts. This leaves 2.77 volts to appear across $R_2$ and $R_3$, as shown in Figure 8–5.

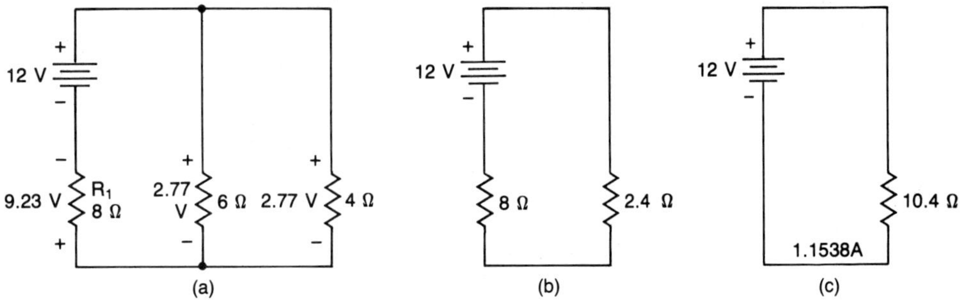


**Figure 8–5**

For the 8-volt source:

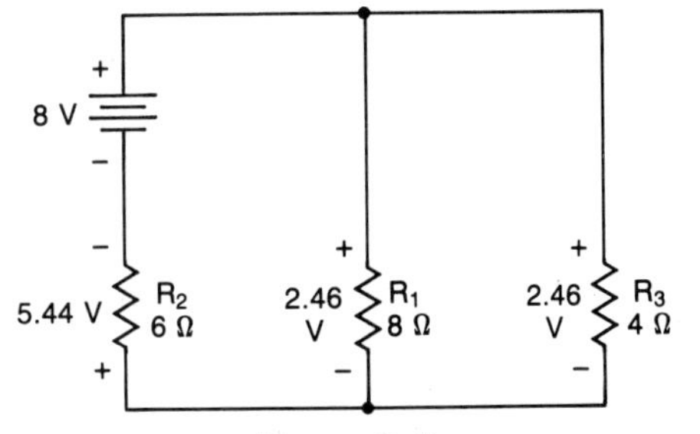


**Figure 8–6**

For the 16-volt source:

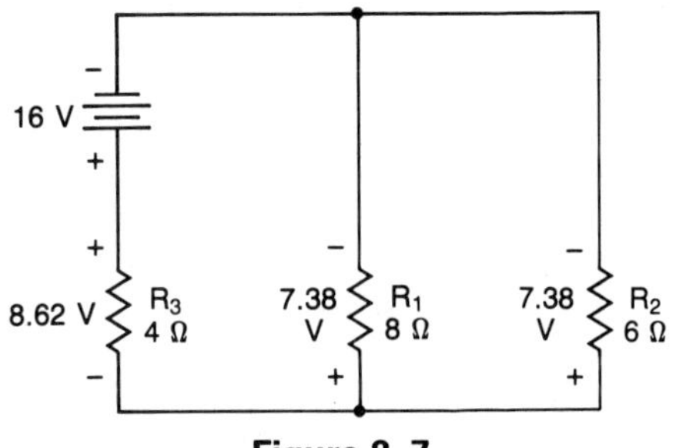


**Figure 8–7**

Combining the voltages on the original circuit:

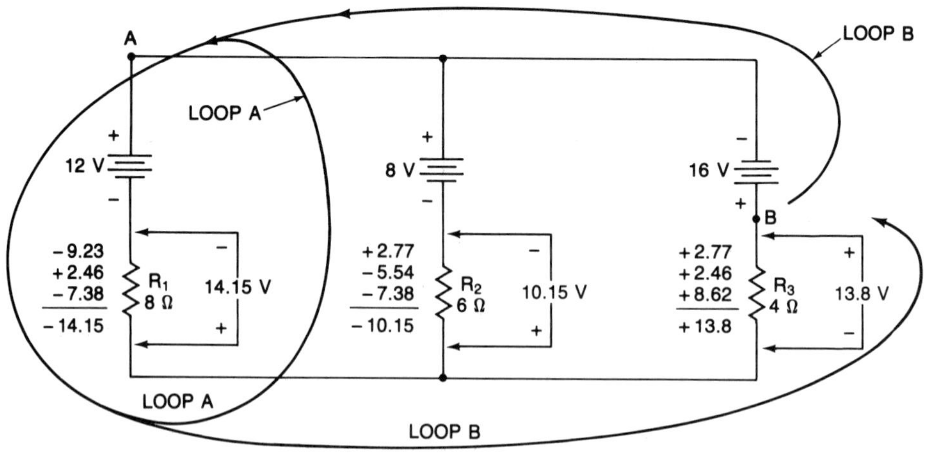


**Figure 8–8**

The problem may be checked by the sum of the voltages around a complete loop.

Loop A $+12 - 14.15 + 10.15 - 8 = 0$
Loop B $+16 + 12 - 14.15 - 13.8 = 0.05$ (zero)

The 0.05 error is due to rounding off numerals beyond the second decimal.

## *THEVENIN'S THEOREM*

This theorem states that *any two terminals in a resistive circuit may be replaced by a single voltage source and a single resistor in series with the source.*

The following are two required conditions:

1. $V_{th}$ is the value of the voltage at the open circuit terminals.
2. $R_{th}$ is the apparent resistance between the two terminals with all voltage sources shorted.

Again, internal battery resistance should be considered, but it is normally so small compared to other resistors that it is ignored.

In the circuit in Figure 8–9, the value of $V_{th}$ and $R_{th}$ for points A and B may be determined in the following manner.

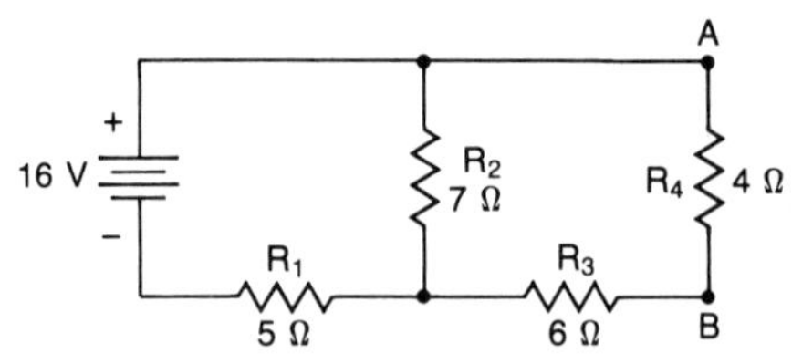


**Figure 8–9**

Open circuit points A and B and solve for $V_{th}$, the voltage across the open terminals.

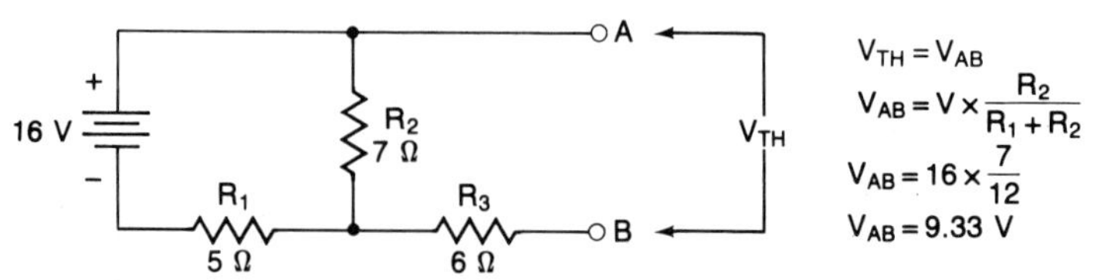


**Figure 8–10**

Solve for $R_{th}$ with the voltage sources shorted.

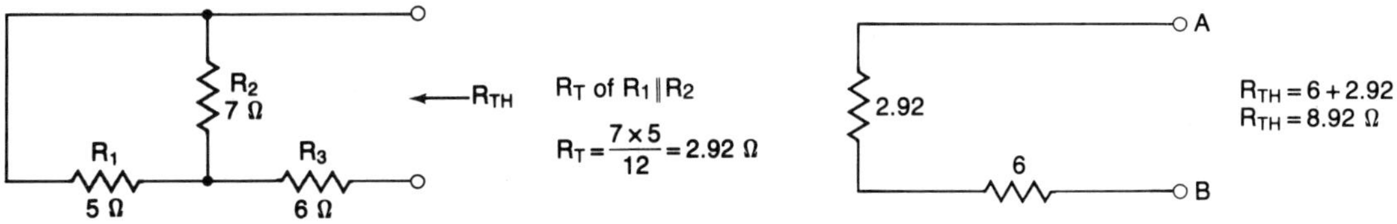


Figure 8–11

The voltage across $R_4$ of the original circuit is:

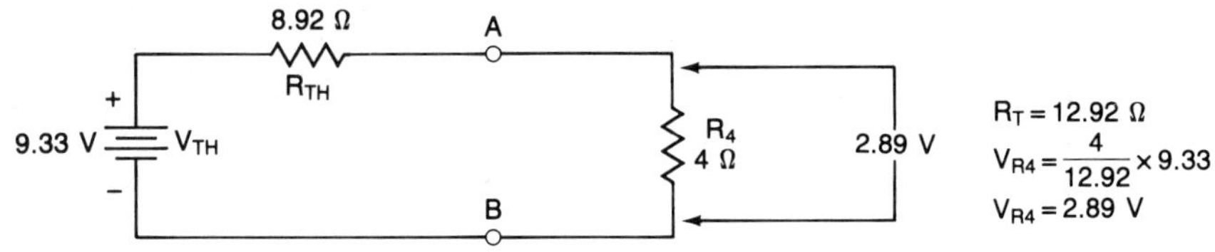


Figure 8–12

### NORTON'S THEOREM

This theorem can be stated thus: *Any two terminals in a resistive network may be replaced by a single current source ($I_N$) and a single resistor ($R_N$) in parallel with the source.*

The following are two required conditions:

1. $I_N$ is the value of current flowing through the short-circuited two terminals.
2. $R_N$ is the resistance looking into the open-circuited terminals with all voltage sources shorted. ($R_N$ is the same value as $R_{th}$ as calculated in Thevenin's problems.)

In the circuit of Figure 8–13, the value of $I_N$ and $R_N$ may be determined in the following manner.

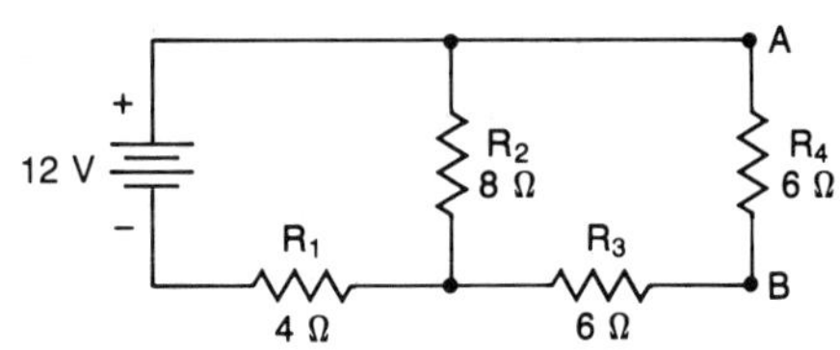


Figure 8–13

Place a short between points A and B and calculate the current through the short, as in Figure 8–14.

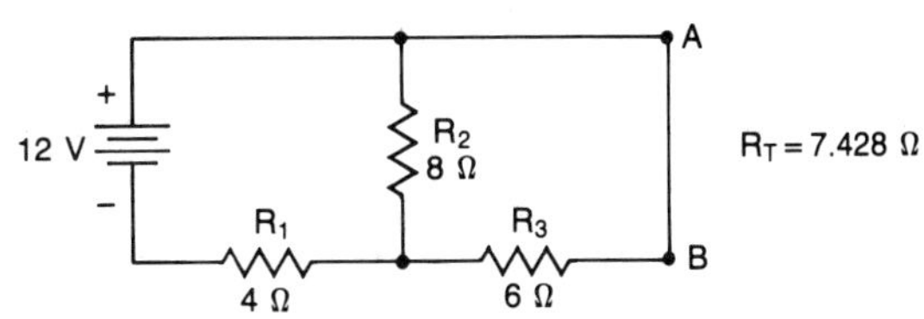


Figure 8–14

Solve the parallel combination of $R_2$ and $R_3$.

$$R_{eq} = \frac{R_2 \times R_3}{R_2 + R_3} = \frac{6 \times 8}{14} = 3.428$$

$$R_{eq} = 3.428$$

$$R_T = R_1 + R_{eq} = 4 + 3.428 = 7.428\Omega$$

$$I_T = \frac{V}{R} = \frac{12V}{7.428} = 1.615\ A$$

$$V_{R3} = I \times R_{eq} = 1.615\ A \times 3.428\Omega$$

$$V_{R3} = 5.538\ V$$

$$I_N = \frac{VR_3}{R_3} = \frac{5.538}{6} = 0.923A$$

For $R_N$, look in the open-circuited terminals and determine resistance with the voltage sources shorted.

Combine $R_1$ and $R_2$

$$R_{eq} = \frac{R_1 \times R_2}{R_1 + R_2} = \frac{32}{12} = 2.66 \quad \text{or} \quad R_{eq} = \frac{1}{\frac{1}{R_1} + \frac{1}{R_2}}$$

$$R_{eq} = 2.66\Omega$$

$$R_N = R_{eq} + R_3 = 2.666\Omega + 6\Omega$$

$$R_N = 8.666\Omega$$

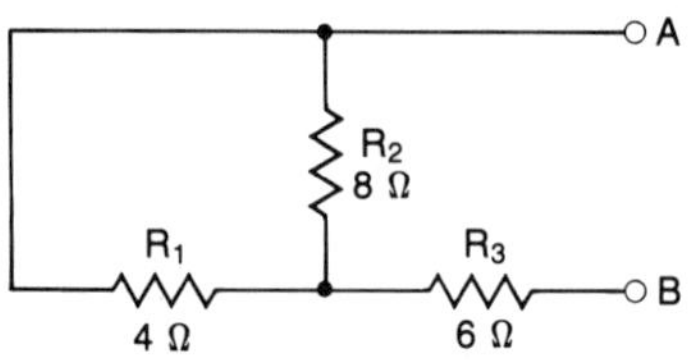


Figure 8–15

Solve for the voltage and current of $R_4$ in the original circuit using a current generator ($I_N$) and the Norton resistance ($R_N$) across the generator.

$$R_{eq} = \frac{8.666 \times 6}{8.666 + 6} = \frac{51.996}{14.666} = 3.545\Omega$$

$$V_{R4} = I \times R_{eq} = 0.923 \times 3.545 = 3.272V$$

$$V_{R4} = 3.272V$$

$$I_{R4} = \frac{3.272V}{6\Omega} = 0.545\ \text{Amp}$$

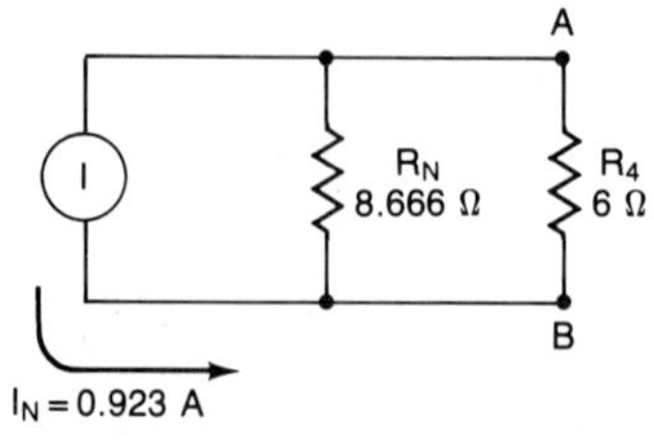


Figure 8–16

## *MILLMAN'S THEOREM*

This theorem is stated thus: *Constant current sources and their associated resistances connected in parallel may be combined to provide a single current source and a parallel resistance where*:

1. $I_M$ is the algebraic sum of the original currents.
2. $R_M$ is the parallel equivalent resistance of the original resistors.

Millman's Theorem relates directly to current sources. In practice, the theorem has great application to voltage sources connected in parallel. Each voltage source must be converted to its Norton equivalent current before applying Millman's Theorem.

---

### EXAMPLE 1

Find the voltage across each resistor in Figure 8–17.

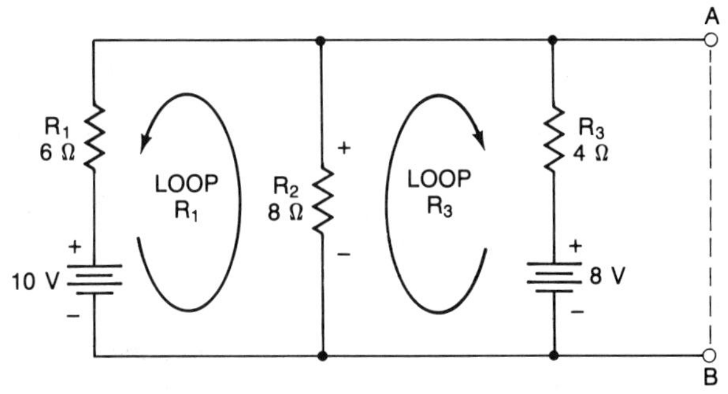


**Figure 8–17**

***Solution:*** To find the Norton or Millman current, place a short from A to B and solve for the individual currents. The sum of the currents is the Norton current, or in this case the Millman current ($I_M$).

$$I_M = \frac{10V}{6\Omega} + \frac{0V}{8\Omega} + \frac{8V}{4\Omega} = 3.667 \text{ Amp}$$

$$I_M = 3.667 \text{ Amp}$$

To find the equivalent resistance ($R_M$), combine the resistances in parallel.

$$R_M = \frac{1}{\frac{1}{R_1} + \frac{1}{R_2} + \frac{1}{R_3}} = \frac{1}{\frac{1}{6} + \frac{1}{8} + \frac{1}{4}}$$

$$R_M = 1.846$$

The voltage A to B is then:

$$V_{AB} = I_M \times R_M = 3.667 \text{ A} \times 1.846\ \Omega$$

$$V_{AB} = 6.769V$$

Kirchhoff's Voltage Law may now be used to find the voltage across each resistor.

Loop of $R_1$ (starting at battery + terminal).

$$+10V - V_{R2} - V_{R1} = 0$$

$$+10V - 6.767V = V_{R1}$$

$$V_{R1} = +3.233V$$

Loop of $R_3$ starting at the battery + terminal.

$$+8V - 6V - V_{R3} = 0$$

$$+8V - 6.767V = V_{R3}$$

$$V_{R3} = +1.233V$$

The operations used to find the voltage A to B could have been combined into the formula:

$$V_{AB} = \frac{\frac{V_1}{R_1} + \frac{V_2}{R_2} + \frac{V_3}{R_3}}{\frac{1}{R_1} + \frac{1}{R_2} + \frac{1}{R_3}}$$

Substituting in the formula

$$V_{AB} = \frac{\frac{10}{6} + \frac{0}{8} + \frac{8}{4}}{\frac{1}{6} + \frac{1}{8} + \frac{1}{4}} = 6.769V$$

---

Millman's Theorem has proven to be very useful in solving for the input voltage to summing amplifiers. (Summing amplifiers will be covered in a later unit.)

In Figure 8–18, a summing amplifier is shown with inputs to both the inverting (−) and noninverting (+) terminals. If the output voltage of the amplifier is to be determined, the voltage at the (+) terminal must first be calculated.

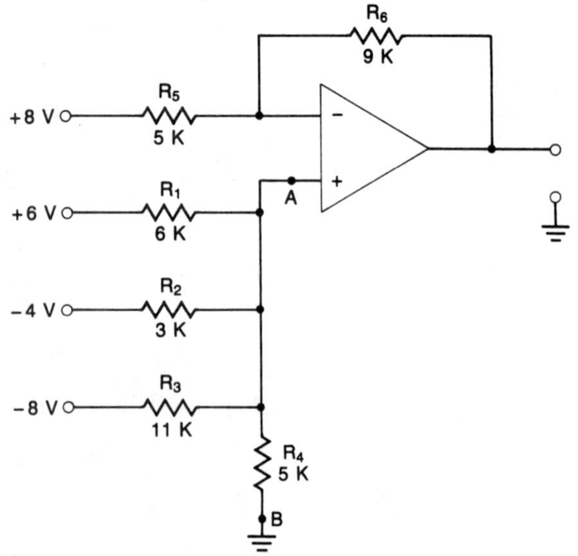


Figure 8–18

Find the Millman current for the (+) terminal to ground (A to B).

$$I_M = \frac{6V}{6k} + \frac{-4V}{3k} + \frac{-8V}{11k} + \frac{0V}{5k} = -1.06 \text{ ma}$$

Find the equivalent resistance:

$$R = \frac{1}{\frac{1}{6k} + \frac{1}{3k} + \frac{1}{11k} + \frac{1}{5k}} = 1.264k$$

Find the voltage at the noninverting input:

$$V = I \times R = 1.06 \text{ ma} \times 1.264k$$
$$V = -1.34V$$

By formula

$$V_{AB} = \frac{\frac{V_1}{R_1} + \frac{V_2}{R_2} + \frac{V_3}{R_3}}{\frac{1}{R_1} + \frac{1}{R_2} + \frac{1}{R_3} + \frac{1}{R_4}}$$

$$V_{AB} = \frac{\frac{6V}{6k} + \frac{-4V}{3k} + \frac{-8V}{11k}}{\frac{1}{6k} + \frac{1}{3k} + \frac{1}{11k} + \frac{1}{5k}}$$

$$V_{AB} = -1.34V$$

## REVIEW QUESTIONS

1. With reference to Figure 8–19, develop a formula for $I_3$ as related to $I_1$ and $I_2$.

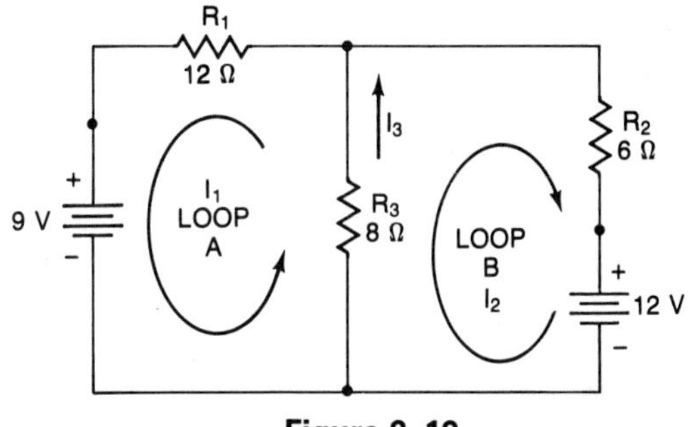


**Figure 8–19**

2. With reference to Figure 8–19, develop a formula for the voltages around loop A starting at the plus terminal of the 9-volt battery.
3. With reference to Figure 8–19, develop a formula for the voltages around loop B starting at the plus terminal of the battery.
4. Using the formulas developed in questions 1, 2, and 3 and Kirchhoff's Voltage Law, determine the voltage across $R_1$, $R_2$, and $R_3$ in Figure 8–19.

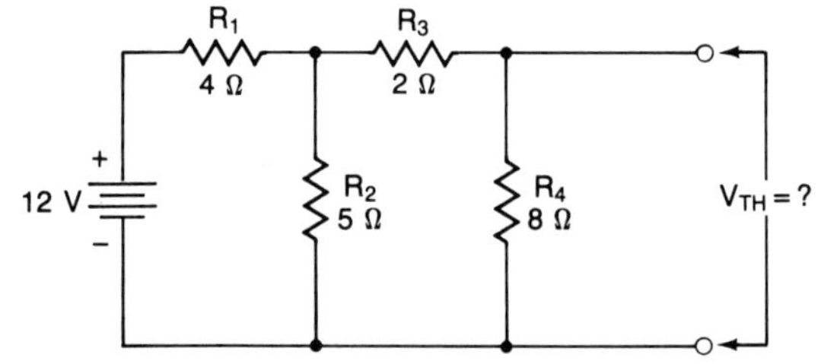


**Figure 8–20**

5. With reference to Figure 8–20, determine Thevenin's voltage. Consider $R_4$ as the load resistor.
6. With reference to Figure 8–20, determine Thevenin's resistance.
7. With reference to Figure 8–20, determine the voltage across $R_4$ if its value were as given:

$$R_4 = 8\Omega\ V_{R4}$$

$$R_4 = 12\Omega\ V_{R4}$$

$$R_4 = 16\Omega\ V_{R4}$$

8. With reference to Figure 8–20, what is the Norton current generator value?
9. Using the Norton current generator and the Norton resistance, determine the voltage across $R_4$ if its value were as given.

$$R_4 = 7\Omega\ V_{R4}$$

$$R_4 = 9\Omega\ V_{R4}$$

$$R_4 = 11\Omega\ V_{R4}$$

10. In Figure 8–21, use Millman's Theorem and determine the voltage from point A to ground.

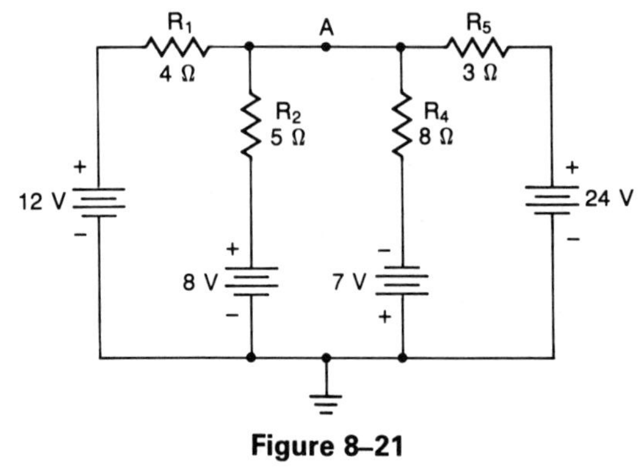


**Figure 8–21**

# UNIT NINE

# *CAPACITANCE AND INDUCTANCE IN DC*

## *CAPACITANCE*

*Capacitance* is the property of any electric circuit that opposes a change in voltage. Electrons must move before the voltage across a capacitor will change. It takes time for electrons to move; therefore, it takes time for the voltage to change. The first hint of capacitive action was given in Unit 1. A generator was shown pulling electrons from a wire connected to the positive terminal of the generator and pushing them out on the wire connected to the negative terminal. A similar situation is shown in Figure 9–1; the generator is running and wires are connected to the positive and negative terminals. In order for a voltage to appear between positive and negative wires, electrons have to be pulled from the positive side and forced out on the negative side.

If a circuit constructed with a sensitive meter were connected as shown in Figure 9–2, the meter would indicate electron movement or current flow during the charging of the wires. The meter needle would move up as the switch is closed and immediately drop back to zero as the wires became charged. The number of free electrons that move would control the meter needle movement.

The number of electrons that have to move in order to ''charge'' the wires could be increased by replacing the external wires with large, flat plates. Increasing the surface area enlarges the ''charge'' capability.

Metal plates with a large surface area separated by some high resistance insulating material make up a capacitor. In Figure 9–3, air is the insulating material.

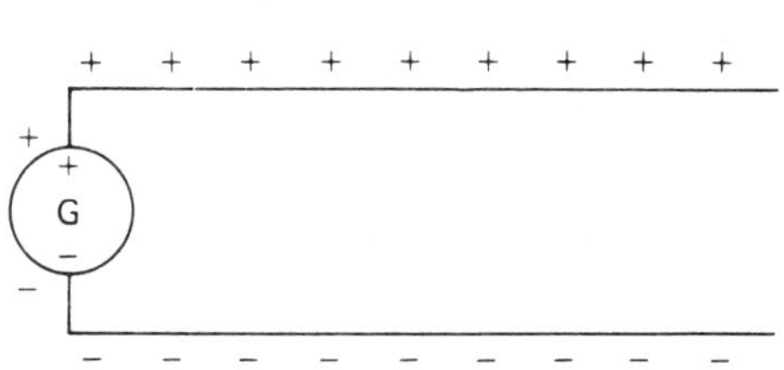


**Figure 9–1**

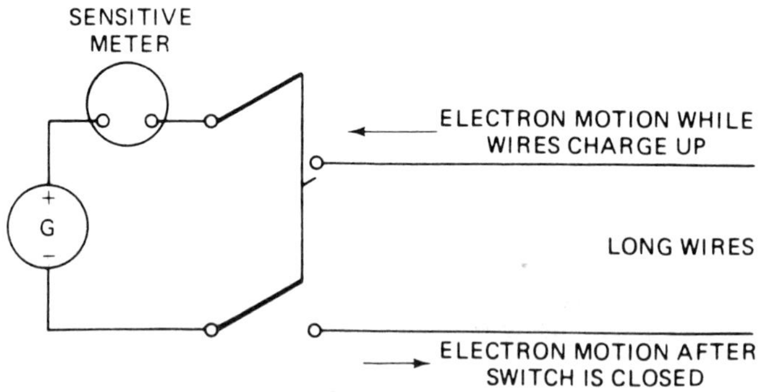


**Figure 9–2**

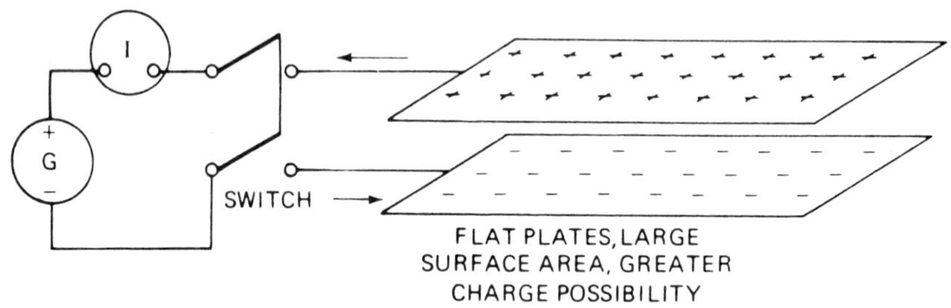


**Figure 9–3**

### *CURRENT FLOW IN CAPACITIVE CIRCUITS (DC)*

It was indicated in the explanation for Figure 9–2 that, after the switch is closed, there is a momentary movement of electrons. With the dc generator connected to the wires, some of the free electrons move from the top wire through the generator and out on the bottom wire. After the wires become "charged," voltage appears between all points on the top and bottom wires. Electron movement then stops.

A similar situation exists if a capacitor is connected across a dc generator or any other dc power source. As the capacitor is connected across the generator, the current jumps up. As the capacitor charges, the current flow drops to zero and the source voltage appears across the capacitor terminals. A graph of voltage and current as related to time is shown in Figure 9–4.

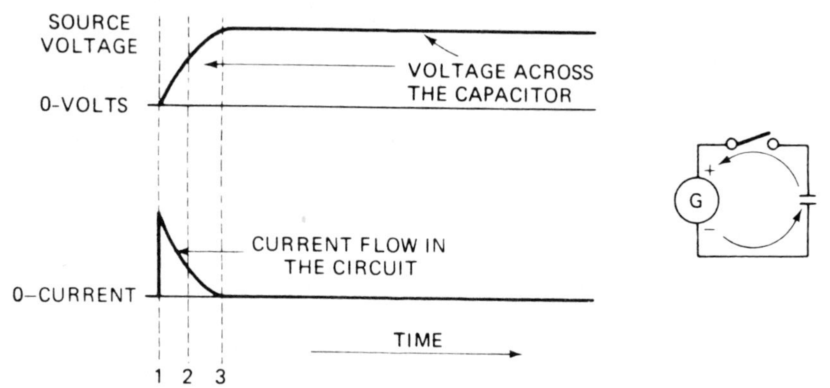


**Figure 9–4**

### Current

1. As the switch is closed at time zero, the current rises immediately to maximum.
2. The current decreases as the capacitor charges.
3. The capacitor is fully charged; the current drops off to 0 amperes.

### Voltage

1. As the switch is closed, the voltage across the capacitor rises rapidly from 0 volts.
2. The voltage rise starts to taper off as the voltage across the capacitor approaches the source (generator) voltage.
3. The voltage across the capacitor reaches the source voltage; the capacitor is fully charged. The voltage across the capacitor remains constant at the source voltage level.

The number of electrons that flow into and charge a capacitor is related to the size of the capacitor in *farads* (F) and the applied voltage. The charge (Q) in *coulombs* is equal to the *capacitance* (C) in farads times the voltage (V).

$$Q = C \times V$$

---

**EXAMPLE 1**

Determine the charge on a 10-microfarad (MF) capacitor that has 30 volts applied to it.

***Solution***

$$Q = C \times V$$

$$Q = 10 \times 10^{-6} \times 30$$

$$Q = 300 \times 10^{-6}$$

How many electrons are stored in the capacitor?

$$e = 3 \times 10^{-4} \times 6.25 \times 10^{18}$$

$$e = 1.875 \times 10^{15}$$

There are $1.875 \times 10^{15}$ electrons stored on the negative plate of the capacitor. There is, of course, an absence of the same number of electrons from the positive plate.

---

This simple formula, Q = CV, will prove helpful in the investigation and understanding of integrated amplifier circuits covered in later units.

## *RC TIME CONSTANT*

In electric circuits containing resistance and capacitance there is a relationship called the *time constant of the circuit* (TC). Any dc resistance-capacitance (RC) circuit will exhibit characteristics similar to those shown in Figure 9–4. The time constant, R × C, provides an indicator of how fast the capacitor will charge.

## EXAMPLE 2

In Figure 9–5, a 1,000,000-ohm (1MΩ) resistor is connected in series with a 2-microfarad (MF) capacitor. What is the time constant of the circuit?

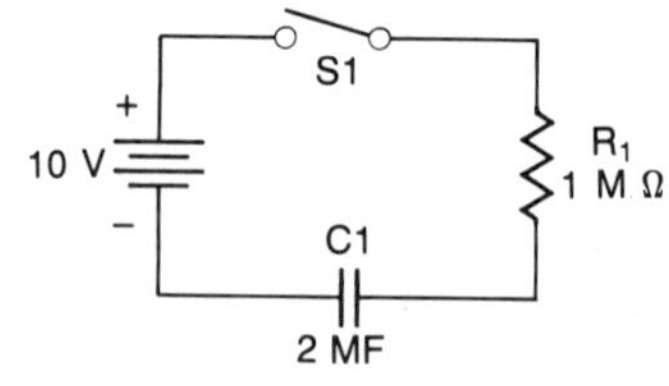


**Figure 9–5**

***Solution***

$$T = RC$$

$$T = 1 \times 10^6 \times 2 \times 10^{-6}$$

$$T = 2 \text{ sec}$$

The capacitor in Figure 9–5 will become fully charged in 5 time constants, or in this case 10 seconds. As $S_1$ is closed, the rate at which the capacitor charges changes. The capacitor charge is shown as a curve in Figure 9–6. This is known as a *universal time-constant curve*.

The capacitor charges 63.21 percent of the voltage left to charge in each time constant. In Figure 9–6, with 10 volts applied, the capacitor will charge to 6.321 volts in the first time constant (2 seconds). This leaves 3.679 volts left to charge. The capacitor will charge 2.325 volts in the next time constant, bringing the voltage across the capacitor to 8.646 volts. Continuing:

BY TIME CONSTANT

| | |
|---|---|
| $0.632 \times 10V = 6.32V$ | charge after the 1st TC |
| $10V - 6.32V = 3.679V$ | volts left to charge |
| $0.632 \times 3.679V = 2.325V$ | change in voltage |
| $6.32V + 2.325V = 8.646V$ | charge after the 2nd TC |
| $10V - 8.646V = 1.354V$ | volts left to charge |
| $0.632 \times 1.354V = 0.8558V$ | change in voltage |
| $8.646V + 0.8558V = 9.50V$ | charge after the 3rd TC |
| $10V - 9.5V = 0.5V$ | volts left to charge |
| $0.632 \times 0.5V = 0.316V$ | change in voltage |
| $9.5V + 0.316V = 9.816V$ | charge after the 4th TC |
| $10V - 9.816V = 0.814V$ | volts left to charge |
| $0.631 \times 0.814V = 0.116V$ | change in voltage |
| $9.816V + 0.116V = 9.932V$ | charge after the 5th RC |

This process could go on forever with 36.79 percent of the voltage always being available to charge in the next time constant.

For practical reasons, the capacitor is said to be fully charged at the end of 5 time constants where the capacitor will be charged to 99.3 percent of the available voltage. Normally, an approximation of the voltage across the capacitor in an RC circuit is all that is necessary. The universal time constant curve can be used to make

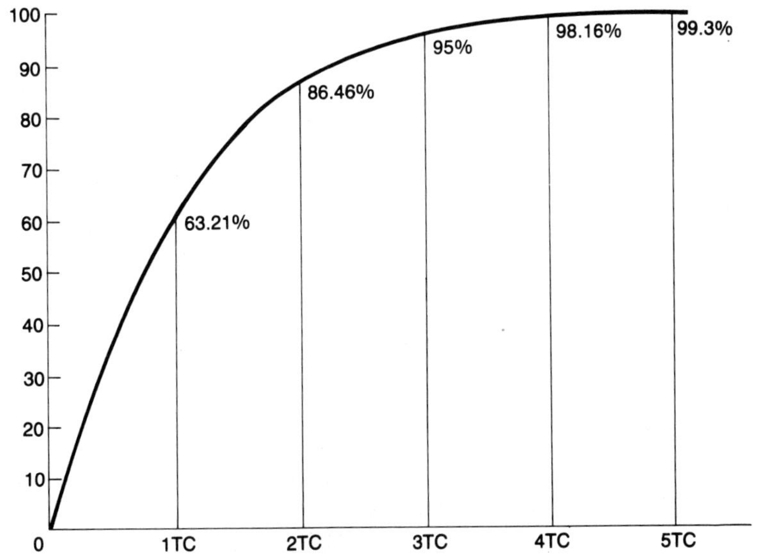


**Figure 9–6**

this determination. The voltage across the resistor equals the supply voltage minus the voltage across the capacitor.

In special cases where the exact voltage across the capacitor is needed, use the formula:

$$V_C = V_S\,(1 - e^{-\frac{t}{TC}})$$

where $V_C$ = voltage across the capacitor
$V_S$ = supply voltage
e = natural logarithm base 2.718
t = time in seconds the capacitor charges
TC = RC time constant

As an example, consider the circuit of Figure 9–5. What is the voltage across the capacitor 1.8 seconds after $S_1$ is closed?

By formula:

$$V_C = V_S\,(1 - e^{-\frac{t}{RC}})$$

$$V_C = 10\,(1 - 2.718^{-\frac{1.8}{2}})$$

$$V_C = 10\,(1 - 2.718^{-0.9})$$

$$V_C = 10\,(1 - 0.407)$$

$$V_C = 10 \times 0.593$$

$$V_C = 5.93\text{V}$$

By formula using a scientific calculator:

$$V_C = V_S\,(1 - e^{-\frac{t}{RC}})$$

| STEP | ENTER | PRESS | DISPLAY |
|---|---|---|---|
| 1 | 2.71828 | $y^x$ | 2.71828 |
| 2 | 0.9 | +/− | −0.9 |
| 3 | | = | 0.40657 |
| 4 | | +/− | −0.40657 |
| 5 | 1 | + | 0.59343 |
| 6 | | × | 0.59343 |
| 7 | 10 | = | 5.9343 |

The voltage across the capacitor will be 5.9343 volts, 1.8 seconds after switch $S_1$ is closed.

One point worth noting concerns the relationship of current to voltage in the capacitor. As shown in Figure 9–4 at (1), the current is maximum when the voltage is zero; at (3) the voltage is maximum when the current is zero. The current is ahead of the voltage.

## *CAPACITOR CONSTRUCTION*

For practical reasons, capacitors are constructed with thin metal sheets, usually aluminum, as the plates. Wax paper, oil-saturated paper, or a chemical electrolyte is used as the high-resistance material between the plates. For air conditioning and refrigeration service, the two common types of capacitors are the oil-filled and the electrolytic capacitors (Figure 9–7).

Oil-filled capacitors are used in series with the start winding of ac motors in order to increase the torque of the motor. Oil-filled capacitors are used when the application requires the capacitor to be in the circuit continuously.

Electrolytic capacitors are used in the start circuits of high-horsepower ac motors. An electrolytic capacitor provides a high capacitance in a small space. Electrolytic capacitors are usually connected for intermittent use.

**Figure 9–7**

## *CAPACITOR RATINGS*

Capacitors are usually rated in capacity and a maximum voltage rating. This capacity unit is the *farad*, which is a very large unit. Capacitors are usually rated in microfarads,

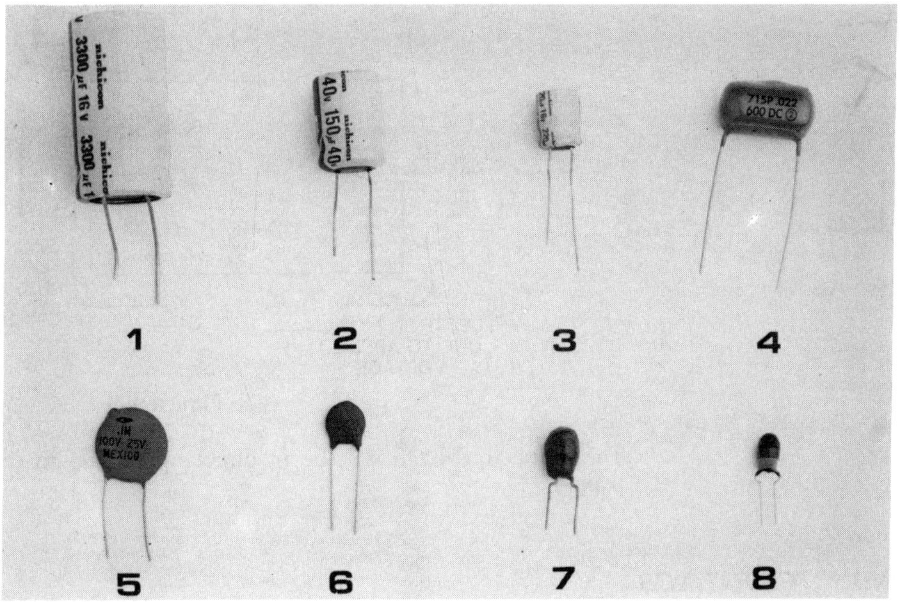


**Figure 9–8**

one-millionth of a farad, which is a practical unit for most applications. Some capacitors are marked 10 MFD, 10mfd, or 10 MF, each of which means ten microfarads. The common abbreviation for microfarad is μf.

Whenever a capacitor is replaced in an electrical device, the capacitance values should be matched as closely as possible. The voltage rating of a capacitor is also an important consideration. The voltage rating of the replacement capacitor should equal or exceed that of the original capacitor.

Standard oil-filled capacitors for continuous use are usually available in capacitance values up to 50 μf with voltage ratings up to 500 volts. Electrolytic capacitors for intermittent use are usually available in capacitance values of 70 μf to 800μf at 125 volts and up to 200 μf at 440 volts.

Another group of capacitors is shown in Figure 9–8. These capacitors are used in electronic circuitry. Capacitors 1, 2, and 3 are aluminum electrolytic capacitors with values of (1) 3,300 MF/17V; (2) 150 MF/40V; (3) 220 MF/16V. Capacitor 4 is a dipped mylar capacitor of 0.022 MF/600V. Capacitors 5 and 6 are dipped ceramic capacitors with values of (5) 0.1 MF/100V and (6) 0.068 MF/25V. Capacitors 7 and 8 are dipped tantalum capacitors with values of (7) 47 MF/20V and (8) 10 MF/30V. *Dipped* simply means that the capacitor after construction is dipped in a plastic to provide a protective cover.

Most electrolytic capacitors are polarized. They have a positive and negative terminal that must be considered when connecting in a circuit. The tantalum capacitors are also polarized. In electronic circuitry, electrolytic capacitors are connected for continuous operation.

## INDUCTANCE

*Inductance* is the property of any electric circuit that opposes changes in current.

In Figure 9–9, a coil of wire is shown connected to a resistor, a switch, and a power source. When the switch $S_1$ is closed, current will start to flow in the circuit. The current flowing through the coil will cause a magnetic field to build up around the coil. The magnetic field cutting through the coil will induce a voltage in the coil.

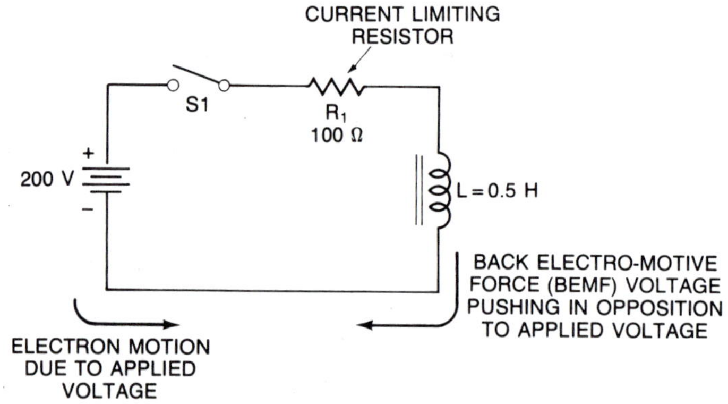


**Figure 9–9**

The induced voltage will be in direct opposition to the applied voltage from the battery.

## *INDUCTOR RATINGS*

Inductors are usually rated in inductance and maximum current-carrying capability. The unit of inductance is the *henry* (H).

A 1-henry coil will have 1 volt induced in it when the current is *changing* at a rate of 1 ampere per second. Practical sizes of coils used in electronics include coils rated in millihenrys (mH) and microhenrys (μH).

## *CURRENT AND VOLTAGE RELATIONSHIP*

The relationship between current and voltage in an inductor is as shown in Figure 9–10. At the instant $S_1$ is closed, the voltage across the coil is maximum while the current is zero. As the voltage across the coil decreases, the current increases (2). When the voltage across the coil decreases, the current increases (2). When the voltage across the coil is minimum, the current is maximum (3). This is exactly opposite the effects that were noted with the capacitor-resistor circuit. In the inductor-resistor (LR) circuit, the current lags behind the voltage.

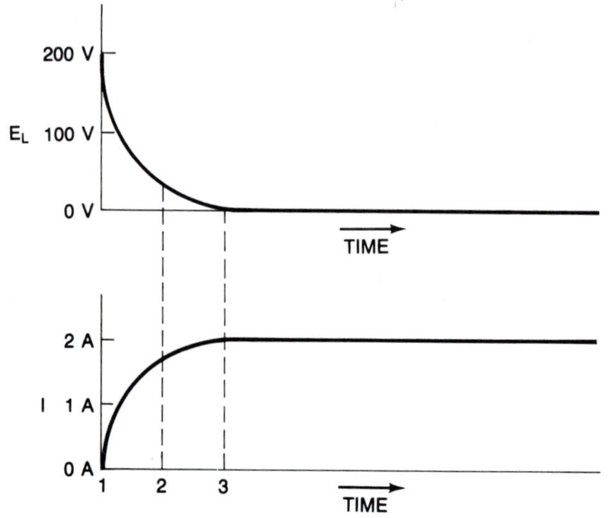


**Figure 9–10**

As a timed operation at point 1, Figure 9–10, the current is rising most rapidly. The magnetic field is changing most rapidly, and the back electro-motive force (BEMF) is greatest as is the voltage across the coil.

At point 2, the rate of change of current has decreased. The BEMF across the coil decreases as does the voltage across the coil.

At point 3, the current has reached its maximum (E applied/R) steady-state value. The magnetic field is not changing. There is no BEMF across the coil. The voltage across the coil is minimum.

## *LR TIME CONSTANT*

As with the RC circuit, there is a direct relation between the values of the inductor and resistor in an LR circuit and the time required for the current to reach its steady-state value. The time constant (TC) in an LR circuit is equal to the inductance divided by the resistance in the circuit.

$$T = \frac{L}{R}$$

The universal time constant curve may be used to determine the current level at any point up to 5 time constants. At 5TC, the current for all practical purposes equals $\frac{V}{R}$.

The current at any instant may also be found using the formula:

$$i = \frac{V_S}{R}(1 - e^{-\frac{t}{TC}})$$

where $i$ = the instantaneous value of current
$V_S$ = the supply voltage
$R$ = the circuit resistance
$e$ = natural logarithm 2.718
$t$ = the time the circuit is connected
$TC$ = the time constant of the circuit; $TC = L/R$

As an example, consider Figure 9–9, the LR circuit. What is the value of the current and the voltage across each component 2 milliseconds after $S_1$ is closed?

$$i = \frac{V_S}{R}(1 - e^{-\frac{t}{TC}})$$

$$i = \frac{200}{100}(1 - 2.718^{-\frac{0.002}{0.005}})$$

$$i = 2\,(1 - 2.718^{-0.4})$$

$$i = 2\,(1 - 0.67)$$

$$i = 2 \times 0.33$$

$$i = 0.66\ A$$

$$V_R = i \times R$$

$$V_R = 0.66A \times 100$$

$$V_R = 66\ V$$

$$V_L = Vs - V_R$$

$$V_L = 200V - 66V$$

$$V_L = 134V$$

For a scientific calculator solution of the problem using the formula, examine the following:

$$i = \frac{V_S}{R}(1 - e^{-\frac{t}{TC}})$$

where e = 2.718

$$\frac{t}{TC} = \frac{0.002}{0.005} = 0.4$$

| STEP | ENTER | PRESS | DISPLAY |
|---|---|---|---|
| 1 | 2.718 | $y^x$ | 2.718 |
| 2 | 0.4 | +/− | −0.4 |
| 3 | | = | 0.67034 |
| 4 | | +/− | −0.67034 |
| 5 | | + | −0.67034 |
| 6 | 1 | = | 0.32966 |
| 7 | | × | 0.32966 |
| 8 | | ( | 0 |
| 9 | 200 | ÷ | 200 |
| 10 | 100 | ) | 2 |
| 11 | | = | 0.6593 |

The current, 0.002 seconds after $S_1$ is closed, will be 0.6593 amperes.

## REVIEW QUESTIONS

Questions 1 through 10 refer to Figure 9–5.

1. If the resistor $R_1$ were reduced to 500kΩ, the capacitor $C_1$ would charge (faster, slower) ________ .
2. What is the voltage across the capacitor $C_1$ four seconds after the switch $S_1$ is closed?
3. What is the voltage across the resistor $R_1$ six seconds after the switch $S_1$ is closed?

4. What is the voltage across the capacitor $C_1$ one minute after the switch $S_1$ is closed?
5. What is the charge in coulombs on the capacitor $C_1$ one minute after the switch $S_1$ is closed?
6. How many excess electrons are stored on the negative plate of the capacitor 2 time constants after $S_1$ is closed?
7. What is the voltage across $C_1$ 2.72 seconds after the switch $S_1$ is closed?
8. What is the voltage across $R_1$ 3.16 seconds after the switch $S_1$ is closed?
9. What is the value of the current in the circuit 2.86 seconds after $S_1$ is closed?
10. How many time constants must pass after $S_1$ is closed before the current is reduced below 50 percent of maximum? (less than 1, 2, 3, 4, or 5 time constants)

Questions 11 through 20 refer to Figure 9–9.

11. Increasing the value of the inductance in the circuit will ______ (increase, decrease) the time required for the current to reach maximum.
12. Increasing the value of the resistance in the circuit will ______ (increase, decrease) the time required for the current to reach maximum.
13. What is the value of voltage across $R_1$, 0.10 milliseconds after $S_1$ is closed?
14. What is the value of the voltage across the coil 0.005 seconds after $S_1$ is closed?
15. What is the value of the current in the circuit 3 time constants after the switch $S_1$ is closed?
16. What is the value of the voltage across the coil 0.0092 seconds after $S_1$ is closed?
17. What is the value of the voltage across $R_1$ 0.012 seconds after $S_1$ is closed?
18. What is the value of the current in the circuit 1.373 time constants after $S_1$ is closed?
19. Inductance in an electric circuit opposes a change in (current, voltage) ______.
20. Capacitance in an electric circuit opposes a change in (current, voltage) ______.

# UNIT TEN

# *ELECTRIC POWER AND ENERGY*

Power is a measure of the rate of doing work. Mechanical power is measured in horsepower (HP), and electrical power is measured in watts (W). The unit of electrical power is named for James Watt (1736–1819), a Scottish engineer who developed the modern condensing steam engine. In the electrical system, power is equal to voltage times current.

$$W = I \times E$$

An electric heat (strip) unit of an air conditioning system may have a hot resistance of about 6 ohms (Figure 10–1). (Cold resistance of this unit would be only slightly lower.) When in use, 9,600 watts of power are produced in the unit. The unit is designed for 240 volts at 40 amperes.

$$I = \frac{E}{R}$$

$$I = \frac{240}{6}$$

$$I = 40 \text{ amps}$$

$$W = I \times E$$

$$W = 40 \times 240$$

$$W = 9{,}600 \text{ watts}$$

This may be seen written on nameplates as 9.6 kW. The k stands for thousand.

A good example of a power-consuming device is an electric bread toaster. In Figure 10–2, an electric toaster is shown plugged into a 120-volt electrical source.

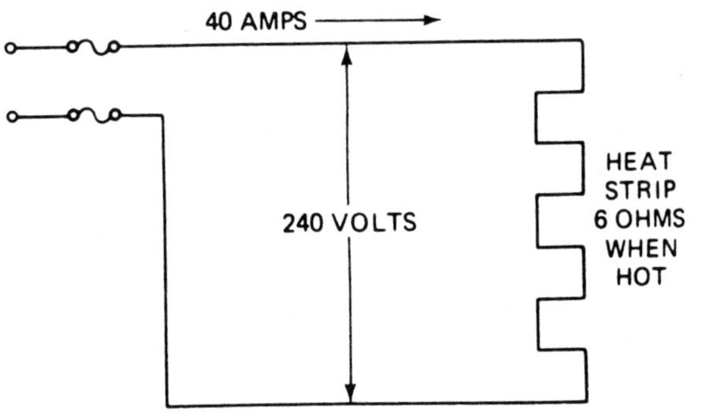


**Figure 10–1**

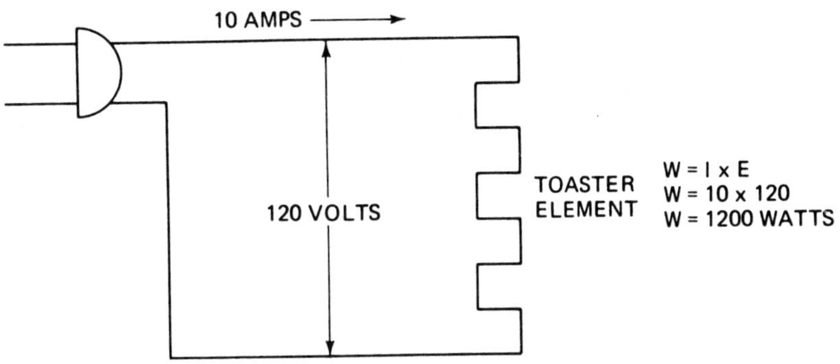


**Figure 10–2**

Ten amperes of current are being drawn from the power source. The power developed in the toaster may be determined by the formula: watts = amperes × volts.

$$W = I \times E$$

$$W = 10 \times 120$$

$$W = 1{,}200 \text{ watts or } 1.2 \text{ kW}$$

Another formula is sometimes used to determine power in electrical circuits.

$$W = I^2R$$

As a statement, the formula is that power (W) is equal to amperes (I) squared times resistance (R). Another way of stating the formula is that power (W) is equal to amperes (I) times amperes (I) times resistance (R).

Since in Ohm's Law $E = I \times R$, the factors $I \times R$ can be substituted for E in the original power formula.

$$W = I \times E$$

substituting $(I \times R)$ for E

$$W = I \times (I \times R)$$

$$W = I \times I \times R$$

$$W = I^2R$$

In the toaster example, the resistance of the toaster heating element, according to Ohm's Law, is:

$$R = \frac{E}{I}$$

$$R = \frac{120}{10}$$

$$R = 12 \text{ ohms}$$

When this factor is used in the formula $W = I^2R$:

$$W = 10 \times 10 \times 12$$

$$W = 100 \times 12$$

$$W = 1{,}200 \text{ watts}$$

This same answer was obtained using the formula $W = I \times E$.

It may be necessary at times to calculate power without information about the current. For example, consider a 30-ohm resistor with 90 volts applied to it. How much power will be developed in the resistor?

$$W = I \times E$$

$$I = \frac{E}{R}$$

$$W = \frac{90}{30} \times 90$$

$$W = 3 \times 90 = 270 \text{ W}$$

The problem could also have been solved using the following method:

$$W = I \times E$$

$$I = \frac{E}{R}$$

$$W = \frac{E}{R} \times E$$

$$W = \frac{E^2}{R}$$

$$W = \frac{90 \times 90}{30}$$

$$W = \frac{8{,}100}{30}$$

$$W = 200$$

$$W = \frac{E^2}{R}$$

This expression shows that the power developed in the circuit is equal to the voltage squared divided by the resistance.

## *HORSEPOWER*

Electric power is required to operate electric motors. In a motor, the desired mechanical output is a rotation of the motor shaft. Some electrical power is continually lost in a motor because of heat produced. A motor is, therefore, not 100 percent efficient.

Consider an electric motor used to drive a refrigeration compressor (Figure 10–3). The compressor is rated as 1 HP. The motor, also rated as 1 HP, is delivering 1-HP mechanical power to the compressor. With 4 amperes of current being drawn from the power line (at 220 volts), 880 watts of electrical power are being delivered to the motor.

$$W = I \times E$$

$$W = 4 \times 220$$

$$W = 880 \text{ watts}$$

If the motor is operating at about 85 percent efficiency, about 134 watts are converted into heat in the motor. The remaining 746 watts are converted into mechanical rotating power at the motor shaft (output).

*The conversion between horsepower and watts is 746 watts equal 1 horsepower.*

Therefore, with 746 watts being converted to rotating power, there is 1 HP being delivered to the motor shaft. Motors are rated in shaft horsepower, that is, output, not power input.

As an aid in remembering the HP/watt conversion factor, remember: Columbus discovered America in 1492; 1492 ÷ 2 = 746; 746 watts = 1 HP.

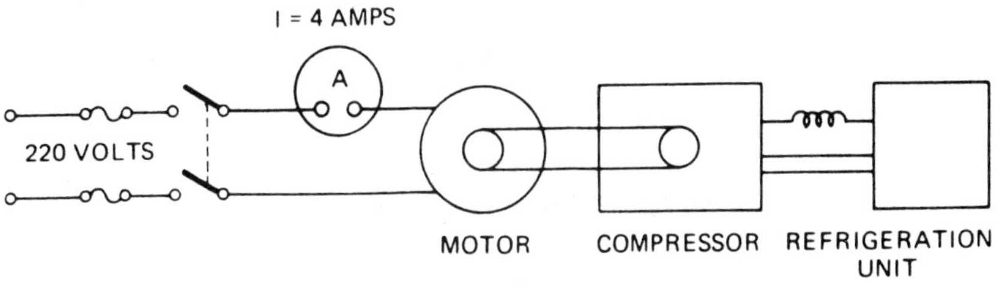


**Figure 10–3**

## *ELECTRIC ENERGY*

Electric power was defined earlier as a measure of the *rate* of doing work. Electric energy relates to the total amount of work done. Stated a different way, the rate of doing work (power) is multiplied by time.

$$\text{Energy} = \text{rate} \times \text{time}$$

The power company charges its customers for the total amount of electrical energy it supplies. Since 1 watt-hour (Wh) is a very small unit of energy, the power company charges per kilowatt hour. A kilowatt hour is equal to 1,000 watts for 1 hour, 500 watts for 2 hours, 2,000 watts for $\frac{1}{2}$ hour, etc.

$$1{,}000 \text{ watt} \times 1 \text{ h} = 1{,}000 \text{ Wh (1 kWh)}$$

$$500 \text{ watt} \times 2 \text{ h} = 1{,}000 \text{ Wh (1kWh)}$$

$$2{,}000 \text{ watt} \times 0.5 \text{ h} = 1{,}000 \text{ Wh (1 kWh)}$$

where "kilo" (k) stands for thousand. The power company bases its charge for providing electrical energy on its cost of production plus a small profit. The cost varies by geographic location and is obtained only from your local company. The following example, based on \$0.06 per kWh, illustrates how cost of operation may be determined when power consumption is known.

---

**EXAMPLE 1**

An electric power company charges its customers 6 cents for 1 kilowatt hour (kWh) of energy. How much does it cost the customer to heat for 8 hours if the heat is obtained from a heat strip of 22 ohms connected to a 220-volt source (Figure 10–4)?

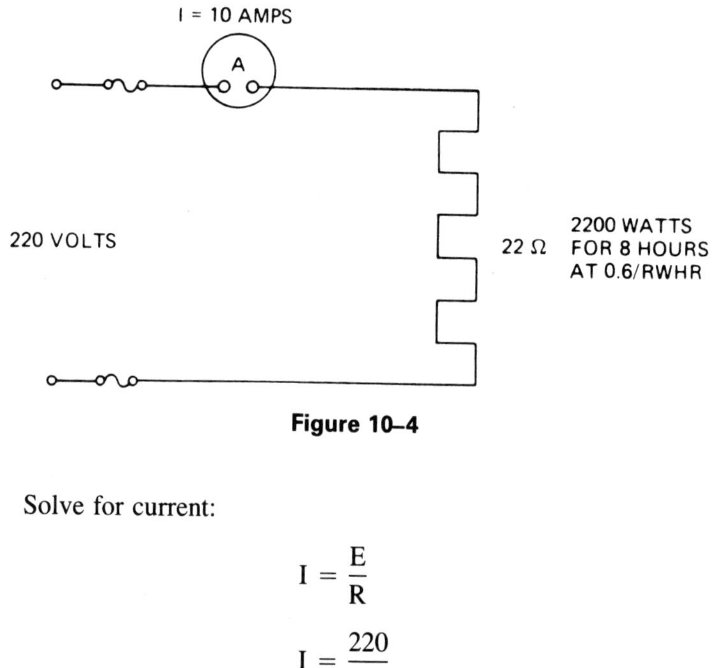


**Figure 10–4**

***Solution:*** Solve for current:

$$I = \frac{E}{R}$$

$$I = \frac{220}{22}$$

$$I = 10 \text{ amperes}$$

Solve for power:

$$W = 1 \times E$$

$$W = 10 \times 220$$

$$W = 2{,}200 \text{ watts}$$

Solve for energy:

$$Wh = W \times \text{time}$$

$$Wh = 2{,}200 \times 8$$

$$Wh = 17{,}600 \text{ watt-hours or } 17.6 \text{ kWh}$$

Solve for cost at 6 cents per kWh:

$$\text{Cost} = \text{kW} \times 0.06$$

$$\text{Cost} = 17.6 \times 0.06$$

$$\text{Cost} = \$1.06$$

In order for the power company to keep track accurately of the total amount of electrical energy being supplied to a customer, a watt-hour meter is usually installed where the electric service is delivered to a building. A watt-hour meter is an electromechanical device that rotates dials, recording the energy delivered to the customer (Figure 10–5). This meter is read monthly by the power company, and the customer is billed for energy used.

**Figure 10–5** Watt-hour meter.

## *CONTROL OF ELECTRIC POWER*

The control of electric power is a problem that must be dealt with every day. Electric power must be turned off and on either manually or automatically as needed for lights, motors, heaters, and other power-consuming devices.

In low-power units, such as lamps, electric drills, and electric toasters, simple switches may be used to provide the required control. In the larger power-consuming devices, such as air conditioning compressor units or electric heat strip units, high power (usually at higher voltages) must be controlled. In these types of units, power must be controlled at a point remote from where it is being consumed. For example, an air conditioning compressor (often outside the house) is controlled by a thermostat in the living area inside the house. A low-voltage signal (usually 24 volts) is fed to a relay (an electrically operated switch) in the air conditioning unit. The relay, when energized, supplies high voltage (usually 220 volts) at high current levels to the compressor motor.

*Relays* are electromechanical devices used as controls for electric power. A relay has a movable metal part called the *armature*. An electrical contact or contacts may be connected to the armature; when the armature moves, the contacts also move. Depending on the application, this action may open some electric circuits while closing others.

In Figure 10–6, the relay is in the deenergized (no coil power applied) position. When no power is supplied to the relay coil, the spring pulls the armature down at the right end. The armature pivots at the point, moving the left end of the armature up. An electric circuit is complete from the normally closed contact through the movable contact to the connecting wire to the terminal output.

In Figure 10–7, the same relay is shown, but electric power has been applied

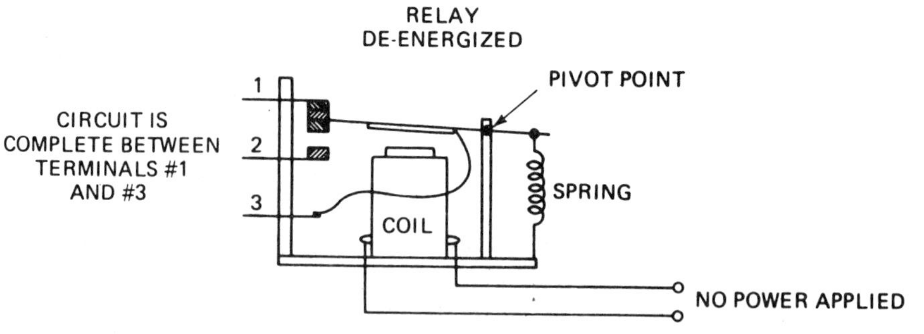


**Figure 10–6**

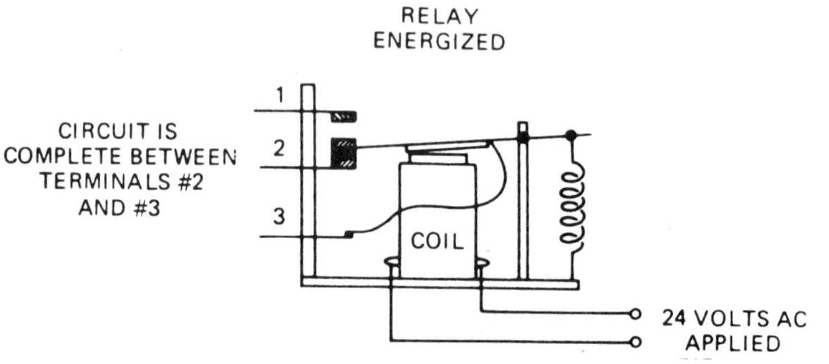


**Figure 10–7**

to the relay coil. The magnetic field developed around the coil attracts the iron in the armature. The armature moves down, making electrical contact between the normally open contact and movable contact. As long as power is supplied to the coil, the relay remains energized. When power is removed from the coil, the spring pulls the armature to its original position, returning the connection to the normally closed contacts.

Relays come in varied shapes and sizes and are designed for special purposes. Sometimes, the name of the device, because of its application, is changed, such as to "contactor." The mechanical construction of a contactor is somewhat different from that of a relay, but it operates on the same principle (Figure 10–8). Low power supplied to the coil of the contactor produces a magnetic field that attracts an armature. When the armature moves, it closes electric contacts that provide for the control of high-power circuits.

**Figure 10–8** Contactor.

## *MEASURING ELECTRIC POWER*

One method of determining electric power in a circuit is to measure the voltage across the circuit and the current through the circuit. The power developed is found by the formula

$$W = I \times E$$

In Figure 10–9, the circuit of an electric heat strip is shown. The voltage across the heat strip is 220 volts as indicated by the voltmeter. The ammeter indicates 43.6 amperes.

$$W = I \times E$$

$$W = 43.6 \times 220$$

$$W = 9{,}600 \text{ watts}$$

Another method of measuring power is to use a wattmeter. A wattmeter is a device that uses both the current in the circuit and the voltage across the circuit to provide an indication of the power developed in the circuit. The heat strip of Figure 10–9 could have a wattmeter connected rather than a voltmeter and an ammeter. The power developed would be read directly from the wattmeter, as shown in Figure 10–10.

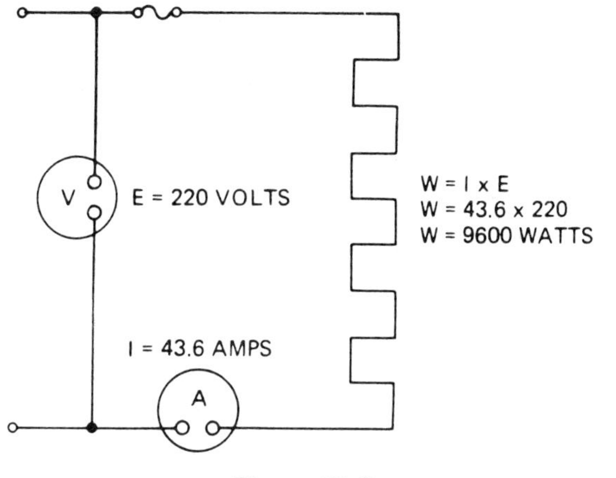


**Figure 10–9**

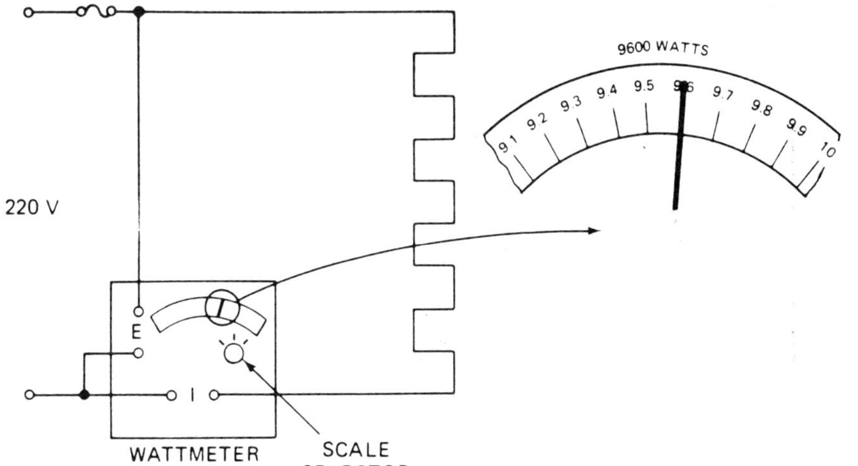


**Figure 10–10**

It is always better to use a wattmeter for determining power, if one is available. Because only one meter reading is needed, the chance of error is lessened. In ac circuits, it is always better to use a wattmeter for power measurement because of other factors having effects on ac. Further information on power in ac circuits will be covered in Unit 12.

## REVIEW QUESTIONS

1. The formula for power using voltage and current is W = ______ .
2. The formula for power using current and resistance is W = ______ .
3. A strip heater draws 30 amperes of current at 220 volts. How much power is developed in the unit?
4. If a strip heater of 8,600 watts is operated for 100 hours, how much electrical energy will be converted to heat?
5. A power company is selling electric energy at $0.04 per kW/h. How much will it cost to operate a 10,000-watt heater for 60 hours?
6. The movable part of a relay is called an ______ .
7. A contactor is a special purpose ______ .
8. In Figure 10–11, if the power developed in the heater is 12,000 watts and the supply voltage is 220 volts, what is the current flow through the heater?
9. In Figure 10–11, the power developed in the heater is 8,000 watts. The power source is 220 volts. What is the resistance of the heater?
10. The circuit of Figure 10–11 is to supply 18,000 watts of heat. The power source is 220 volts. What is the resistance value of the heater?

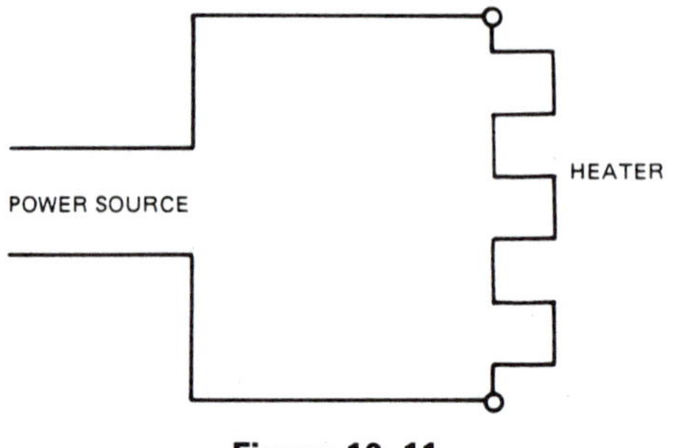


**Figure 10–11**

# UNIT ELEVEN

# *ALTERNATING CURRENT*

In the previous discussion of electrical current, the implication was that electrons always move in one direction. Actually, the most common form of electrical energy generated today is *alternating current*, ac, which changes direction 60 times each second.

Alternating current is the most common form of electrical energy because it is the natural form of a generated current. Alternating current may be delivered over great distances with much less loss in power than direct current (dc). For these reasons, alternating current is generally less expensive than direct current for most applications.

## *GENERATION OF AC VOLTAGE*

In Unit 2, there was a discussion of the fact that an electric current flowing in a wire produces a magnetic field about the wire. Another relationship between magnetic fields and electricity is:

> *Whenever a wire passes through a magnetic field, a voltage is induced in the wire.*

The amount of voltage that will be induced in the wire is dependent on two factors: the strength of the magnetic field and the speed at which the wire cuts through the field.

In generating electric voltages:

Stronger magnetic field = higher voltage
Higher speed = higher voltage

### Generated Electromotive Force (voltage)

A *generator* is a device that converts mechanical energy into electrical energy by mechanically rotating coils of wire in a magnetic field. An *alternator* is a device that converts mechanical energy into electrical energy by rotating magnetic fields past coils of wire.

In Figure 11–1, a piece of wire is shown moving down through a uniform magnetic field. In position 1, the wire is not cutting through the field, and no voltage will be generated. As the wire enters the magnetic field, position 2, a level of voltage is established. As the wire moves through the uniform field, position 3, at a constant speed, the level of voltage remains constant. At position 4, the wire leaves the magnetic field, and the voltage drops to zero. In position 5, the wire does not cut through the the magnetic field, and the voltage remains at zero.

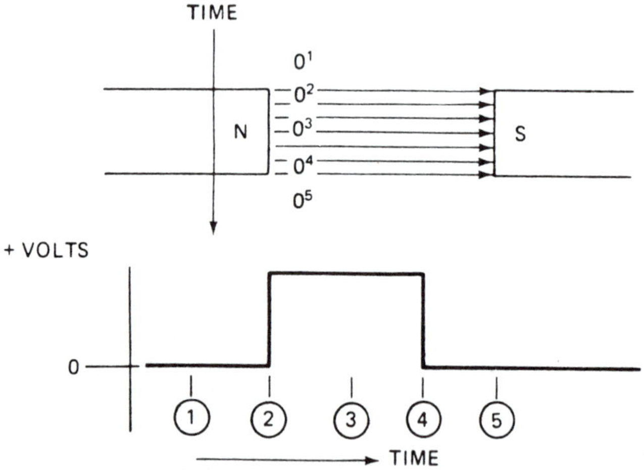

**Figure 11–1**

### Direction of Generated Voltage

When a piece of wire passes through a magnetic field, some of the free electrons in the wire are forced to move. The direction in which the electrons move, and therefore the polarity of the voltage, is dependent on two things:

1. The direction in which the wire moves through the magnetic field.
2. The direction of the magnetic field.

### Left-Hand Rule

There is a rule associated with generator action, namely, the left-hand rule. The rule states: Receive the lines of force in the palm of the left hand with the thumb pointed in the direction of wire motion. The extended fingers will point in the direction of electron movement. (Remember, lines of force travel from the N pole to the S pole.)

Figure 11–2 shows an example of the rule. The direction of electron movement is indicated in Figure 11–2 with a (+) at the tail of an arrow. The electrons move from the (+) end of the wire to the other end, creating a voltage. The (+) end is positive since electrons move away from this end of the wire. The other end of the wire is negative because the electrons crowd in.

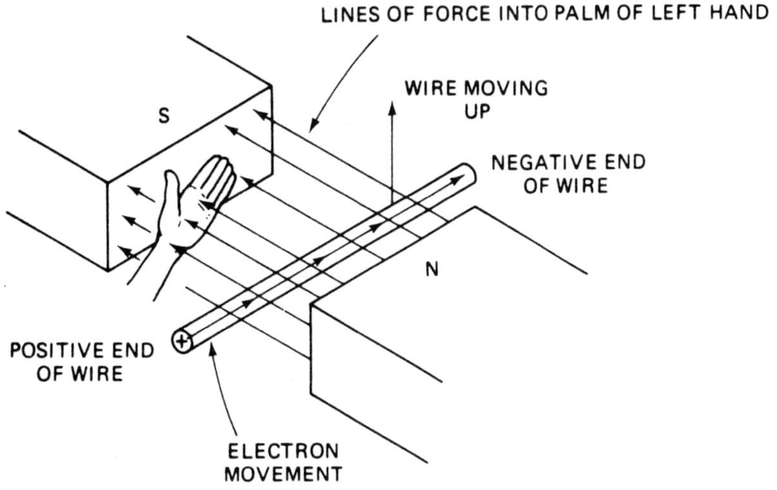


**Figure 11–2**

## Rotating Coils

For practical mechanical reasons, generators are made up of coils rotating in a magnetic field. When a coil rotates in a magnetic field, an alternating voltage is produced. In Figure 11–3, an example of a coil section, one wire is shown. As the wire rotates through a complete revolution, a complete cycle of alternating voltage is produced. The complete revolution is shown, divided into twelve equal parts.

In position 0, the wire is moving parallel to the magnetic lines. The wire does not cut through them. No voltage is produced. As the wire moves from position 0 to position 1, it cuts through a few magnetic lines. The (+) indicates the direction of electron movement and produced voltage. (The graph, Figure 11–4, indicates the level of voltage.) As the wire moves from 1 to 2, it cuts through an increasing number of lines. A corresponding increase in voltage is indicated on the graph. At position 3, the wire is cutting directly down through the magnetic lines. The maximum voltage is generated, as indicated in the graph. As the wire moves to positions 4 and 5, it cuts through fewer lines, until at position 6, it is again moving parallel to the magnetic

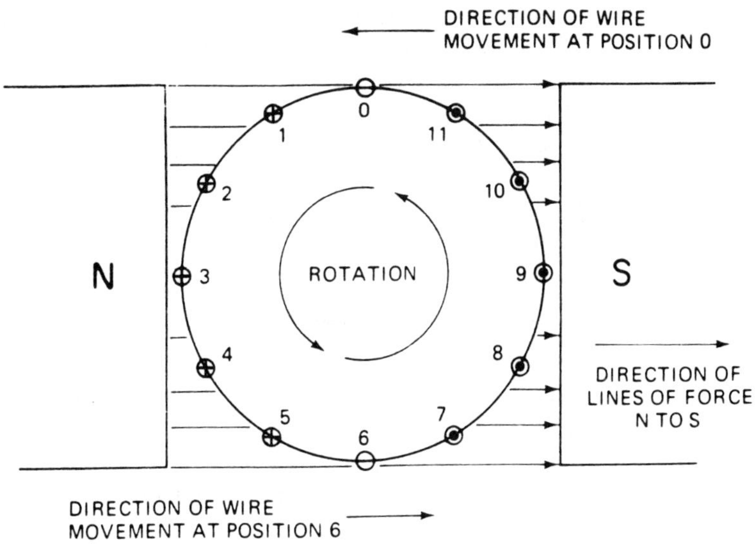


**Figure 11–3**

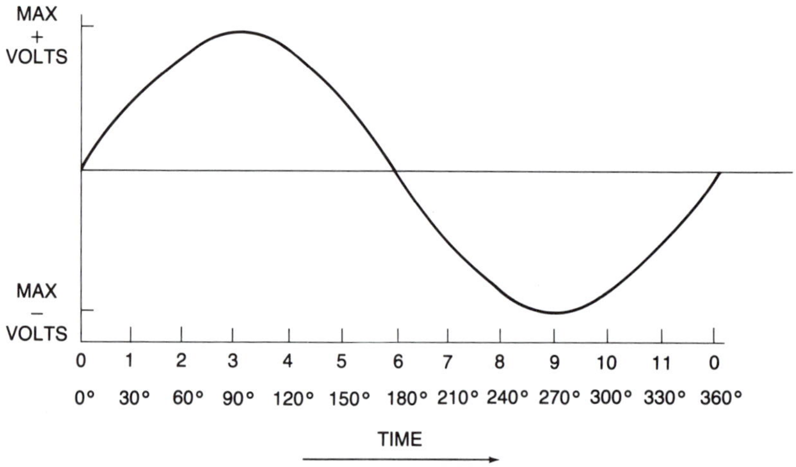


**Figure 11–4**

lines. In the graph, this is shown as decreasing voltage to a zero value at position 6. As the wire moves from position 6 to position 7, the direction of movement through the magnetic lines changes.

The level of voltage starting at 7 is increasingly negative, reaching a maximum at position 9. The voltage decreases at 10 and 11, falling to zero volts at position 0. This completes one revolution and one complete cycle of voltage. As the wire continues to rotate, another cycle starts.

According to the left-hand rule for positions 1, 2, 3, 4, and 5, electron motion would be into the page as the wire moves down. In positions 7, 8, 9, 10, and 11, the wire is moving up. The left-hand rule provides for electron motion out from the page, as indicated by a dot in the center of the wire.

The voltage is called *alternating voltage* because the output of the generator will first be positive and then negative at each of the output terminals.

### Sine Wave

The output voltage varies in the form of a *sine wave*. The generator winding cuts through the magnetic field at various rates for each position of rotation. As a measure of rotation, the circle is divided into 360 degrees. This is shown in Figure 11–4.

Most power companies in the United States generate electricity with a frequency of 60 hertz (Hz) or 60 cycles per second (60 cps). There are sixty complete cycles of positive and negative alternations during each second. Equipment designed to operate with voltage at this frequency is marked on the nameplate at 60 cps or 60 Hz. Many foreign countries produce voltages and operate equipment at 50 cycles per second (50 cps or 50 Hz).

### Peak and Effective (RMS) Values of Voltage

When an ac voltage is applied to a resistor, the power developed in the resistor varies instant by instant. The instantaneous powers developed in the resistor are equal to $E^2/R$. The average power will be less than the power developed at the peak voltage. Accordingly, the effect of the voltage in producing power is not equal to the voltage at the peak. The effective value is equal to 0.707 of the peak value.

This means that a sine wave of voltage with a peak value of 100 volts has the same heating effect as 70.7 volts dc.

The following table shows the development of voltage and power over a quarter cycle of a sine wave. Each quarter cycle would contain the same values. The information concerns an ac voltage of 100 volts peak applied to a 100-ohm resistor. The calculations that follow show how the effect of the voltage is determined. This results in the effective, or RMS, voltage value. Power is calculated using the formula $W = E^2/R$, $W = E^2/100$.

| *Degree rotation* | *Instant voltage* | *Voltage squared* | *Watts power* |
|---|---|---|---|
| 0° = 100V sin 0° = | 0.00 V | 0 | 0.00 |
| 10° = 100V sin 10° = | 17.36 V | 301 | 3.01 |
| 20° = 100V sin 20° = | 34.20 V | 1,170 | 11.70 |
| 30° = 100V sin 30° = | 50.00 V | 2,500 | 25.00 |
| 40° = 100V sin 40° = | 64.28 V | 4,130 | 41.30 |
| 50° = 100V sin 50° = | 76.60 V | 5,868 | 58.68 |
| 60° = 100V sin 60° = | 86.60 V | 7,500 | 75.00 |
| 70° = 100V sin 70° = | 93.97 V | 8,830 | 88.30 |
| 80° = 100V sin 80° = | 98.48 V | 9,698 | 96.98 |
| 90° = 100V sin 90° = | 100.00 V | 10,000 | 100.00 |
| | | 49,997 V² | 499.97 W |

Find the average power over the 10 positions of rotation:

$$W\text{ avg.} = \frac{Wt}{10}$$

$$W\text{ avg.} = \frac{499.97}{10} = 49.99 \text{ or } 50 \text{ watts}$$

Find the voltage required to produce the average power:

$$W = \frac{V^2}{R}$$

$$V^2 = W \times R$$

$$V = \sqrt{W \times R}$$

$$V = \sqrt{49.997 \times 100}$$

$$V = \sqrt{4999.7}$$

$$V = 70.7 \text{ volts}$$

One hundred volts peak equals 70.7 volts effective, or RMS. Almost all electrical devices are rated in effective (RMS) voltage. An appliance rated at 120 volts means 120 volts effective ac voltage.

Electrical instruments are designed to indicate effective or RMS voltage. If a standard voltmeter were to be connected to a wall outlet, the meter would indicate the effective value of the voltage, roughly, 120 volts. The peak value of the voltage is 1.414 times the effective value, or 169.68 volts. The peak-to-peak value is 339.36 volts.

Standard voltmeters are designed to provide the RMS value of sine waves only. Other voltage waveform measurements should be made using an oscilloscope. (The oscilloscope will be covered in a later unit.)

## INSTANTANEOUS VOLTAGE VALUES

The instantaneous value of voltage at any point along the sine waves may be determined by multiplying the peak value by the sine of the angle at the point of concern.

---

**EXAMPLE 1**

If the sine wave in Figure 11–4 has a peak or maximum value of 120 volts, what is the instantaneous value at 45°?

***Solution***

$$V = \text{Vpeak} \times \sin \measuredangle$$

$$V = 120 \times \sin 45°$$

$$V = 120 \times 0.707$$

$$V = 84.85$$

Scientific calculator entry requires either the use of parentheses or the entry of the angle first.

$$V = 120 \times (45°\sin) = 84.85 \text{ or}$$

$$V = 45° \sin \times 120 = 84.85$$

---

**EXAMPLE 2**

With the same 120V peak voltage, what is the instantaneous voltage value at 283°?

***Solution***

$$V = \text{Vpeak} \times \sin 283°$$

$$V = 120 \times -0.97437$$

$$V = -116.9 \text{ volts}$$

By scientific calculator:

$$V = 120 \times (283° \sin) = -116.9 \text{ volts} \quad \text{or}$$

$$V = -283° \sin \times 120 = -116.9 \text{ volts}$$

---

## FREQUENCY AND TIME

The speed of rotation of a generator determines the frequency of the voltage output. The higher the rotational speed, the higher the frequency.

If the generator shown in Figure 11–3 takes one second to complete one revolution, the frequency of the output would be one cycle per second. If the speed of the generator is increased to 10 revolutions per second, the frequency would be 10 cycles per second. The time or period of one cycle would be 1/10 second.

To convert from frequency to time or time to frequency, use the formula:

$$f = \frac{1}{t} \quad \text{or} \quad t = \frac{1}{f}$$

where $f$ = frequency in cycles/second (hertz)
$t$ = time for one cycle

---

**EXAMPLE 3**

What is the length of time for one cycle of a 2328-hertz sine wave?

***Solution***

$$t = \frac{1}{f}$$

$$t = \frac{1}{2328}$$

$$t = 0.0004295 \text{ sec, or } 429.5 \text{ microseconds}$$

What is the frequency of a sine wave with a period for one cycle of 1327 microseconds?

***Solution***

$$f = \frac{1}{t}$$

$$f = \frac{1}{1327 \times 10^{-6}}$$

$$f = 753.6 \text{ hertz}$$

---

## *THREE-PHASE POWER*

Most of the electricity generated today is in the form of three-phase alternating current. Producing power as three-phase is the most economical method of generating and distributing electricity.

There are two main systems of electrical connection of three-phase power. They are the *delta* and *Wye* connections. The circuits are shown in Figure 11–5. When considering the operation of a generator or alternator, it is the connection that determines output voltage and currents.

In the delta connection, the line voltage is equal to the coil voltage. The line current is equal to the coil current times the square root of three (1.732). The line current is the vector sum of the currents of the two coils joined at the line connection. A vector addition is necessary as the currents are 120 degrees out of phase with each other.

In the Wye connection, the line current is equal to the coil current. The line voltage is equal to the coil voltage times the square root of three. The line voltage is the vector sum of the coil voltages between two lines.

The description of the delta and Wye systems was made considering a balanced load on the three-phase system. The voltage and current distribution to delta and Wye

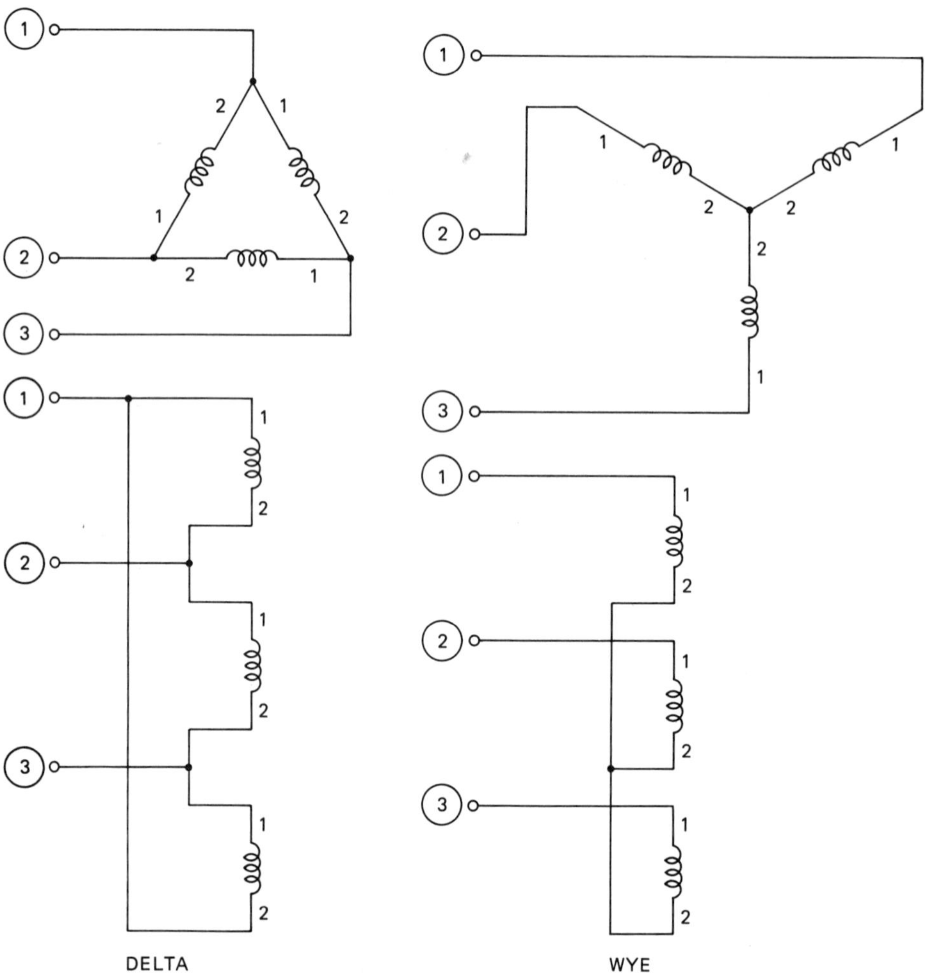


**Figure 11–5**

loads follows the same pattern as with the generator. This distribution is shown in Figures 11–6a and 11–6b.

### Power in Three-Phase

The total power developed in a three-phase system is the sum of the powers developed in the individual loads. In Figure 11–6c, a three-phase delta load is shown.

If a load contains a reactive component, the power developed is found by the formula:

$$W = I \times V \times pf$$

In three-phase motor circuits, the standard formula for power is:

$$W = I \times V \times pf \times \sqrt{3}$$

Because a three-phase motor circuit is nearly balanced, only the current and voltage of one line need be considered. The factor in the formula compensates for the power of the two other lines.

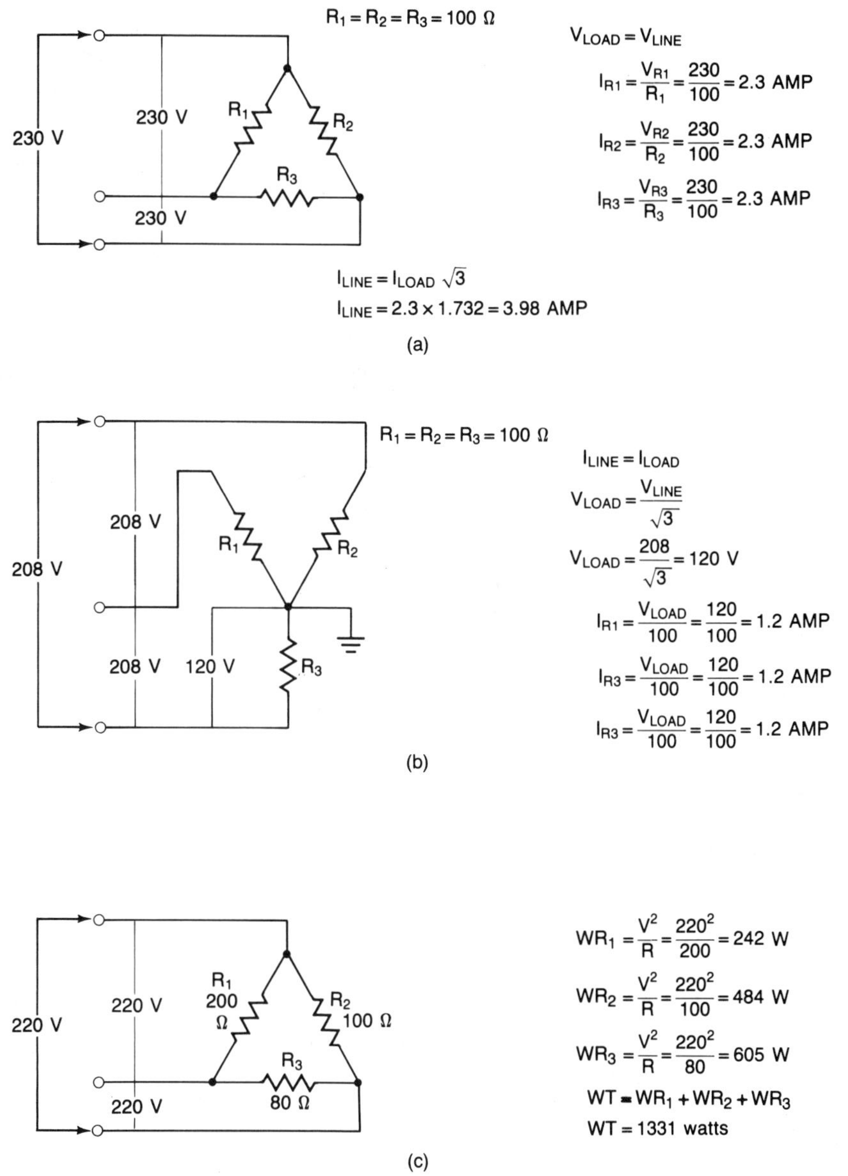


**Figure 11–6**

## REVIEW QUESTIONS

1. What two factors determine the magnitude of a voltage induced in a wire traveling through a magnetic field?
2. What two factors determine the polarity of the voltage developed in a wire passing through a magnetic field?
3. In Figure 11–3, is the rate of cutting through lines of force greater at position 1 or position 8?
4. In Figure 11–4, if the value of voltage at position 5 (150°) is 100 volts, what is the peak value?
5. In Figure 11–4, if the value of voltage at position 10 (300°) is $-100$ volts, what is the voltage value at position 2 (60°)?
6. What is the frequency of most of the power generated in the U.S. today?

7. A standard ac voltmeter indicates the (average, effective, peak, peak-to-peak) _______ value of voltage.
8. A standard ac voltmeter will correctly indicate the effective value of both sine waves and square waves. True _______ False _______.
9. If the rotational speed of a generator is increased, the frequency will (increase, decrease) _______.
10. The higher the frequency, the greater the time for one cycle. True _______ False _______.
11. The effective value of an ac sine wave of voltage is 88 volts. What is the peak value of the voltage?
12. The peak-to-peak value of a sine wave of voltage is 263 volts. What is the effective value of the voltage?
13. The peak value of a sine wave of voltage is 96 volts. What is the instantaneous value of the voltage at 121 degrees?
14. The frequency of a certain sine wave is 327 hertz. What is the time for one cycle?
15. The time for one cycle of a sine wave is 83 microseconds. What is the frequency of the waveform?

# UNIT TWELVE

# *PHASORS AND POWER FACTOR*

### *POLYPHASE GENERATORS AND ALTERNATORS*

Generators that produce a single output, alternating voltage, are called *single-phase generators* (Figure 12–1). The voltage completes one cycle (360°) in 1/60 of a second.

A second generator could be connected to the same drive source and produce an output voltage (Figure 12–2). If the armatures of the two generators are exactly aligned, as in Figure 12–3, the output voltages will be in phase with each other.

Another alignment of the armatures could be made which would not provide in-phase voltages at the outputs. In Figure 12–4, the armature of generator B has been rotated clockwise 90°. The output of generator B is 90° out of phase with the output of generator A. The two outputs of the generator could be connected together (Figure 12–5). The voltages available would then be as given in Figure 12–5.

The output between the two generators is not the sum of the two voltages because the voltages are not in phase and cannot be added directly. Since the voltages are

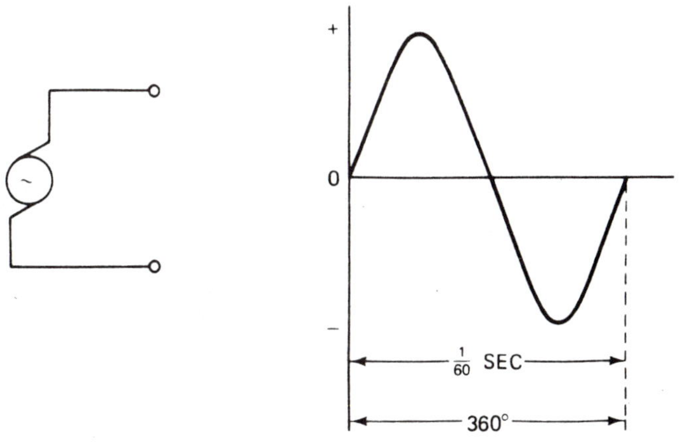


**Figure 12–1**

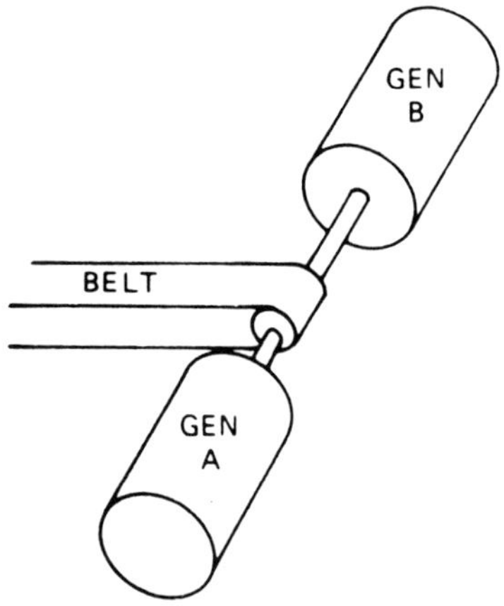


Figure 12–2

90° out of phase with each other, the output between A and B will be 100 times the square root of 2 ($\sqrt{2} = 1.414$).

$$100 \text{ volts} \times 1.414 = 141.4 \text{ volts}$$

When adding voltages that are not in phase with each other, the instantaneous values must be used. For example, in Figure 12–6a, the two voltages will be added at 0°, 45°, 90°, 135°, 180°, 225°, 270°, 315°, and 360°. The resultant voltage is shown in Figure 12–6b. Refer the following to Figure 12–6a.

At 0 degrees, phase A is 0 volts while phase B is 100 volts.

$$0 + 100 = 100$$

At 45 degrees, phase A is +70.7 volts and phase B is +70.7 volts.

$$70.7 + 70.7 = 141.4$$

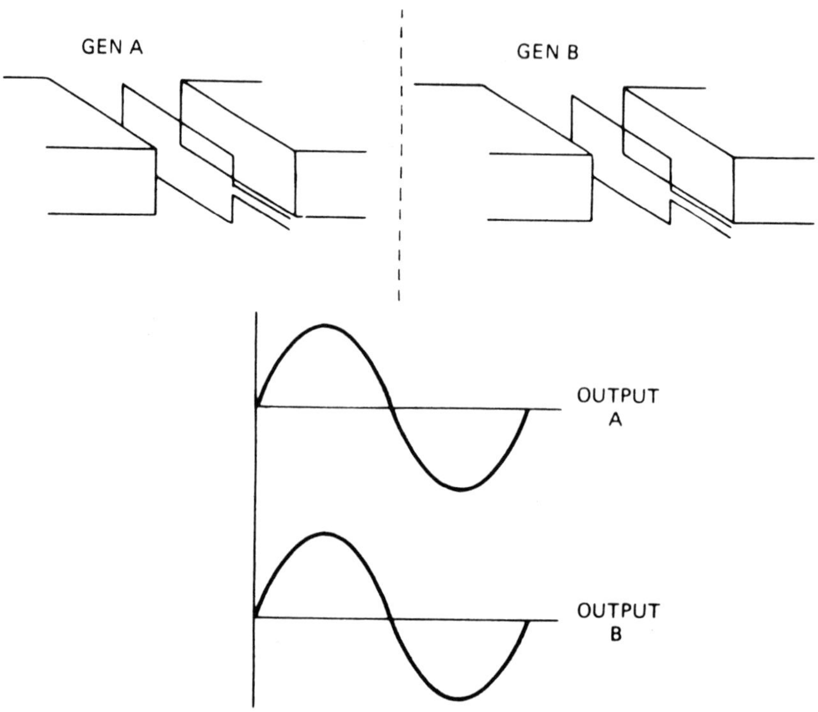


Figure 12–3

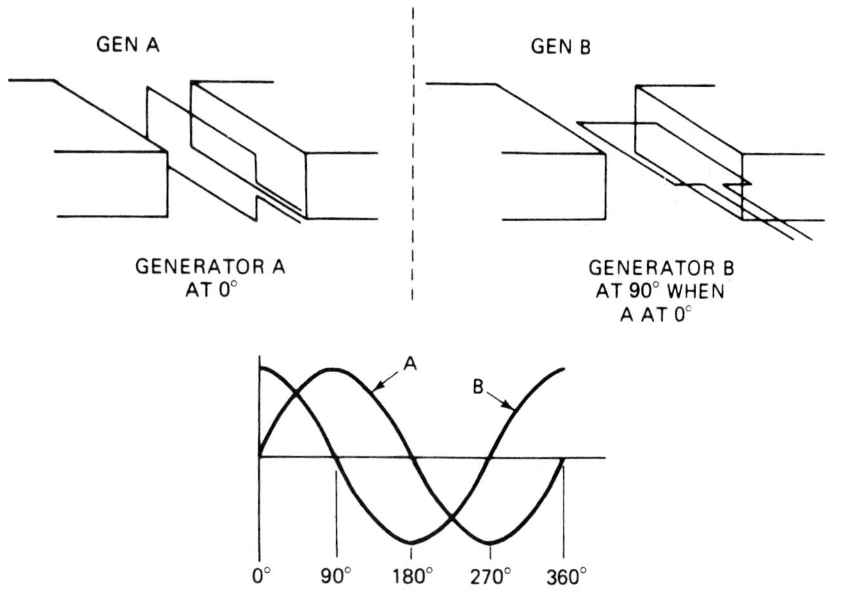


**Figure 12–4**

At 90 degrees, phase A is +100 volts and phase B is 0 volts.

$$100 + 0 = 100$$

At 135 degrees, phase A is +70.7 volts and phase B is −70.7 volts.

$$+70.7 - 70.7 = 0$$

At 180 degrees, phase A is 0 volts and phase B is −100 volts.

$$0 - 100 = -100$$

At 225 degrees, phase A is −70.7 volts and phase B is −70.7 volts.

$$-70.7 - 70.7 = -141.4$$

At 270 degrees, phase A is −100 volts and phase B is 0 volts.

$$-100 + 0 = -100$$

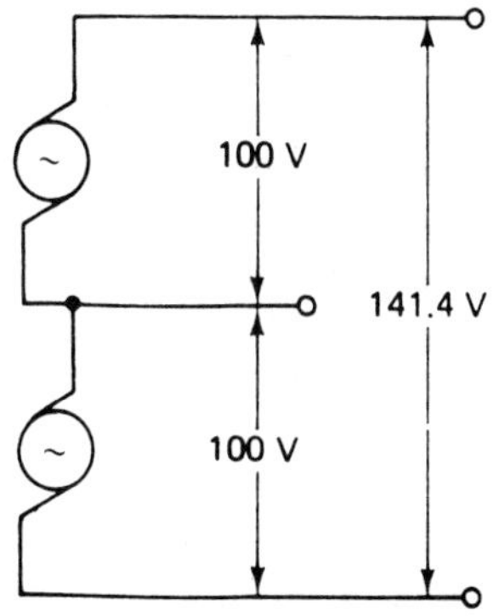


**Figure 12–5**

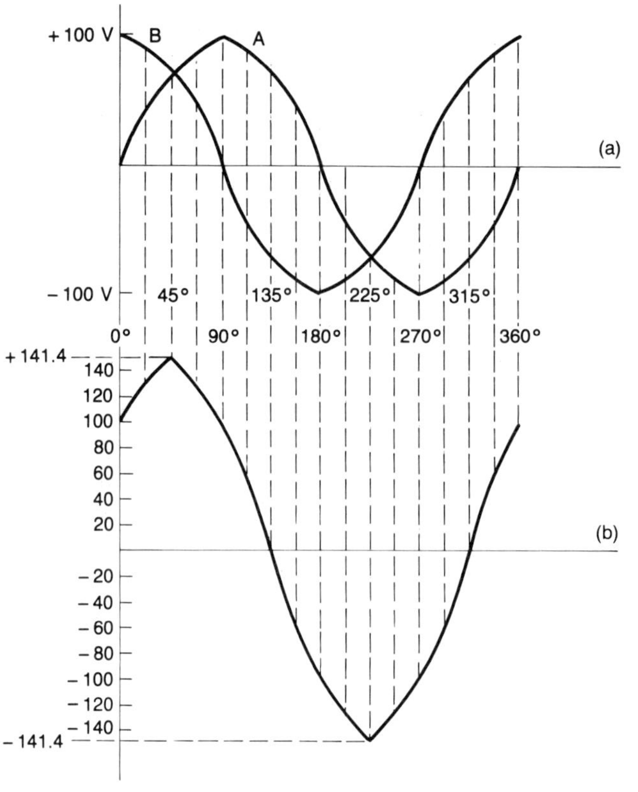


**Figure 12–6**

At 315 degrees, phase A is −70.7 volts and phase B is +70.7 volts.

$$-70.7 + 70.7 = 0$$

The 360-degree (0°) point is the starting point, and the waveforms are ready to repeat themselves.

The resultant sine wave has a peak amplitude of 141.4 volts and is 45 degrees out of phase with both of the original waveforms. The resultant leads A by 45 degrees and lags B by 45 degrees.

It is obvious when adding the instantaneous values that the peak voltages do not occur at the same time. The addition of voltages out of phase with each other therefore provides a result that is less than the algebraic sum of the peaks.

## VECTORS

Another method that may be used to determine the result of adding voltages (or currents) that are not in phase with each other is to make a *vector* diagram. A vector is a line that indicates magnitude and direction. A vector diagram shows the relation of two or more factors as the magnitude and direction.

For example, the two-phase voltages of Figure 12–6a may be shown as a vector diagram. As in Figure 12–6a, voltage B is leading voltage A by 90 degrees. All vectors are considered to rotate counterclockwise; therefore, B is ahead of A in Figure 12–7.

To add the vectors, complete the parallelogram and determine the resultant, as shown in Figure 12–8.

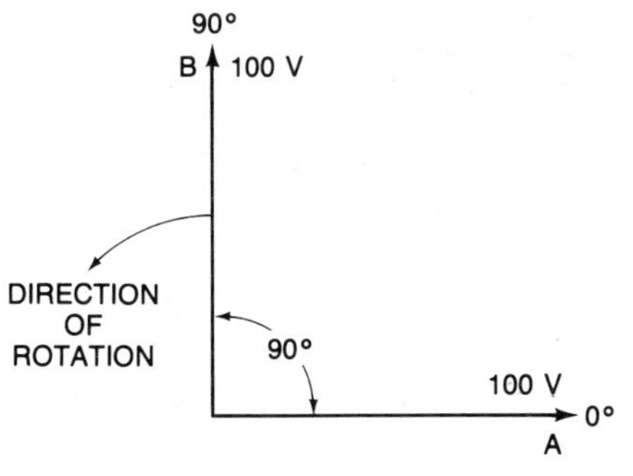


Figure 12–7

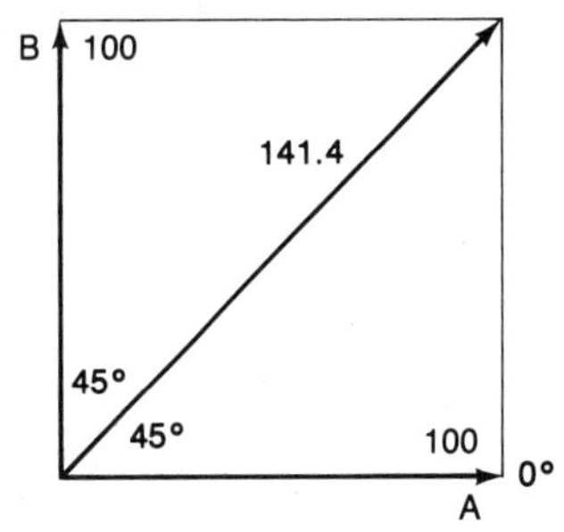


Figure 12–8

Using the Pythagorean theorem

$$C = \sqrt{A^2 + B^2}$$

$$C = \sqrt{100^2 + 100^2}$$

$$C = 141.4$$

Because the resultant bisects the original 90-degree angle, it is obvious that the resultant voltage leads voltage A by 45 degrees and lags voltage B by 45 degrees. This agrees with the conclusion of adding the instantaneous voltages in both magnitude and direction.

In practice, it is much easier to work with vector diagrams than with adding the instantaneous values.

## *RECTANGULAR AND POLAR COORDINATES*

In order to more easily solve problems, it is sometimes necessary to change a vector to its components at 0 degrees and 90 degrees.

If it were necessary to convert the 141.4 volts at 45 degrees to its rectangular coordinate components, they are 100 volts at 0 degrees and 100 volts at 90 degrees as shown in Figure 12–8.

In some cases it is necessary to change voltages (or currents) from polar coordinates (any angle 0° to 360°) to rectangular coordinates, 0 degrees and 90 degrees, 180 degrees, and 360 degrees.

For example, convert a voltage of 120 volts at 37 degrees, as shown in Figure 12–9, to its components at 0 degrees and 90 degrees.

Draw perpendiculars to the 0-degree and 90-degree lines as shown in Figure 12–10.

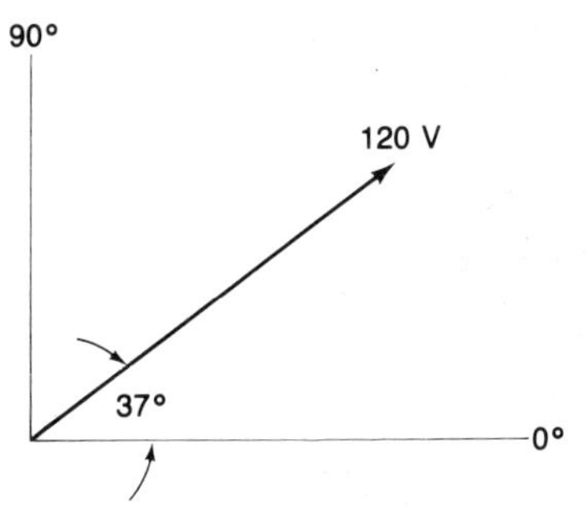


Figure 12–9

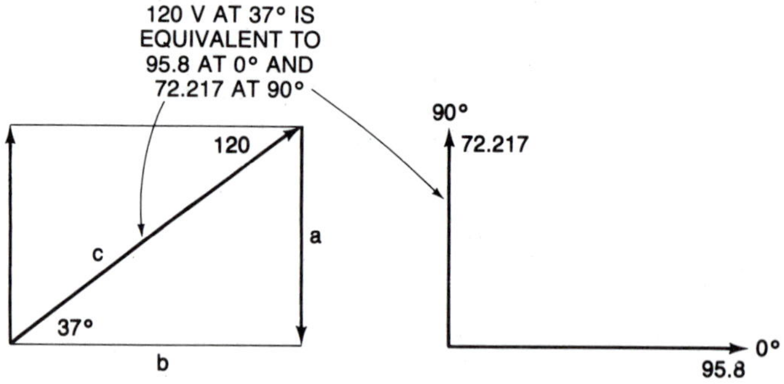


**Figure 12–10**

Solve the right-angle trigonometry problem for side a and side b.

$$\sin = \frac{\text{opp}}{\text{hyp}}$$

$$\cos = \frac{\text{adj}}{\text{hyp}}$$

$$37^\circ \sin = \frac{a}{120}$$

$$a = 37^\circ \sin \times 120$$

$$a = 72.217 \text{ V}$$

$$37^\circ \cos = \frac{b}{120}$$

$$b = 37^\circ \cos \times 120$$

$$b = 95.8$$

***INDUCTANCE IN AC***

The effect of inductance in an electric circuit is to oppose the change in current. In an ac circuit, the current is always changing.

It is reasonable, then, that there will be a continuous opposition to current from inductance in an ac circuit. The opposition is called *inductive reactance*. The symbol $X_L$ is used to indicate inductive reactance.

Two factors determine the inductive reactance of a coil. They are the *inductance* (L) of the coil and the *frequency* (f) (rate of change) of the applied voltage and current.

By formula:

$$X_L = 2\pi fL$$

---

**EXAMPLE 1**

An inductor of 8 henrys is connected in a circuit where a voltage of 200 volts at 80 hertz is applied. What is the effect of the inductor?

***Solution:*** See Figure 12–11.

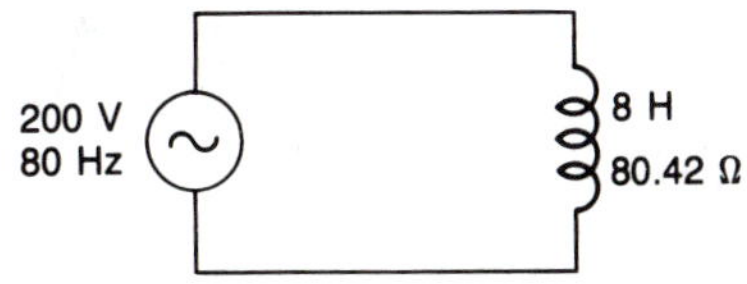


$X_L = 2\pi fL$

$X_L = 6.28 \times 80 \times 8$

$X_L = 8042\ \Omega$

$I = \frac{V}{X_L}$

$I = \frac{200}{8042}$

$I = 24.87$ ma

Figure 12–11

## *CAPACITANCE IN AC*

The effect of capacitance in an electric circuit is to oppose a change in voltage. In an ac circuit, the voltage is always changing.

There will be a continuous opposition to the voltage in an ac circuit. The opposition is known as *capacitive reactance*. The symbol for capacitive reactance is $X_C$.

The two factors controlling $X_C$ are frequency and capacitance of the capacitor. By formula:

$$X_C = \frac{1}{2\pi fC}$$

### EXAMPLE 2

How much current will flow in a circuit containing a 10-μf capacitor supplied by a generator producing 500 V at 80 hertz?

***Solution*:** See Figure 12–12.

$X_C = \frac{1}{2\pi fc}$

$X_C = \frac{1}{6.28 \times 80 \times 10 \times 10^{-6}}$

$X_C = 199\ \Omega$

$I = \frac{V}{X_C}$

$I = \frac{500}{199}$

$I = 2.51$ Amps

Figure 12–12

### Phase Shift

The effect of resistors in ac circuits is the same as in dc circuits. The resistor controls the amount of current. The effect of capacitors and inductors in ac circuits is quite different from their effects in dc circuits. Or at least it appears that way.

1. In a dc circuit, a capacitor opposes a change in voltage. But in an ac circuit, the voltage is always changing.
2. In a dc circuit, an inductor opposes a change in current. But in an ac circuit, the current is always changing.

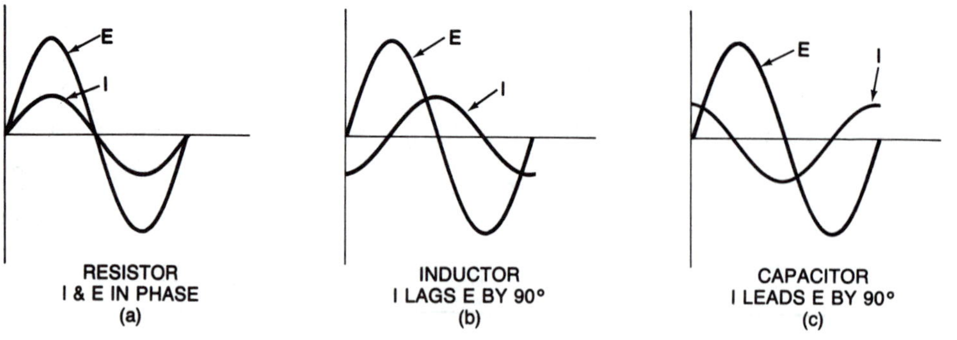


**Figure 12–13**

These two considerations provide the background for an approach to ac circuit problems. Before looking at these problems, three other aspects of components should be restated.

1. In dc circuits with inductors, it was shown (page 86) that the current lags the voltage, that is, the current was zero when the voltage was maximum. The same condition exists in ac circuits. (See Figure 12–13b.)
2. In dc circuits with capacitors, it was shown (page 80) that the current led the voltage, that is, the current was maximum when the voltage was changing most rapidly. This took place at 0 volts. The same condition exists in ac circuits. (See Figure 12–13c.)
3. The current and voltage in a resistor are always in phase with each other. Figure 12–13a shows this relationship graphically.

## *AC CIRCUITS*

In ac circuits, the effects of current and voltage in resistors, inductors, and capacitors are not the same. The current and voltage have different phase relationships depending on the component.

Because there are phase differences between voltages and between currents, when voltages are added or currents are added, the addition must be vectorial. (Refer to the paragraphs relating to two-phase voltages on page 112.)

### Series AC Circuits

If a combination of components (resistors, inductors, and capacitors) is connected in a series circuit, the total effect opposing current will be based on the values of the individual components. The values of the components may not simply be added together, for they do not act in phase with each other. In order to take into account the phase angle of the components, vector addition is used. The result is called *impedance* and is symbolized by the letter Z.

## *SERIES RL CIRCUITS*

**EXAMPLE 3**

Find the total current flow in a circuit containing a 2-henry inductor and an 800-ohm resistor. Determine the voltage across each component across a generator producing 440 volts at 60 hertz.

***Solution:*** See Figure 12–14.

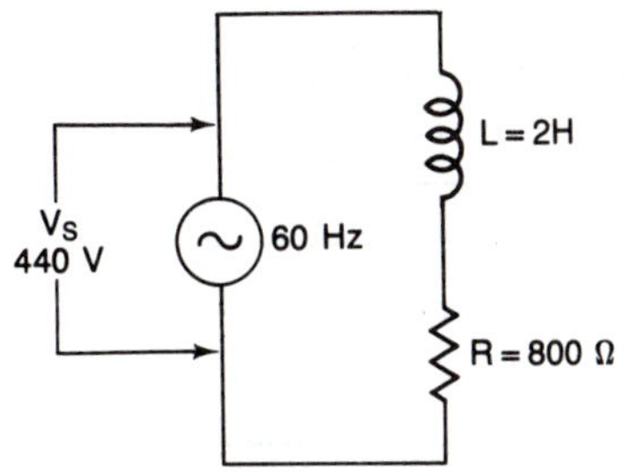


$X_L = 2\pi fL$

$X_L = 6.28 \times 60 \times 2$

$X_L = 753.6\ \Omega$

$Z = \sqrt{R^2 + X_L^2}$

$Z = \sqrt{800^2 + 753.6^2}$

$Z = 1099\ \Omega$

$I = \frac{V}{Z} = \frac{440\ V}{1099\ \Omega} = 0.4$ Amp

$V_R = I \times R = 0.4 \times 800 = 320$ V

$V_L = I \times X_L = 0.4 \times 753.6 = 301$ V

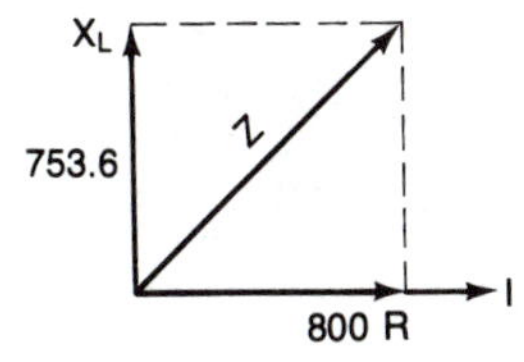


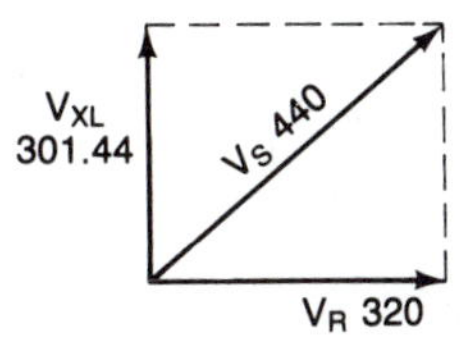


$V_S = \sqrt{V_R^2 + V_X^2}$

$V_S = \sqrt{320^2 + 301.44^2}$

$V_S = 439.6$ or 440 V

**Figure 12–14**

---

The sum of the voltages across the components must equal the supply voltage. The addition is vector addition. The total voltage in a series circuit is the sum of the voltages across the individual components.

***Remember*:** Problems in ac are solved just as dc problems were solved with the exception that all addition and subtraction must be done vectorially.

The basis for the vector diagram in Example 3 is that the circuit is in series. Current is the common factor in a series circuit and is used as the reference. The current vector is drawn at 0 degrees.

The current is in phase with the resistive component and the resistive component is also shown at 0 degrees.

The current lags the voltage across the inductive component. The inductive reactance is shown at 90 degrees. Assuming vectors rotate in a counterclockwise direction, the current and resistance vectors are lagging behind the inductive component. The current lags in an inductive circuit.

## *SERIES RC CIRCUITS*

The current leads the voltage in a capacitor by 90 degrees. Consider the series circuit shown in Figure 12–15. A 10-μf capacitor and 200-ohm resistor are connected in series across a generator producing 200 volts at 100 Hz.

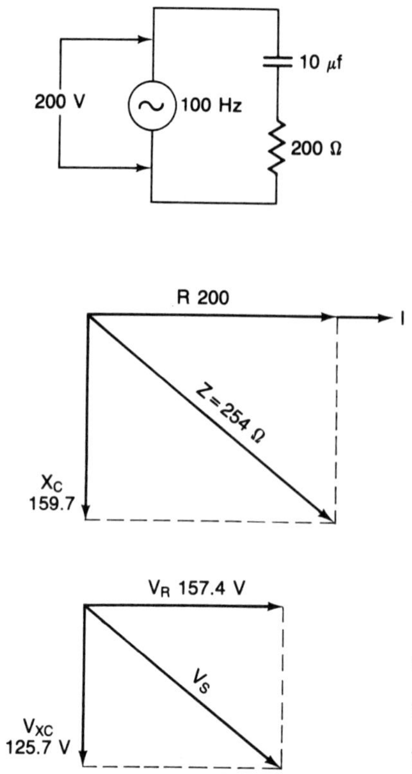


$$X_C = \frac{1}{2\pi fC}$$

$$X_C = \frac{1}{6.28 \times 100 \times 10 \times 10^{-6}}$$

$$X_C = 159.7\ \Omega$$

$$Z = \sqrt{R^2 + X_C^2}$$

$$Z = \sqrt{200^2 + 159.7^2}$$

$$Z = 254\ \Omega$$

$$V_S = \sqrt{V_R^2 + V_C^2}$$

$$V_S = \sqrt{157.4^2 + 125.7^2}$$

$$V_S = 201 \text{ or } 200\ \text{V}$$

**Figure 12–15**

---

**EXAMPLE 4**

Determine the current of the circuit of Figure 12–15 and the voltage across each component.

***Solution:*** The total impedance of the circuit is 254 ohms. Find the current.

$$I = V/R = 200/254 = 0.787 \text{ amp}$$

$$V_R = I \times R = 0.787 \times 200 = 157.4\ \text{V}$$

$$V_C = I \times X_C = 0.787 \times 159.7 = 125.7\ \text{V}$$

Check that the sum of the voltage is equal to the supply voltage. (See Figure 12–15.)

---

Note that in the vector diagram, the current as reference is at 0 degrees. The resistance and the voltage across the resistor are shown at 0 degrees in phase with the current. The capacitive reactance and the voltage across the capacitive reactance are shown with the reference current leading by 90 degrees.

## *SERIES RCL CIRCUITS*

When inductance and capacitance are both connected in a series circuit, the effects of the two components directly oppose each other. This relationship is shown in Figure 12–16 where the vector diagram shows $X_L$ directly opposite $X_C$.

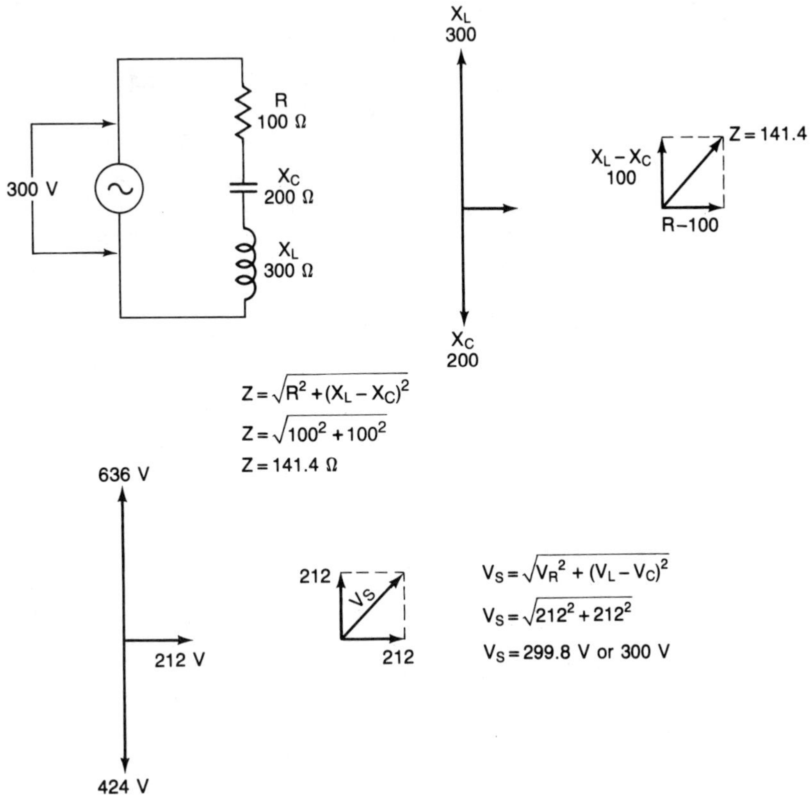


**Figure 12–16**

Vectorial adding of $X_L$ and $X_C$ provides a resultant of 100 in the inductive direction (300 − 200 = 100). The total impedance is then:

$$Z = \sqrt{R^2 + (X_L - X_C)^2} \qquad Z = \sqrt{100^2 + 100^2}$$

$$Z = \sqrt{R^2 + X^2} \qquad Z = \sqrt{20{,}000}$$

$$Z = 141.4\ \Omega$$

Solve for the current in the circuit:

$$I = E/Z$$

$$I = 300/141.4$$

$$I = 2.12 \text{ amps}$$

Solve for the voltage across each component:

$$E_R = I \times R \qquad E_C = I \times X_C \qquad E_L = I \times X_L$$

$$E_R = 2.12 \times 100 \qquad E_C = 2.12 \times 200 \qquad E_L = 2.12 \times 300$$

$$E_R = 212 \text{ V} \qquad E_C = 424 \text{ V} \qquad E_L = 636 \text{ V}$$

Note that the voltages across the inductance coil and the capacitor are both higher than the applied voltage. Even so, vectorially, the voltages do add up to the supply voltage of 300 volts.

## *SERIES RESONANT CIRCUITS*

*Resonant circuits* are a special class of reactance circuits where inductive reactance $X_L$ is equal to $X_C$. In such circuits, the reactances cancel each other. The resistance in the circuit is the only factor controlling current flow.

Circuits containing both inductance and capacitance are frequency sensitive. Consider the effect of frequency on individual components of a series RCL circuit. As shown graphically in Figure 12–17, frequency does not affect resistance. The 100-ohm resistor is 100 ohms at all frequencies.

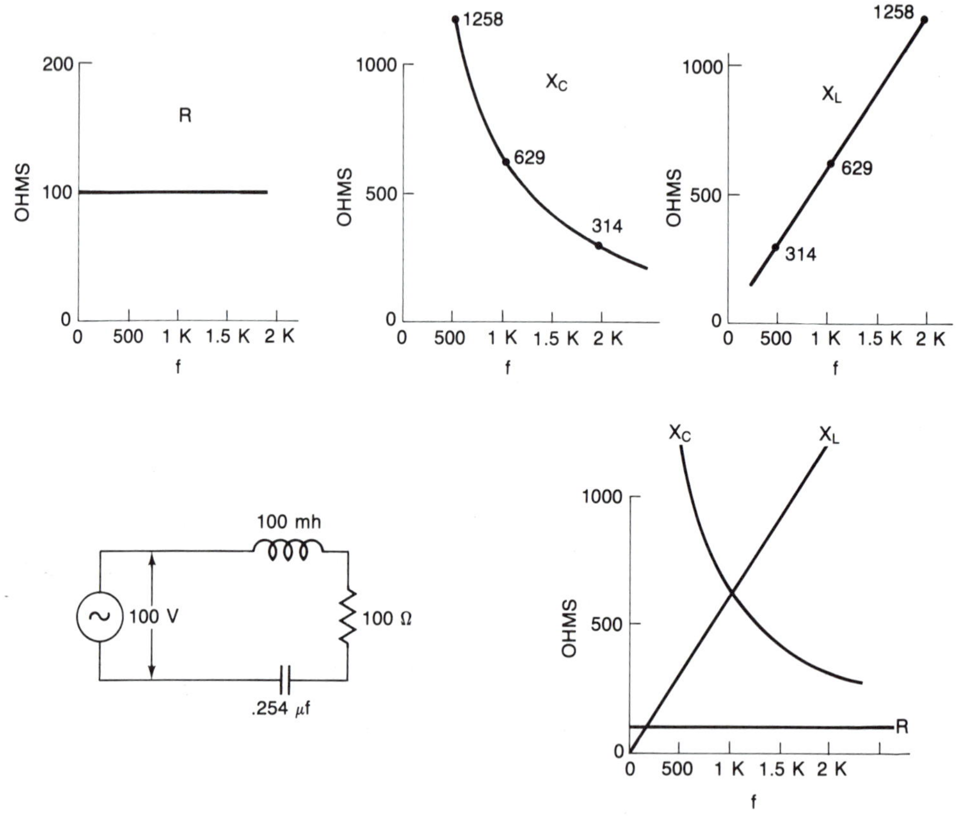


**Figure 12–17**

Frequency *does* affect capacitive reactance. Capacitive reactance decreases as frequency increases. (Figure 12–17.)

Frequency also affects inductive reactance. Inductive reactance increases with frequency. (Figure 12–17.)

The changing reactances with frequencies are shown together in Figure 12–17. Note the resonant frequency at 1,000 Hz where $X_C$ and $X_L$ are equal. What is a definition of resonance? It is where inductive reactance and capacitive reactance are equal and cancel each other. A formula may be derived for resonance.

Resonance:

$$X_L = X_C$$

$$2\pi fL = \frac{1}{2\pi fC} \qquad \text{Multiply both sides by } 2\pi fC$$

$$4\pi^2 f^2 LC = \frac{1}{1} \qquad \text{Divide both sides by } 4\pi^2 LC$$

$$f^2 = \frac{1}{4\pi^2 LC} \qquad \text{Take square root of both sides}$$

$$f = \frac{1}{2\pi\sqrt{LC}}$$

The frequency of resonance is by formula:

$$f = \frac{1}{2\pi\sqrt{LC}}$$

In Figure 12–17, the reactance of the components changes with frequency. These changes are shown in Table 12–1. Also shown is the difference $X_L - X_C$ or $X_C - X_L$ depending on which one is the larger.

Finally, in Figure 12–18, a plot of the current against frequency is shown. Note the peak current of 1 ampere at resonance. Also note that the current drops off rapidly on either side of resonance.

Because the current is maximum at resonance, the impedance must be minimum at this frequency. The impedance is minimum at resonance and is equal to the resistance of 100 in the circuit. An impedance plot is also shown in Figure 12–18.

The ability of the LC circuit to distinguish between different frequencies makes the circuit useful as a filter. The series LC circuit may be used to select a band of frequencies while rejecting other frequencies.

**TABLE 12–1**

Resonance Chart

The effect of frequency on the circuit of Figure 12–17:

| *Freq.* Hz | $2\pi fL$ $X_L$ | $\frac{1}{2\pi fC}$ $X_C$ | $X$ $X_L - X_C$ | $\sqrt{R^2 + X^2}$ $Z$ | $V/Z$ $I$ |
|---|---|---|---|---|---|
| 100 | 62.8 | 6,269 | 6,206 | 6,207 | 0.016 |
| 200 | 125 | 3,134 | 3,009 | 3,010 | 0.033 |
| 300 | 188 | 2,089 | 1,901 | 1,903 | 0.053 |
| 400 | 259 | 1,567 | 1,308 | 1,312 | 0.076 |
| 500 | 314 | 1,254 | 940 | 945 | 0.105 |
| 600 | 376 | 1,045 | 669 | 676 | 0.148 |
| 700 | 439 | 895 | 456 | 467 | 0.214 |
| 800 | 502 | 784 | 282 | 299 | 0.334 |
| 900 | 565 | 697 | 132 | 166 | 0.624 |
| 1,000 | 628 | 628 | 0 | 100 | 1.0 |
| 1,100 | 691 | 570 | 121 | 156 | 0.641 |
| 1,200 | 753 | 522 | 231 | 252 | 0.397 |
| 1,300 | 816 | 482 | 334 | 348 | 0.287 |
| 1,400 | 879 | 447 | 432 | 443 | 0.226 |
| 1,500 | 942 | 417 | 525 | 534 | 0.187 |
| 1,600 | 1,004 | 392 | 612 | 620 | 0.161 |
| 1,700 | 1,067 | 368 | 699 | 706 | 0.142 |
| 1,800 | 1,130 | 348 | 782 | 788 | 0.127 |
| 1,900 | 1,193 | 330 | 863 | 868 | 0.115 |
| 2,000 | 1,256 | 313 | 943 | 948 | 0.105 |
| 3,000 | 1,884 | 209 | 1,675 | 1,677 | 0.060 |
| 5,000 | 3,140 | 125 | 3,015 | 3,016 | 0.033 |
| 10,000 | 6,280 | 62.7 | 6,217 | 6,218 | 0.016 |

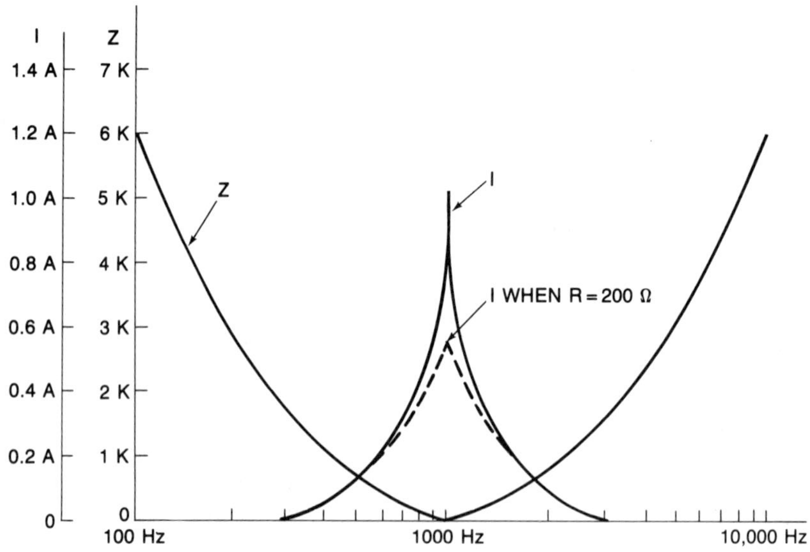


**Figure 12–18**

## *HALF-POWER POINTS AND BANDWIDTH*

A rating system has been devised to provide technical people with a means of categorizing the characteristics of circuits. With regard to filters, this is called the *half-power point*. At this point, the voltage or current is down to 0.707 of the maximum. It indicates when the power will be down to 0.5, or one-half. If the voltage is down 0.707, the current will be down 0.707. Accordingly, 0.707 E × 0.707 I will provide 0.5 of the original power.

In the circuit of Figure 12–17, there is a peak current of 1 ampere at 1,000 Hz. The current is down to 0.707 at frequencies of about 925 Hz and 1,080 Hz. These two frequencies are at the half-power points.

The bandwidth of the circuit is obtained by subtracting the frequency at the lower half-power point from the frequency at the higher half-power point.

$$1{,}078 \text{ Hz} - 922 \text{ Hz} = 156 \text{ Hz}$$

Another measurement of circuit performance is the quality, or Q, of the circuit. The Q of a circuit is a number without units. In the case of the circuit of Figure 12–17, the Q is about 6.25. The Q is found by dividing the center frequency by the bandwidth of the circuit.

Another method of determining Q is by the formula:

$$Q = X_L/R$$

$$628\Omega/100\Omega = 6.28$$

$X_L$ is used to determine Q rather than $X_C$ because in many circuits external resistors are not used. The most active resistance in the circuit is the resistance of the inductance coil. As resistance in the circuit is increased, the Q decreases. This effect can be seen in Figure 12–18. The broken line curve shows current if a series resistance of 200 ohms were used. The current would decrease to 0.5 amp. Q would decrease to 3.14.

### *PARALLEL RL CIRCUITS*

In parallel circuits, voltage is the common factor. Voltage is therefore used as the reference when solving a parallel RL circuit by vector diagram.

Consider the circuit of Figure 12–19. Find the line current and total circuit impedance.

$$I = V/R \qquad 1{,}000/1{,}000 = 1 \text{ amp}$$

$$I = V/X_L \qquad 1{,}000/600 = 1.666 \text{ amps}$$

The voltage is the same across each component. In order to determine line current, the branch currents must be added. (ac addition is vectorial.)

$$I_L = \sqrt{I_R^2 + I_L^2}$$

$$I_L = \sqrt{1^2 + 1.666^2}$$

$$I_L = 1.939 \text{ amps}$$

The total impedance and then the current could have been calculated by use of the reciprocal formula. (Note again vector addition.)

$$Z = \frac{1}{\sqrt{\left(\frac{1}{R}\right)^2 + \left(\frac{1}{X_L}\right)^2}}$$

The following is a scientific calculator solution of the problem associated with Figure 12–19.

| STEP | ENTER | PRESS | DISPLAY |
|---|---|---|---|
| 1 | 1000 | 1/x | 0.001 |
| 2 | | $x^2$ | 0.000001 |
| 3 | | + | 0.000001 |
| 4 | 600 | 1/x | 0.001666 |
| 5 | | $x^2$ | 0.00000277 |
| 6 | | = | 0.00000377 |
| 7 | | $\sqrt{\ }$ | 0.001943 |
| 8 | | 1/x | 514.5 |

The answer is 514.5 ohms, which checks with the vector diagram solution shown in Figure 12–19. The 1.2-ohm error is due to the dropping of decimal places in the vector diagram solution.

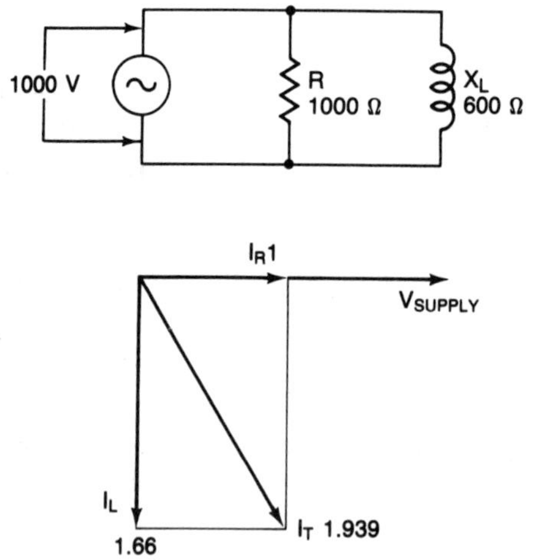


$I_R = \frac{V}{R} = \frac{1000}{1000} = 1 \text{ Amp}$

$I_L = \frac{V}{X_L} = \frac{1000}{600} = 1.66 \text{ Amp}$

$I_L = \sqrt{I_R^2 + I_L^2}$

$I_L = \sqrt{1^2 + 1.66^2}$

$I_L = \sqrt{3.777}$

$I_L = 1.939 \text{ Amp}$

$Z = \frac{V}{I} = \frac{1000 \text{ V}}{1.939 \text{ A}}$

$Z = 515.7\ \Omega$

**Figure 12–19**

### *PARALLEL RC CIRCUITS*

The solution of parallel RC circuits is similar to an RL circuit with the exception that the capacitive current must be drawn leading. Consider the problem in Figure 12–20. Find the total line current and the circuit impedance.

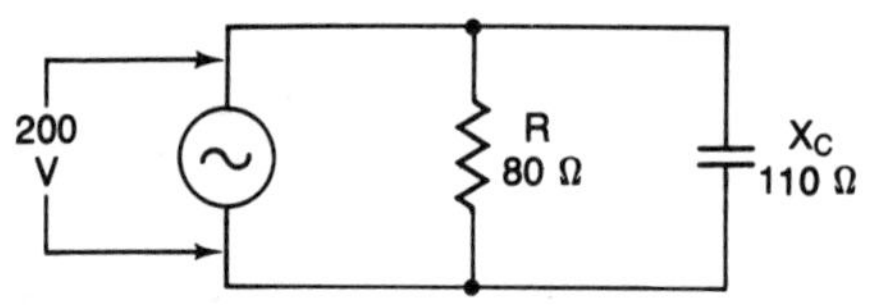


$I_R = \frac{V}{R} = \frac{200}{80} = 2.5 \text{ A}$

$I_C = \frac{V}{X_C} = \frac{200}{110} = 1.818 \text{ A}$

$I_T = \sqrt{I_R^2 + I_C^2}$

$I_T = \sqrt{2.5^2 + 1.818^2}$

$I_T = 3.09 \text{ Amps}$

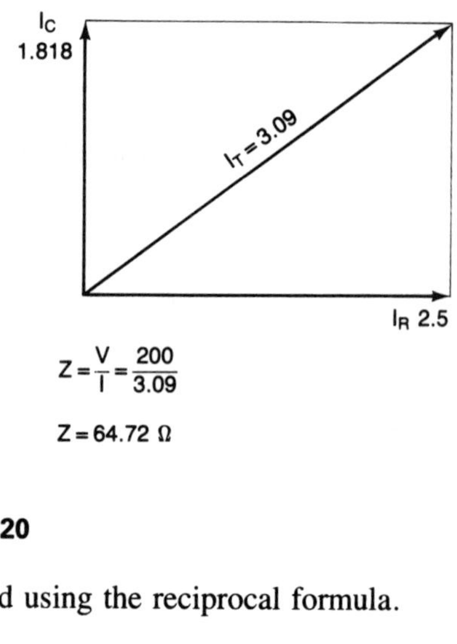


$Z = \frac{V}{I} = \frac{200}{3.09}$

$Z = 64.72\ \Omega$

**Figure 12–20**

The problem could also have been solved using the reciprocal formula.

$$Z = \frac{1}{\sqrt{\left(\frac{1}{R}\right)^2 + \left(\frac{1}{X_C}\right)^2}}$$

### Parallel RCL Circuits

The parallel RCL circuit is also solved using the current-vector diagram. Consider the circuit of Figure 12–21.

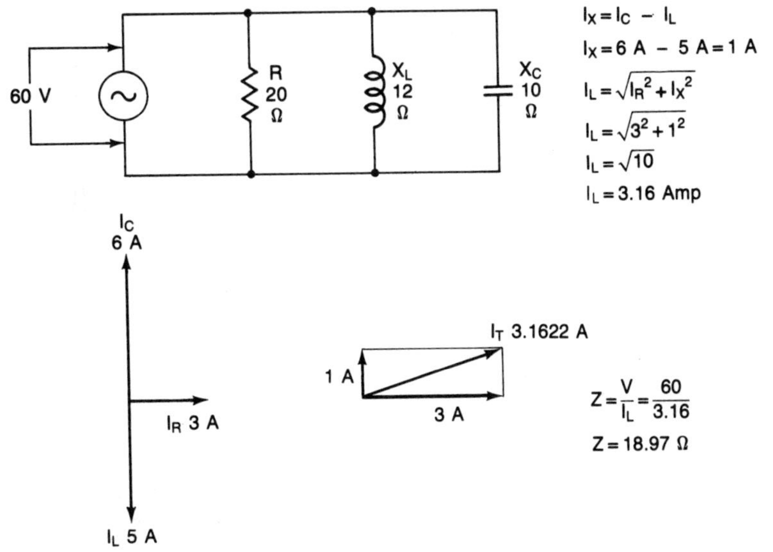


**Figure 12–21**

## *PARALLEL RESONANCE*

A parallel LC circuit operating at the frequency given by the formula

$$f_R = \frac{1}{2\pi\sqrt{LC}}$$

will be resonant at that frequency. The line current will be minimum while the current circulating between the inductance coil and capacitor will be maximum. (See Figure 12–22.)

If it were possible to wind a coil with wire without resistance, the coil and capacitor currents would be equal to:

$$I = V/X_L \qquad (X_L = X_C \text{ at resonance})$$

The line current supplied by the generator would be zero. The impedance of the circuit would be infinite.

In the practical situation, it is not possible to obtain perfect coils or capacitors. There will be some line current at the resonant frequency. The impedance of the circuit is

$$Z = \frac{V}{I_{LINE}}$$

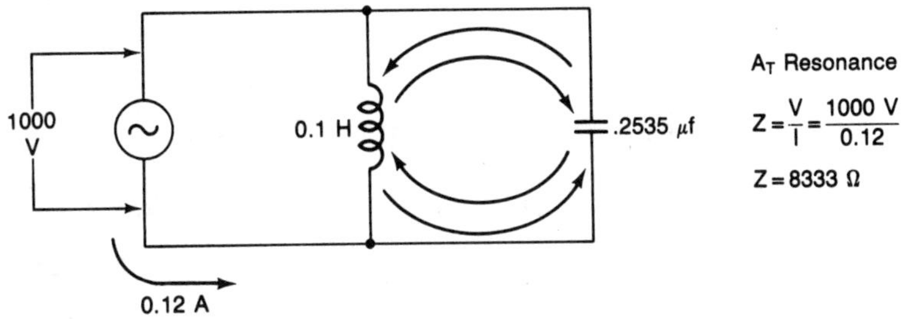


**Figure 12–22**

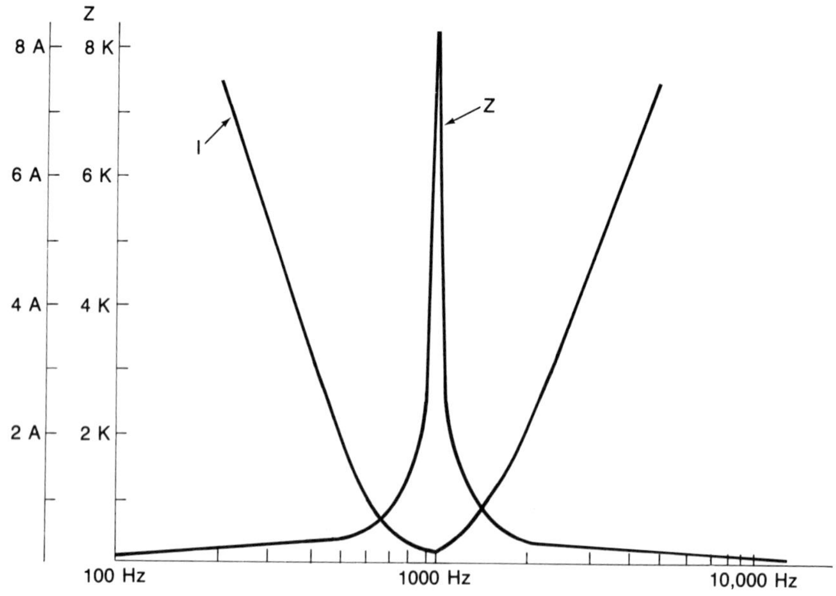


**Figure 12–23**

The impedance of the circuit may be plotted for frequencies below and above resonance. (See Figure 12–23.) With the values given for L and C in Figure 12–22 the resonant frequency will be

$$f = \frac{1}{2\pi\sqrt{LC}}$$

$$f = \frac{1}{6.28\sqrt{0.1 \times 0.2535 \times 10^{-6}}}$$

$$f = 1{,}000 \text{ Hz}$$

As indicated in Figure 12–22 the line current at resonance is 0.12 amp. Both the coil and capacitor have reactances of about 628 ohms at the resonant frequency.

$$X_L = 2\pi fL$$

$$X_L = 6.28 \times 1{,}000 \times 0.1$$

$$X_L = 628$$

$$X_C = \frac{1}{2\pi fC}$$

$$X_C = \frac{1}{6.28 \times 1{,}000 \times 0.2535 \times 10}$$

$$X_C = 628$$

The circulating current is 1.59 amperes.

$$I = V/X$$

$$I = 1{,}000/628$$

$$I = 1.59 \text{ amps}$$

In such a situation, the impedance of the circuit at resonance is

$$Z = \frac{V}{I_{LINE}}$$

$$Z = \frac{1{,}000V}{0.12 \text{ amp}}$$

$$Z = 8{,}333\ \Omega$$

If the circuit drew a higher line current at resonance 0.5 amp (lower Q), the shape of the curve would change. The peak of the impedance curve would be lower at 2,000Ω. The minimum line current dip would be to 0.5 amp.

The Q of the circuit is given by the formula:

$$Q = \frac{I \text{ circulation}}{I_{LINE}}$$

$$Q = \frac{1.59}{0.12}$$

$$Q = 13.25$$

Since Q is also equal to center frequency divided by bandwidth (BW), the approximate bandwidth may be determined by:

$$Q = CF/BW$$

$$BW = CF/Q$$

$$BW = 1{,}000/13.25$$

$$BW = 75.5 \text{ Hz}$$

At frequencies below and above resonance, the impedance of the circuit will be lower than 8,333 ohms, and the line current will be higher. Figure 12–23 shows this relation.

Some of the data in the resonance chart for the series circuit (Table 12–1) apply to this parallel resonant circuit. For example, $X_L$ and $X_C$ are frequency dependent; they do not change because of series or parallel connection. An approximation of the impedance of the parallel resonant circuit may be found using the data in the resonance chart. Keep in mind that the maximum impedance of 8,333 ohms has already been determined. At each frequency other than resonance, the approximate impedance may be found by the formula:

$$Z = \frac{1}{1/X_L - 1/X_C}$$

The line current is determined by $I = V/Z$ for the impedance at each frequency.

The series and parallel resonant circuits are often used as filters in radio circuitry at the higher frequencies. At lower frequencies, the required inductance coils are large and heavy. Quite often, filters at these lower frequencies are constructed using resistors and capacitors.

## RC FILTERS

A *filter* is used either to allow a signal to pass or to reject it. A combination of the *pass* and *reject* types is sometimes used for specific purposes.

### Low-Pass RC Filters

A low-pass filter is one that allows certain low frequencies to pass but attenuates (makes smaller) frequencies above a cutoff point. The cutoff frequency is described as the frequency where the output is down 3 db (decibels), or 0.707 of the maximum.

Consider the circuit in Figure 12–24. This is a low-pass filter. The output voltage is taken across the capacitor. At low frequency, $X_C$ will be high. Almost all of the generated voltage will appear across the capacitor and therefore at the output. As frequency is increased, $X_C$ will decrease. So will the voltage across the capacitor. At about 1,000 Hz, $X_C$ will be 10 k and $V_C$ will be 0.707 of voltage applied. $V_{OUT}$ will be 0.707 of voltage applied. The output will be down to 0.707 of voltage maximum, and this is called the *cutoff frequency*. All higher frequencies produce lower output voltages.

### High-Pass RC Filters

Interchanging the components provides a high-pass filter. The reactance of the capacitor is high at the low frequencies, and with the capacitor in series the low frequencies will be blocked. (See Figure 12–25.) The cutoff frequency is the same,

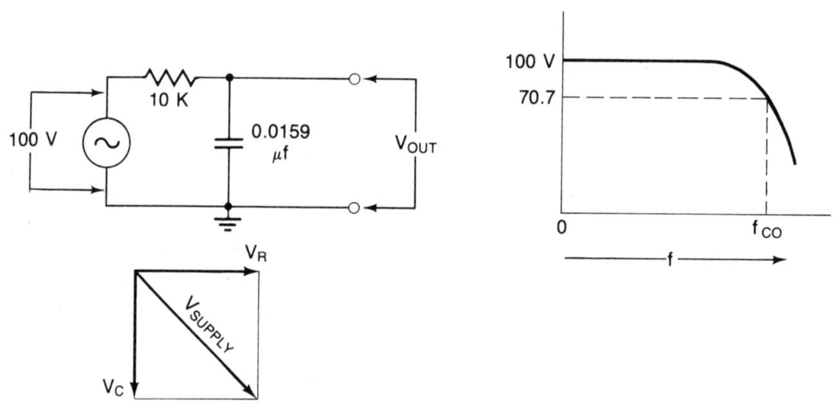


Figure 12–24

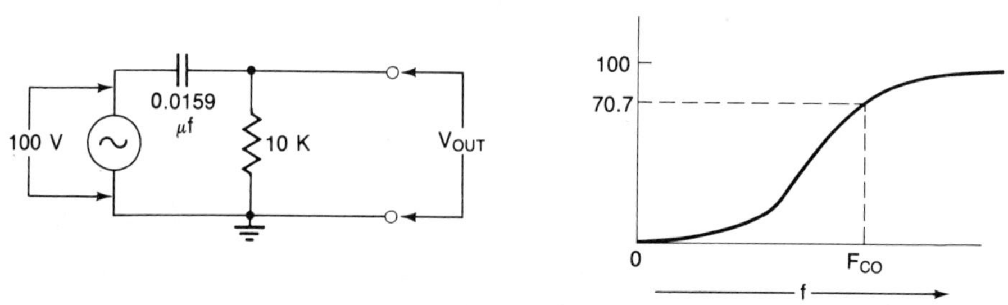


Figure 12–25

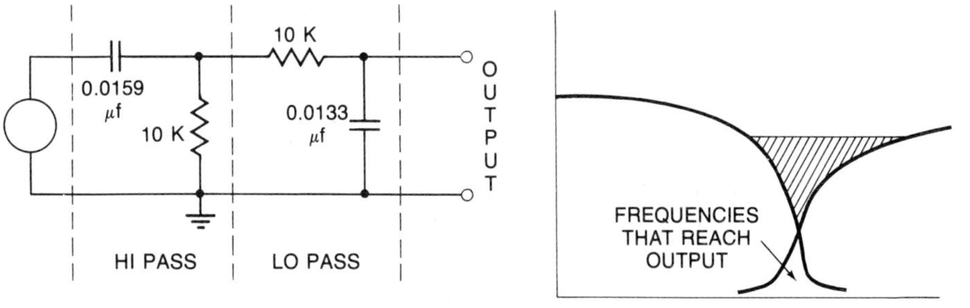


**Figure 12–26**

1,000 Hz, because the same components were used. In this circuit, only the frequencies above 1 kHz will be passed to the output. Keep in mind that in RC filters, the cutoff frequency is always at the point where $X_C = R$.

### Band-Pass RC Filters

If a low-pass filter and a high-pass filter are connected in series, a band-pass filter may be provided, as in Figure 12–26. The high-pass filter of Figure 12–25 and another filter setup for low-pass operations are connected together. The reactance of C is 10 k at 1,200 Hz. Along with R it will function to pass all frequencies below 1,200 Hz. The C-R combination will pass all frequencies above 1,000 Hz.

The curve of Figure 12–26 shows the results of the circuit connection. Only the signals at frequencies between 1,000 Hz and 1,200 Hz will pass through the filter.

## *LC FILTERS*

There are two main types of LC filters that are used in electrical-electronic circuits. They are the *band-pass* and *band-reject* filters. The circuits used to make up the filters are the same for both filters. They are the series-resonant and parallel-resonant circuits. It is the connection of the circuit in the system that determines the passing or rejecting of a band of frequencies.

For example, the series-resonant circuit allows a band of frequencies to pass through the circuit with low impedance. If connected as in Figure 12–27a, a small band of frequencies will pass to the load. It is a band-pass filter. If connected as in Figure 12–27b, the same small band of frequencies will be shorted to ground. All frequencies other than the one the circuit is tuned to will reach the load. It is a band-reject filter.

A parallel-resonant circuit provides high impedance to frequencies around the center frequency. If placed in series with the line as in Figure 12–27d, it will reject those frequencies. If placed from the line to ground, it will allow the band around the center frequency to reach the load while shorting others to ground. (Figure 12–27c.)

Combinations of series-resonant and parallel-resonant circuits tuned to the same frequency provide sharper, more narrow bandwidth to the filters. Examples are shown in Figures 12–27e and 12–27f.

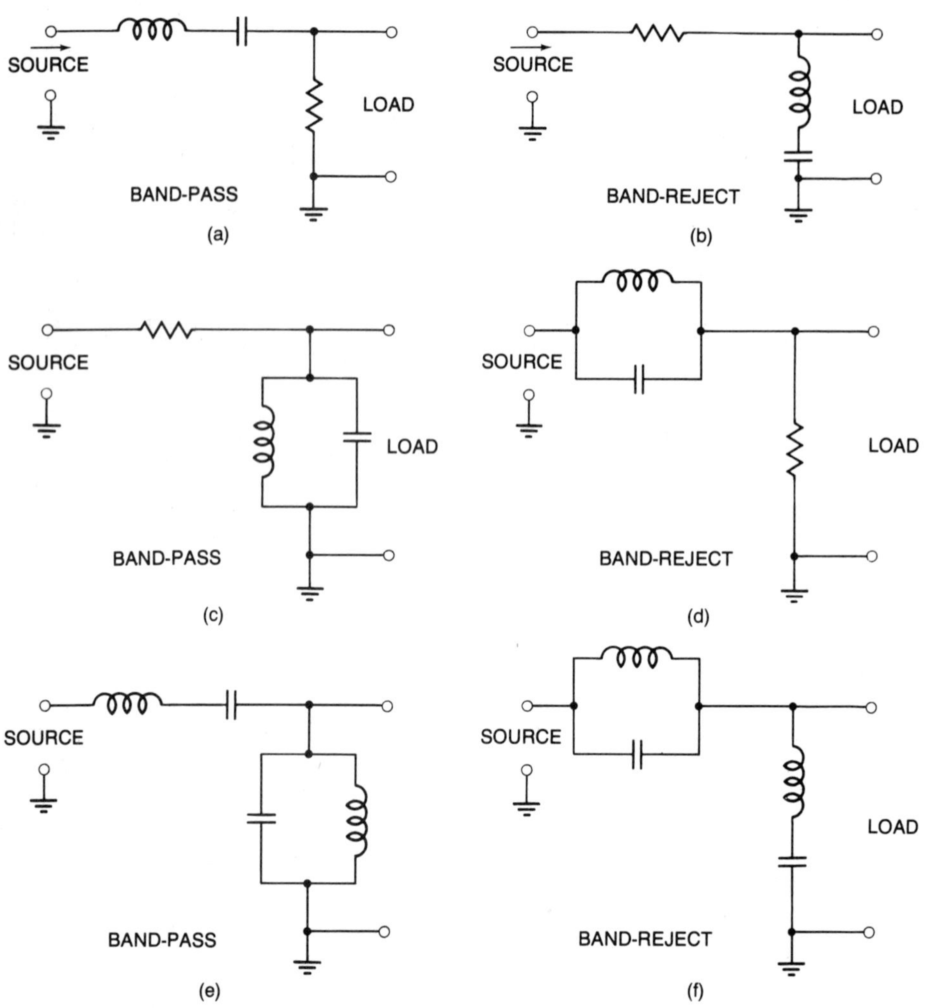


**Figure 12–27**

## *RC FILTER APPLICATIONS*

RC filters have many practical applications. One of the common uses of this type filter is in the Modulator-Demodulators (MODEM) associated with telephone hookup to computers. The RC filter along with special amplifier circuits provides a simple and inexpensive means of picking out the frequencies that represent different numbers.

Another common use of RC filters is in communication transmitters. Federal Communication Commission (FCC) regulations limit the bandwidth of the signal that

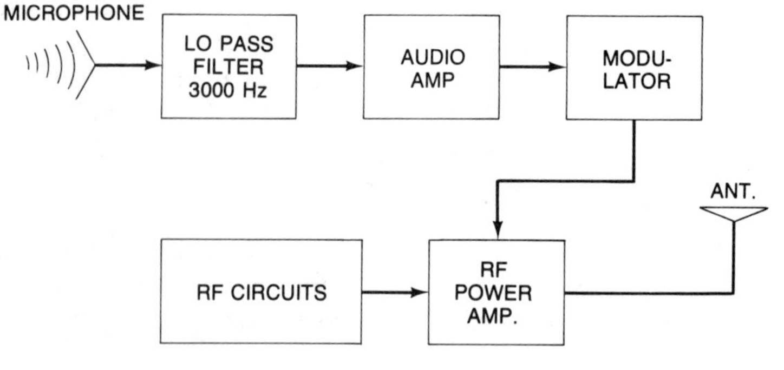


**Figure 12–28**

may be transmitted. One method of achieving this limitation of frequencies is to include an RC filter with the audio amplifier used in the system. A block diagram of such a system is shown in Figure 12–28.

The low-pass filter normally would have a cutoff frequency somewhere slightly above 3,000 Hz. Higher frequencies would be eliminated, but good voice transmission, one well recognized, would result.

## AC POWER AND POWER FACTOR

Current and voltage are not always in phase or 90 degrees out of phase with each other in ac circuits. Power then is not, as in dc, simply the multiplying of current by voltage. The phase-angle between zero and 90 degrees must be considered.

One method of determining power would be as follows. Take the instantaneous value of voltage and current for a number of points throughout a cycle and multiply them together. Figure 12–29a shows the result of multiplying instantaneous values of current and voltage when current and voltage are in phase. Figure 12–29b shows the result of the same multiplication when current and voltage are not in phase.

Figure 12–29a shows positive voltage times positive current providing positive power. This is true algebraically. In Figure 12–29b, the voltage and current are not always of the same polarity. When voltage and current are of opposite polarities, negative power will be produced. Negative power is power that is supplied to the circuit but not used in the circuit. The negative power is returned to the source.

An example of a circuit exhibiting these characteristics may be observed in Figure 12–30, which shows a motor drawing 60 amperes from a 220-volt power source. Further investigation shows the current lagging the voltage by 22 degrees (motors contain inductance coils).

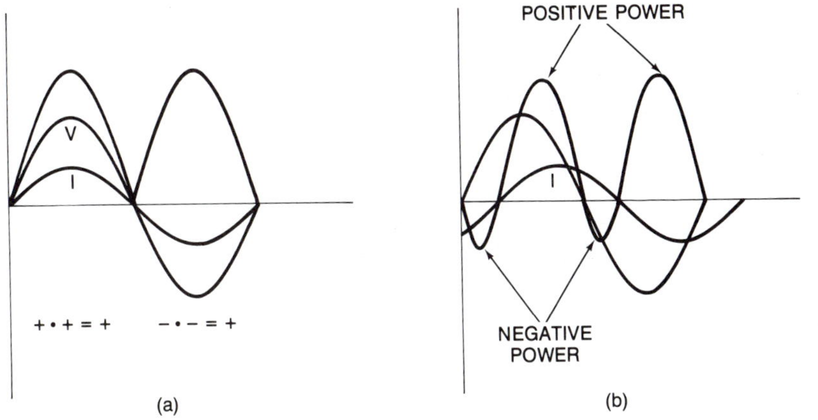


**Figure 12–29**

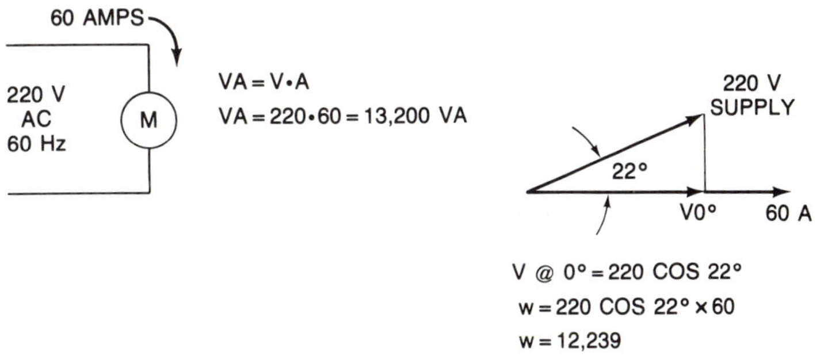


**Figure 12–30**

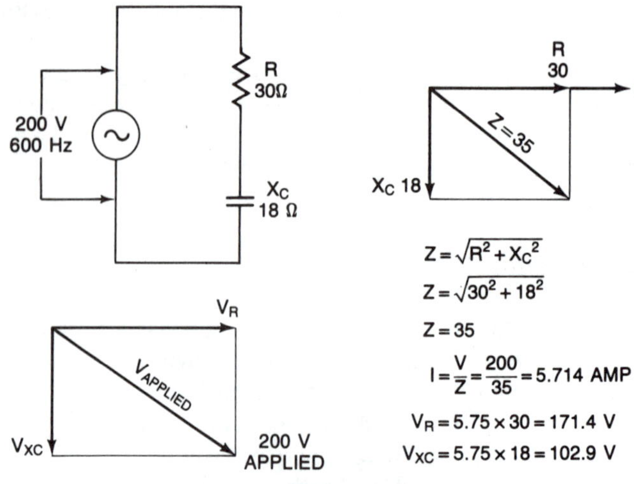


**Figure 12–31**

The power developed in the motor is the current times the voltage resolved to 0 degrees. Simply stated, the power developed in an ac circuit is equal to the current multiplied by the voltage multiplied by the cosine of the angle between the current and voltage. The cosine of the angle is called the *power factor*. The formula for ac power becomes:

$$W = V \times I \times \text{pf or}$$

$$W = V \times I \times \cos \measuredangle$$

Simple RC and RL circuits may be looked at in another light. For example, in Figure 12–31, determine the power developed in the circuit.

1. The current vector I is in phase with the resistance vector R.
2. The supply voltage is applied to the complete circuit, which is the total impedance Z.
3. The phase-angle between current and supply voltage is the same as the angle between R and Z.

$$\text{pf} = \cos \text{ of angle}$$
$$\cos = \text{Adj/Hyp} = R/Z \qquad \text{Power} = I \times V \times \text{pf}$$
$$\cos = 30/35 \qquad W = 5.71 \times 200 \times \cos 31$$
$$\cos = 0.8571 \qquad W = 5.71 \times 200 \times 0.8571$$
$$\measuredangle = 31 \text{ degrees} \qquad W = 978 \text{ watts}$$

Observe in the voltage-vector diagram that if the applied voltage were multiplied by the cosine of the angle, the resultant is the voltage across the resistor.

The formula $W = I \times V \times \text{pf}$ is really nothing more than an indication that power is developed in the resistive, not the reactive component. For the circuit of Figure 12–31:

$$W = I \times V_R$$

$$W = 5.71 \times 171.4$$

$$W = 978 \text{ watts}$$

Since all the power is developed in the resistive component, the formulas

$$W = I^2 R$$

$$W = E^2/R$$

are suitable for use in CR, LC, LR, and LCR circuits.

## REVIEW QUESTIONS

1. What is the most common form of electrical energy in the United States?
2. What is the frequency of the ac produced in the United States?
3. What are the two factors that determine the voltage induced in a wire?
4. What is the left-hand rule for generator action?
5. The abbreviation dc stands for ______.
6. The abbreviation ac stands for ______.
7. In the two-phase system, the voltages are out of phase by ______ degrees.
8. In the three-phase system, the voltages are out of phase by ______ degrees.
9. In the Wye system with a winding voltage of 100 volts, the line voltage is ______.
10. In the delta system with a line current of 34.4 amperes, the winding current is ______ amperes.
11. An RC circuit consists of a 10-kΩ resistor and a capacitive reactance of 14 kΩ connected in series. If the applied voltage is 100 volts, what is the value of current in the circuit?
12. In the circuit of question 11, what is the voltage across each component?

    $V_R =$ $\qquad$ $V_{XC} =$

13. In the circuit of question 11, what is the phase-angle between the current and the supply voltage?
14. A certain low-pass RC filter has an output of 12 volts well below the cutoff frequency. What will be the value of voltage at the cutoff frequency?
15. An LC band-pass filter is resonant at 318 kHz. If the Q of the circuit is 23, what is the bandwidth of the filter at the half-power points?
16. A two-phase generator (90°) has a common connection. The voltage between phases is 282.8 volts. What is the voltage across one of the phases?
17. A three-phase Wye-connecting generator has an output of 208 volts between phases. If the common connection of the Wye system is grounded, what is the voltage from any phase line to ground?
18. A balanced delta load of 200 ohms on each leg is connected to a three-phase source of 220 volts. What is the line current?
19. A balanced Wye load of 200 ohms on each leg is connected to a three-phase source of 220 volts. What is the line current?
20. A three-phase motor is connected to a 220-volt source. The balanced line current is 43.6 amps at a phase-angle of 27 degrees. How much power is being developed in the motor?

# UNIT THIRTEEN

# *ELECTRICAL SAFETY*

Electrical shock—possibly causing serious injury or death—is a hazard that all service technicians must be aware of. In this unit an explanation will be given of

1. Effects of current flow through the body.
2. Factors affecting the amount of current flow.
3. Dangerous conditions and equipment.
4. First aid for shock victims.

## *EFFECT OF CURRENT FLOW THROUGH THE BODY*

Many tests have been made attempting to determine the effect of current levels on the human body. The general effects are shown in Table 13–1. Keep in mind that individuals will have different responses to current flow through their bodies. Table 13–1, then, shows the approximate amount of current causing a given body response.

Besides the responses given in Table 13–1, there are secondary reactions to electrical shock that often are the really serious part of the encounter. Any electrical shock that causes an involuntary response can cause a serious accident.

For example, a technician might be on a ladder or scaffolding while using a drill. If he or she were to receive a shock, the technician might involuntarily drop the drill. This could cause a serious accident to anyone who happened to be below the drill.

Another example is a technician working on a roof of a building. If the technician were to receive a strong electrical shock, a combination of voluntary and/or involuntary muscle contractions could cause the person to jump (be knocked off) the roof. One might live through the shock, but the fall from the roof might be another matter.

It is important that the technician always be aware of electrical dangers.

**TABLE 13–1**

| *Amount of current* | *Response* |
|---|---|
| 0 to 0.5 milliampere | no response |
| 0.5 to 2 milliampere | slight tingling to mild shock; quick withdrawal from body contact |
| 2 to 10 milliampere | mild to heavy shock; muscular tightening |
| 10 to 50 milliampere | painful shock; cannot let go |
| 50 to 100 milliampere | severe shock; breathing difficulties |
| 100 to 200 milliampere | heart convulsion; death |
| Over 200 milliamperes | severe burns; breathing stops |

## *FACTORS AFFECTING BODY CURRENT FLOW*

It was stated earlier in this textbook that Ohm's Law is true for the whole circuit or any part of the circuit. When the human body becomes the circuit (the path for current flow), Ohm's Law is true. The factors needed in determining the amount of current are the applied voltage and the body resistance. The applied voltage is the voltage with which the body comes in contact. The resistance of the body is the variable factor.

If the resistance of the human body were high enough, for example, 1 million (1,000,000) ohms, electrical shock would not be a problem.

### EXAMPLE 1

Air conditioning and refrigeration service technicians normally work with ac voltages under 600 volts. If the body resistance were 1 million ohms, the current flow by Ohm's Law would be

$$I = \frac{E}{R}$$

$$I = \frac{600}{1{,}000{,}000} \qquad \text{600 microamperes or 0.6 milliamperes}$$

The 0.6 milliampere of current is barely enough to cause a human to respond or even recognize that a shock was received.

The problem is the resistance of the human body is not a fixed value. In addition, the resistance of the human body is mainly the skin resistance. Internally, the body is mostly water, and salty at that. Salt water is a good conductor of electricity. It has low resistance.

Because it is the skin resistance that controls the overall body resistance, let's take a closer look at skin resistance. Dry skin has a resistance of about 200,000–700,000 ohms. A resistance at this level would be sufficiently high, keeping current flow below a value that would cause a problem.

Wet skin is a different situation. The resistance of skin wet with perspiration or other moisture can go as low as 300 ohms. Any contact with line voltage when your skin is wet can cause serious injury or death. Remember Ohm's Law is always true.

$$I = \frac{E}{R}$$

$$I = \frac{120 \text{ volts}}{300 \text{ ohms}}$$

$$I = 400 \text{ milliamperes}$$

According to Table 13–1, a current flow of 400 milliamperes through the body can cause death.

---

Consider the conditions in which a service technician might be working. One area in which a technician is often required to work is a kitchen. Here the high temperature can easily cause a body to perspire. In a short time, the body as well as shirt, pants, socks, and shoes could become damp with perspiration. Body resistance as well as contact resistance through clothes would go down rapidly. Any contact with normal working voltages could result in dangerous shock and burns.

Keep in mind that dry clothes provide good insulation. Clothes wet with perspiration or other moisture are usually low-resistance paths for electricity.

When you are wet it is dangerous to come in contact with line voltages.

## *DANGEROUS WORKING CONDITIONS AND EQUIPMENT*

As stated earlier, a dry body in dry clothes and dry shoes has sufficient resistance so that contact with a live wire should not cause a dangerous shock. The major difficulty is that it is nearly impossible to maintain these conditions when working. The dangerous working condition is to work around *live* line voltages whenever your body is damp with perspiration or other moisture. It is also dangerous to work around wet equipment, where a better contact to the body may be made through wet metal surfaces. Whenever water is present, whether it be a pool of water on a roof or floor or dampness due to perspiration or moisture on metal cabinets, consider the situation dangerous from an electrical standpoint.

## *HAND TOOLS*

The service technician often encounters problems where power hand tools must be used. Whenever a power hand tool is picked up, line voltage (120 V ac) is close to the gripped hand. This should always be considered a potentially dangerous situation.

The first rule in using power hand tools is a matter of common sense but often ignored until someone receives a dangerous shock.

1. If anyone receives a slight tingling shock while using a power hand tool, the tool should be discarded *immediately*. Do not suggest that the job be finished and then the tool replaced. Stop using the tool at the first sign of a shock.

The second rule regarding power hand tools is an extension of the first rule.

2. When discarding electrically dangerous power hand tools, destroy the hand tool completely, thus ensuring that some unsuspecting individual does not find access to the tool and receive a dangerous or fatal shock while attempting to use it.

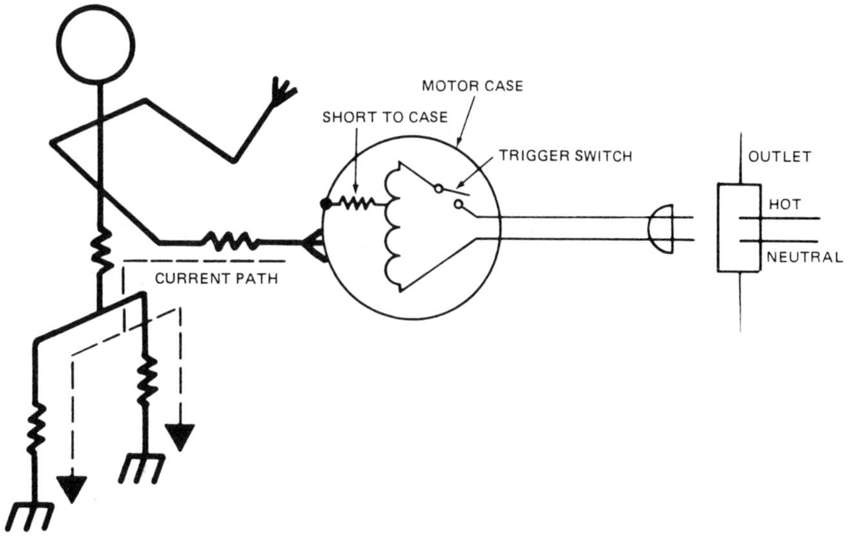


**Figure 13–1**

The electrical safety problem with power hand tools is always present when the older model, two-wire, metal-case power tool is used. The same problem exists when the three-wire power hand tool is used where a good ground does not exist. When a shock occurs, the electrical circuit causing the possibly dangerous shock will first be investigated. The lack of good grounding with the three-wire, grounded case tool will then be looked into.

Observe Figure 13–1 where the example of a hand-drill motor with a motor winding short to metal case is indicated. A stick diagram of the individual holding the tool is also shown.

In Figure 13–1 the diagram of a dangerous situation is shown. A number of different results could be obtained from this situation. Let's consider the two extremes.

1. The work location is dry; the technician is dry; this is the first task of the day. The technician's hand resistance to the drill case is high, and grounding through the technician's left hand and feet is through high resistance.

This situation could result in the technician not feeling any shock at all or perhaps a slight tingling shock. The technician would be lucky if he or she felt the slight tingling shock and would therefore know that the drill was dangerous. One could then discard the drill. Otherwise the drill might be continued in use, resulting in situation 2 at some later time.

2. The work location is damp; the technician is wet with perspiration. The work site is cramped between two damp metal cabinets, one of which needs a hole drilled through its case. Because the metal case of the cabinet to be drilled is thick, pressure will be needed on the drill. The technician braces his or her back against the case of the second cabinet. If the technician presses the hand trigger switch on the shorted drill, the chances are he or she will *die* right there.

It is obvious that if the drill were repaired after the first indication of a tingling shock, the technician would not have been killed in the second situation.

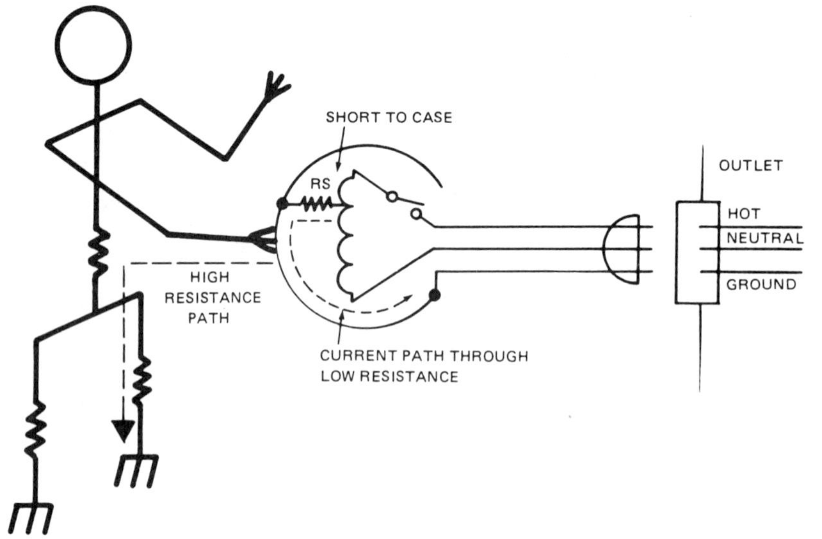


**Figure 13–2**

## Grounding of Tool Housing

If the metal case of a three-wire hand tool is properly grounded, a short within the tool will most likely result in a blown fuse. Even if the fuse does not blow, the possibility of a dangerous electrical shock is lessened to a great extent since the case of the tool will be at close-to-ground potential, as shown in Figure 13–2.

One of the main points of electrical safety when using hand tools with three-wire power cords is to ensure that the ground connection is complete to ground. One problem encountered with three-wire power cords is to find the grounding prong broken off. This usually comes about when someone has previously used the hand tool where only the old type, two-wire receptacle was available. The hand tool should not be used until a new three-wire plug or complete cord has been installed.

Cheater plugs adapting the old two-hole receptacles to the three-wire plug are available. The ground pigtail must be connected to a good ground if a safe situation is to exist. Do not be satisfied with a connection to a screw holding the outlet cover plate. Make sure the outlet box is grounded. The use of double-insulated tools is making inroads in the air conditioning field. Since the tools use the two-prong plug, the need for the three-prong grounding receptacle is eliminated. The problems of the ungrounded three-prong receptical are also eliminated.

Because safety engineers have not yet come to a definite conclusion as to double-insulated tools, it would be well to keep abreast of this changing situation.

Panel taps are often used on job sites as a source of 120-volt power whenever standard outlets are not present. An example might be a rooftop installation. An example of a panel tap is given in Figure 13–3.

The panel tap is made from a standard three-wire extension cord with a molded receptacle at one end. The plug is removed from the opposite end and the wires separated for approximately two feet. The clips are installed at the wire ends as shown in Figure 13–3. An insulated inline fuse cartridge is installed in the red line. Maximum fuse size is 15 amperes. The panel tap is practical if properly constructed and installed.

First, inspect the panel tap to ensure the clips are properly fastened and color coded. Check the panel power with a voltmeter for a source of 110 V to 120 V to neutral/ground at the panel box. Do not assume the voltage is 120 V to neutral/ground. You may have a 440-V or 660-V power source.

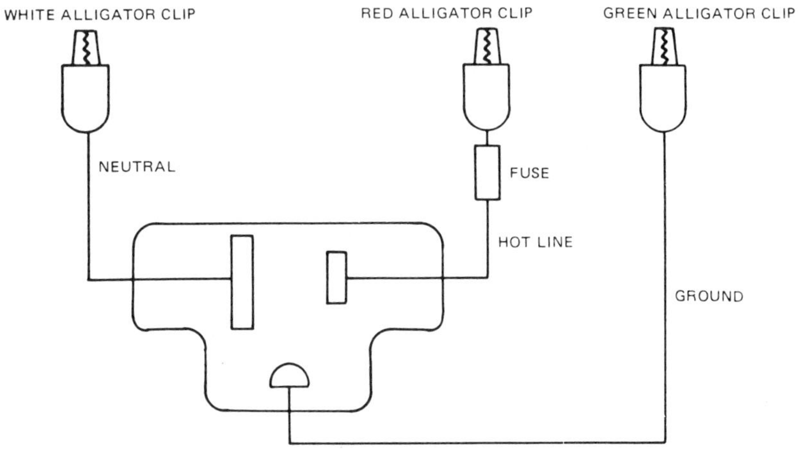


**Figure 13–3**

When the nominal 120-volt source has been located, panel tap connections may be made. The panel tap extension cord should be tied off, providing strain relief at the power box. Movement of the cord at the receptacle end should not cause movement at the clip end. Next, make the ground connection. If a ground bus bar is available, the green ground clip should be connected to it. If there is no ground bus bar, the *green* clip should be connected to the metal case of the box or metal conduit. Make sure that paint does not isolate this connection.

The second connection to be made is the neutral *white* clip. This clip should also be connected to the neutral/ground bus if available or to the metal case of the power box. The *red* clip may then be carefully connected to the hot line terminal. If practical, these connections should be made in a power-off situation.

The technician has created a somewhat unsafe condition when using a panel tap. It should be recognized as such and used only when absolutely necessary.

## *FIRST AID IN ELECTRIC SHOCK*

In cases of electric shock, the first and most important step is to disconnect the victim from the electric power source. If possible, turn the power off. If this is not possible, use a dry board or other nonconducting material to separate the victim from the electric connection.

Start artificial respiration as soon as it is safe to touch the victim. Speed is essential when dealing with victims of electric shock. Do not stop the administration of the artificial respiration even if the victim appears dead. Long time periods up to 8 hours have elapsed before some victims have responded. Only a physician should determine if the victim has died.

Most importantly, technicians should take the official first-aid course offered in their areas. It is extremely important to know the best methods of administrating first aid to shock victims.

Keep yourself aware of accidents as they happen by reading your trade periodicals and newspapers. A recent accident reported in a trade paper described a situation where a technician was working in the attic of a manufacturing plant. Being safety conscious, he had disconnected power from the air conditioning system he was working on. The attic was not well lighted; it was hot; he was perspiring; and his clothes were damp. Along with the electrical conduit feeding the air conditioning systems were commercial gas lines providing service to the building.

The technician had just finished placing a blower motor in the air distribution system. He raised himself up and possibly stretched to relieve the tension in his back. At his head level behind him was an open junction box. The cover had been left off by a previous worker. A wire nut had loosened and fallen off a pigtail connection or perhaps it had never been installed.

As the technician stretched back, his head came in contact with the hot pigtail. The pigtail was the junction in a 278-volt lighting system. The technician had his hand and back against the gas line, which is an excellent ground.

The attempts to revive him were not successful.

Be aware of what you are doing, where you are, and what is around you. Be safety conscious at all times. Do not be satisfied with a half-hearted approach to an understanding of safe working conditions.

## REVIEW QUESTIONS

T ______ F ______ 1. The only electrical shocks that are dangerous are the ones that cause harsh burns.

T ______ F ______ 2. A victim of electric shock is always knocked away from the electric connection by involuntary muscle contractions.

T ______ F ______ 3. Most shocks are the result of carelessness.

T ______ F ______ 4. No voltage should be considered safe since it is the current that causes damage, and body resistance is changeable.

T ______ F ______ 5. Hand power tools are not dangerous if grasped with only one hand.

T ______ F ______ 6. A slight tingling shock from a hand tool is not important.

T ______ F ______ 7. A technician can feel safe as long as a three-wire plug is on his hand drill.

T ______ F ______ 8. Two wire connections are needed on a panel tap.

T ______ F ______ 9. The voltage between the black wire in a panel box and ground is 120 volts.

T ______ F ______ 10. In case of electric shock, the first thing to do is remove the victim from the electrical connection.

# UNIT FOURTEEN

# *ELECTRIC MOTORS*

A common device used in many pieces of equipment is the electric motor. The most common type of motor is the induction motor. The cost of the induction motor is relatively low. This is due to the simplicity of the motor's rotor.

An induction motor operates on a basic principle of electricity: An induced current opposes the field that produced it. This is covered in Unit 16 where the secondary current of a transformer opposes the field of the primary. The back EMF of the primary tends to be reduced, allowing for more current in the primary. In this way, power in the secondary is provided for by increased power in the primary. Refer to Figure 16–4 in Unit 16.

An induction electric motor is actually nothing more than a special type of transformer. The primary of the motor transformer is wound on the stationary parts, called the *stator*. The secondary is made up of copper bars shorted together on the rotating part of the motor, called the *rotor*.

The primary windings on the stator look very much like most transformer windings. The secondary winding on the rotor consists of copper bars on an iron core. The copper bars are all connected together. This essentially provides a combination of single-turn windings, short circuited. Figure 14–1 shows a typical rotor in which the core is made up of die-cut laminations. These laminations serve to reduce eddy currents in the iron core of the rotor.

Whenever an ac voltage is connected to the primary of a transformer, a voltage is induced in the secondary. The secondary induced voltage is in direct opposition to the primary supply voltage. This opposition usually results in a rotation of the motor shaft.

Without some electrical form of direction, the rotor may turn in either direction, clockwise (CW) or counterclockwise (CCW) (Figure 14–2). If the rotor were not started in some direction, the motor would sit like a transformer with a shorted secondary—it hums but does not move. In part of alternation time period 1, current flows through the stator to provide for north and south poles. The induced voltage and the current flow in the rotor produce a magnetic field in the secondary winding

**Figure 14–1** Rotor.

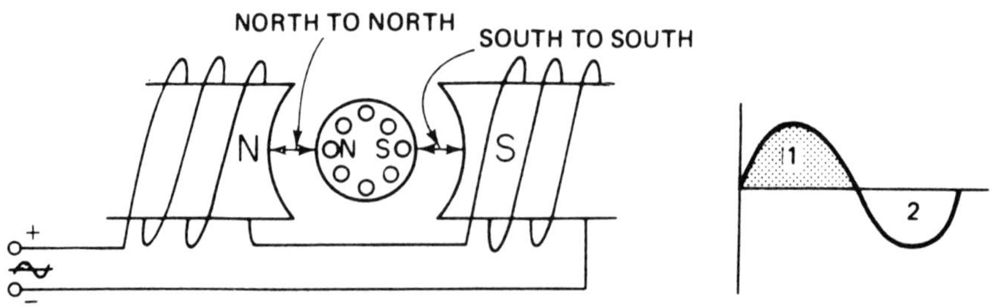


**Figure 14–2**

in opposition to the magnetic field in the primary winding. The system is in balance; the rotor does not rotate. Some direction action is needed.

In part of the second alternation of the applied voltage time period 2, the current flow and the polarity of the magnetic field of the primary change. The current flow and magnetic polarity in the secondary also change (Figure 14–3). Although the magnetic field of the stator and rotor change, the rotor still does not turn since the opposition of magnetic poles is direct, trying to force the rotor toward the shaft.

When the motor shaft is started in either direction, a time delay imbalance is created, allowing the motor to continue to turn in the direction in which it was started. The motor continues to turn after it is started because of the repulsion between the stator poles and the poles set up in the rotor (Figure 14–4). As long as electric power is applied to the stator winding of the motor, it continues to rotate in the direction in which it was started. A continuous shift in the magnetic field about the rotor takes place. Current flow in the rotor results from two actions: transformer action of the primary as previously discussed and generator action as the copper bars in the rotor cut through the primary magnetic field.

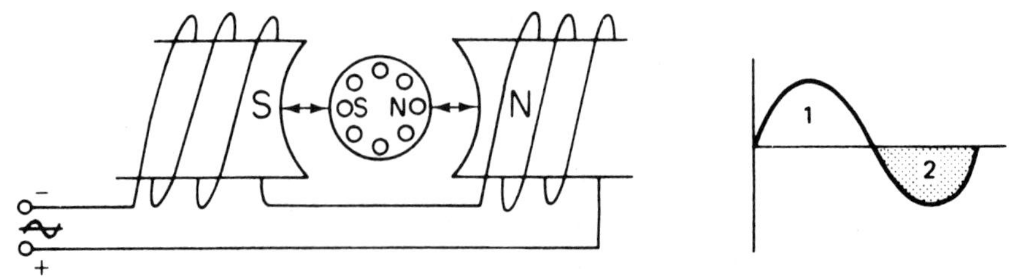


**Figure 14–3**

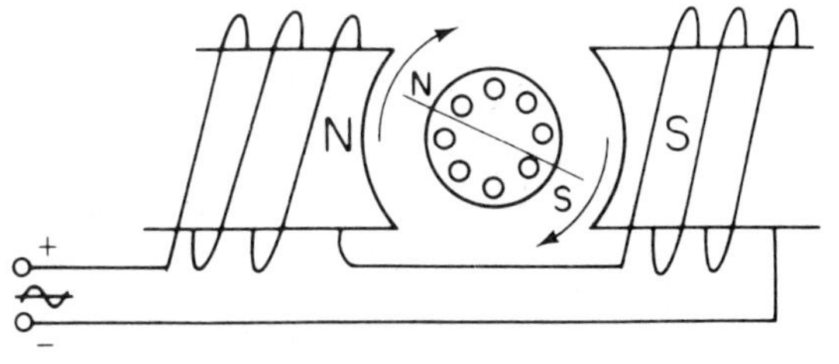


Figure 14–4

### *SHADED-POLE MOTOR THEORY*

One method used to get small motors ($\frac{1}{4}$ HP or less) started is to provide shaded poles in the motor (Figure 14–5). There is a slotted area in the two pole pieces with shorted turns of copper wire in the slots. It is the transformer action involving these shorted turns that starts the motor in the desired direction.

The magnetic polarity relationships of the rotor and stator of a shaded-pole motor are shown in Figure 14–6. This relationship exists just after the current flow in the motor coil passes through zero.

The current flow in the shaded pole (shorted turn) sets up a magnetic field in opposition to the magnetic field that produced it. The magnetic field set up by the shaded pole is of the same polarity as the magnetic field set up by the rotor currents. The magnetic field of the shorted turn is not as strong as the magnetic field of the main coil, but it is sufficient to cause a delay in the magnetic field set up in the shaded area.

The end result is a rotating magnetic field about the stator of the motor. The relationship is shown in Figure 14–7 and is as follows:

1. The current in the main winding passes through zero and starts to increase. A north pole is started in the main pole area. The opposing effect in the shaded area produces a south pole at the shaded pole area.

Figure 14–5 Shaded-pole motor.

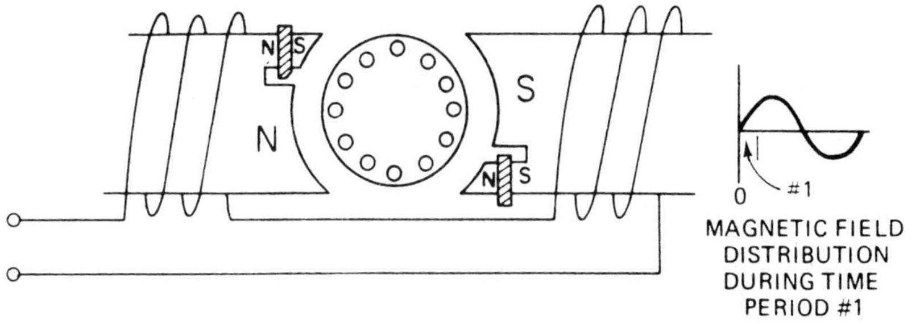


Figure 14–6

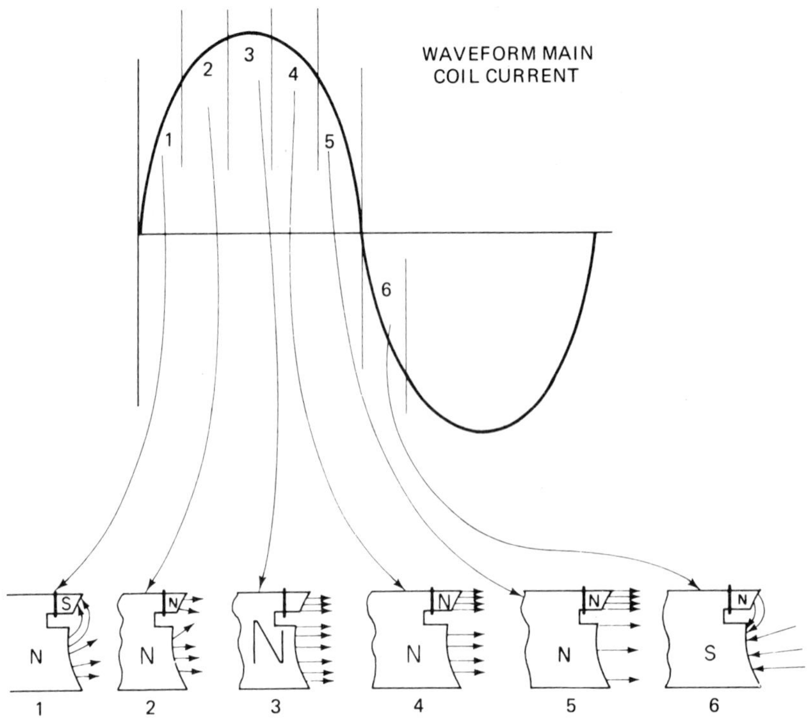


**Figure 14–7**

2. As the main field builds up towards a maximum north pole, the shaded area becomes a weak north pole. The effect of the shaded pole is decreasing.
3. The main pole is at maximum. The effect of the shaded pole is zero.
4. The strength of the main pole decreases from a maximum north pole. The shaded area becomes a north pole.
5. The main pole decreases in strength to zero. The shaded area stays a north pole.
6. The main pole changes polarity to a south pole. The shaded area remains a north pole.

As long as the magnetic field is rotating around the stator, the resulting current flow in the rotor causes the rotor to follow. The direction of rotor rotation is always in the direction of the shaded pole (Figure 14–8).

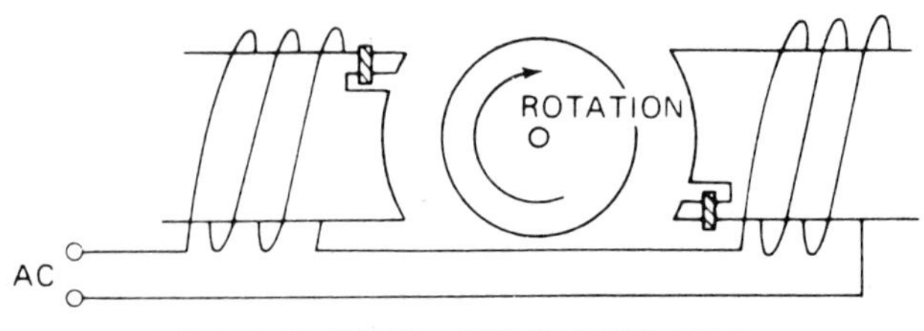


**Figure 14–8**

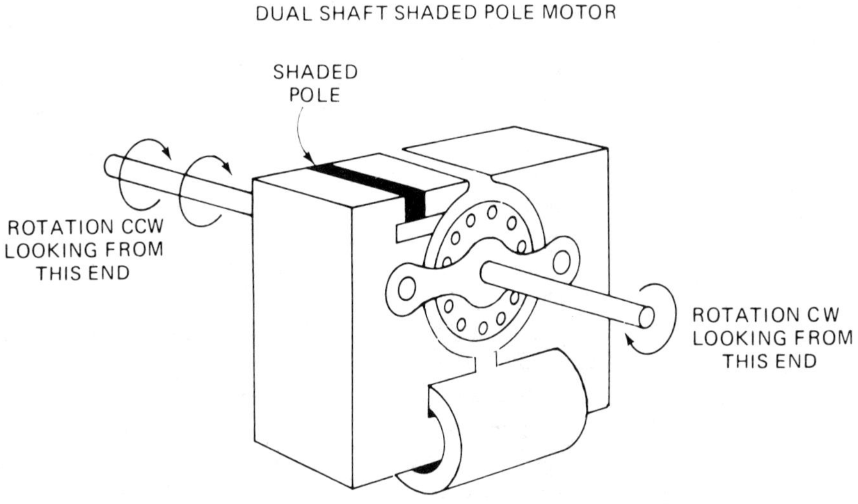


**Figure 14–9**

### Motor Direction (Shaded-Pole)

Many shaded-pole motors have a shaft extending out of end bells on both sides of the rotor. With a double-shaft motor, either clockwise or counterclockwise rotation is obtained, depending on which shaft is used to drive the "load" (Figure 14–9). In many cases, it is possible to obtain reversed rotation of a shaded-pole motor with a single shaft (extending from only one end). If the end bells of the motor are interchangeable, the motor may be reversed simply by reversing the rotor. The end bells are removed, and the motor is reassembled with the shaft extending from the opposite end of the motor (Figure 14–10).

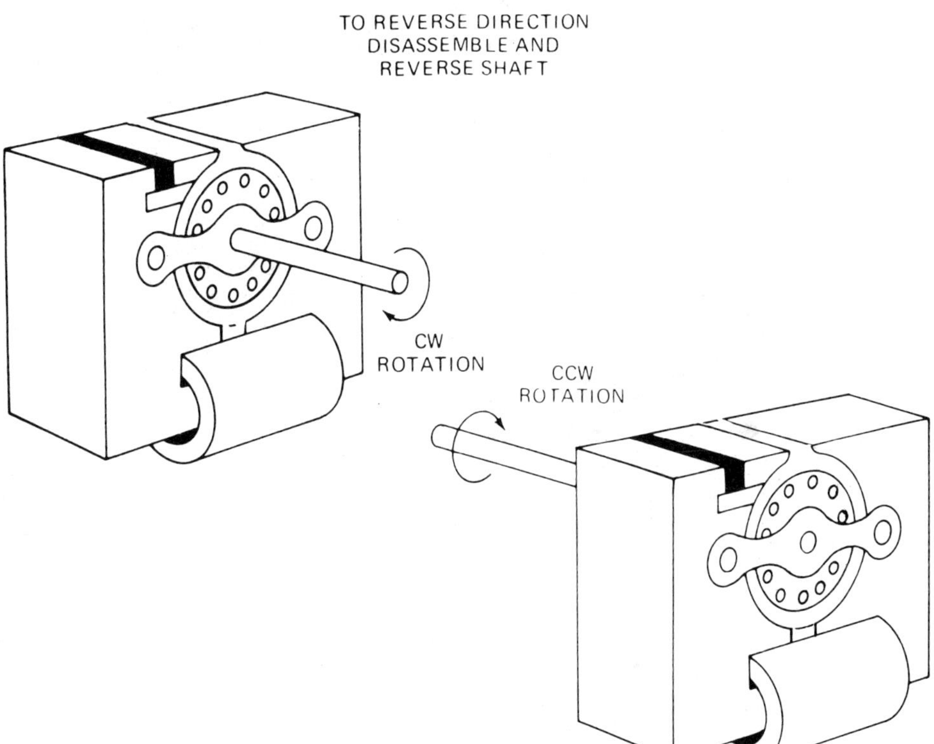


**Figure 14–10**

## *SPLIT-PHASE MOTORS*

Split-phase motors are usually used for applications requiring from $\frac{1}{6}$ to $\frac{3}{4}$ horsepower. The starting torque of split-phase motors is low to moderate compared with other motor types, except shaded-pole motors. Split-phase motors are usually found in service for driving fans, centrifugal pumps, conveyors, and refrigerator compressors.

The rotating magnetic field needed to start split-phase motors is obtained through the use of two stator windings in the motor, a main winding and a start winding. Each of these windings has different inductance and resistance values. Current flows in the two windings are not in phase with each other (Figure 14–11). The current in the main winding (coil A) lags the current in the start winding (coil B). A rotating magnetic field is set up in the stator, and the motor rotates in the direction in which the field rotates.

After a split-phase motor has started its rotation and as its speed approaches normal operating speed, the start winding is no longer needed. A mechanical centrifugal switch is often used to disconnect electrically the start winding from the power source, usually at about $\frac{2}{3}$ of top speed (Figure 14–12). The motor continues to rotate after the start winding has been electrically disconnected.

### Motor Direction (Split-Phase)

The direction of rotation of most split-phase motors may be reversed by interchanging the connections of the windings. The motor windings are marked $S_1$ and $S_2$ for the

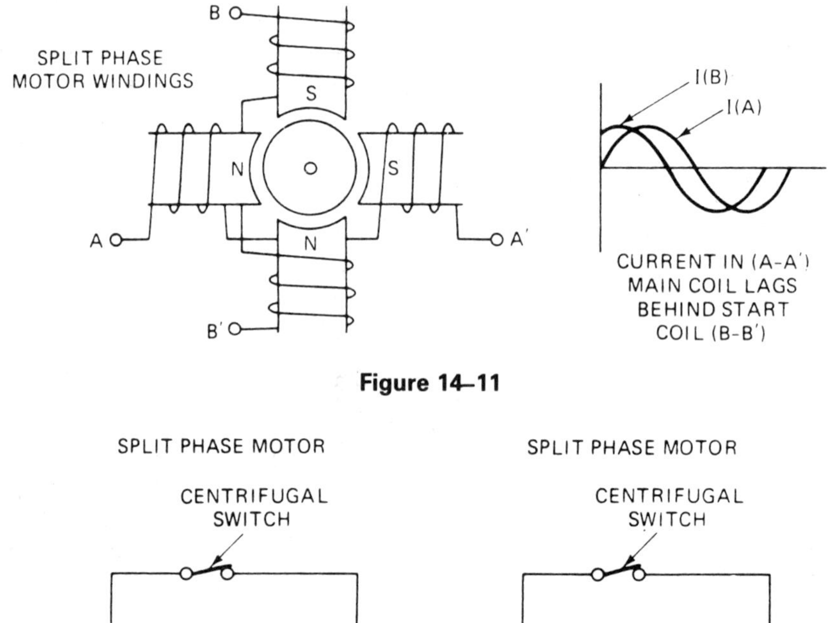


**Figure 14–11**

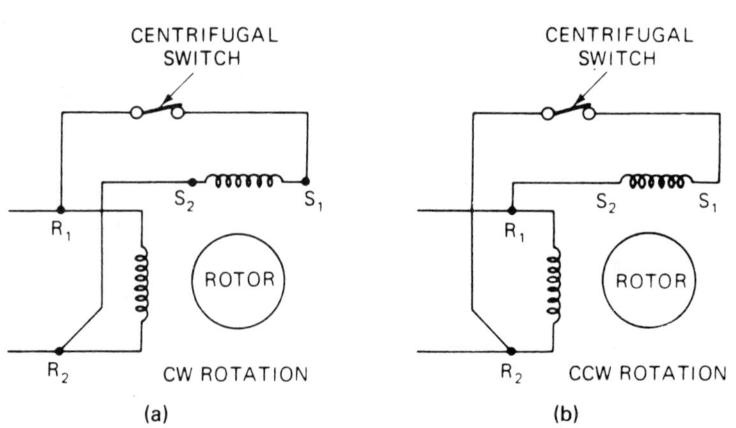


**Figure 14–12**

**Figure 14–13** Motor stator/centrifugal switch.

start winding and $R_1$ and $R_2$ for the run winding. In Figure 14–12a, the winding connection is $S_1$ to $R_1$ and $S_2$ to $R_2$; the motor rotates clockwise. In Figure 14–12b, the connection is $S_1$ to $R_2$ and $S_2$ to $R_1$; the motor rotation is counterclockwise.

The centrifugal switch is in series with the (S) winding of the motor (Figure 14–13). Although bearings often wear out in motors using centrifugal switches, the switch and switch mechanism have proven to be a common source of trouble. Contacts of the switch may become burned and pitted, preventing good electrical contact. Dirt and grit may interfere with the mechanical operation of the switch mechanism, causing the switch to remain open after the motor has been stopped.

If the switch contacts are open and an attempt is made to start the motor, the shaft will not rotate. The motor will hum but will not turn. If this should happen, it is necessary to disassemble the motor and to clean the mechanical portion of the centrifugal switch. The contact surfaces of the electrical switch should be inspected for excessive pitting. Minor pitting may be removed and the contact surface returned to operating condition with fine sandpaper. Care must be taken to remove all sand and sanded contact material from the motor. If in doubt, the centrifugal switch should be replaced as an assembly.

## *CAPACITOR MOTORS (SPLIT-PHASE)*

Capacitor-type, split-phase motors are used in applications where higher starting torque and/or higher horsepower ratings are required. Common capacitor-type, split-phase motors are available from $\frac{1}{3}$ to 10 horsepower.

There are three different types of capacitor split-phase motors in common use today.

1. Permanent split-capacitor (PSC) motor
2. Capacitor-start motor
3. Capacitor-start, capacitor-run motor

### Permanent Split-Capacitor Motors

The permanent split-capacitor motor is usually of the fractional horsepower size, $\frac{1}{70}$ to $\frac{1}{2}$ horsepower. These motors are often used in place of shaded-pole motors because they are somewhat more efficient. The capacitor remains in series with the auxiliary winding of the motor at all times. An electrical circuit of this motor is shown in Figure 14–14.

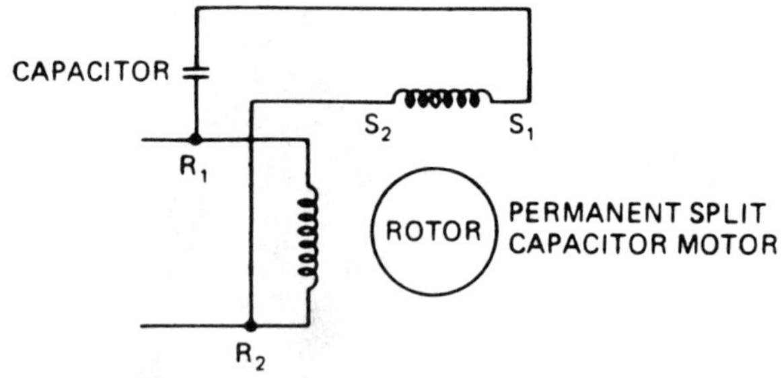


**Figure 14–14**

### Capacitor-Start Motors

Capacitor-start motors are readily available from $\frac{1}{4}$ to 10 horsepower. They are most often used in applications in which high starting torque is required.

Capacitor-start motor operation is very similar to that of the split-phase induction motor with one exception. The capacitor-start motor has an electrolytic capacitor in series with the start winding, sometimes called the auxiliary winding. A centrifugal switch or some other control device is used to remove electrically the capacitor and the start winding from the circuit after the motor reaches approximately $\frac{3}{4}$ of operating speed. The diagram of a capacitor-start motor, Figure 14–15, is very similar to that of the split-phase motor, Figure 14–12; the only difference is the addition of the capacitor.

*Motor Direction (Capacitor-Start).* The direction of rotation of capacitor-start motors may be changed by reversing the electrical relationship of the two windings. If the motor connections in Figure 14–15 are taken as standard, then the motor direction is reversed by the motor connections shown in Figure 14–16. The method that the technician uses to reverse motor direction is dependent on what access is available to the winding connections.

A common trouble area in capacitor-start motors is the starting mechanism, which may be a centrifugal switch or some other control device such as a relay. A second trouble area is the start capacitor. Failure of the start capacitor, usually an electrolytic one, may be due to an open or short.

### Capacitor-Start, Capacitor-Run Motors

The capacitor-start, capacitor-run motor is essentially a combination of the permanent split-capacitor motor and the capacitor-start motor. The capacitor-start, capacitor-run motor provides both the high starting torque of the capacitor-start motor and the efficiency (as well as good power factor) of the permanent split-capacitor motor.

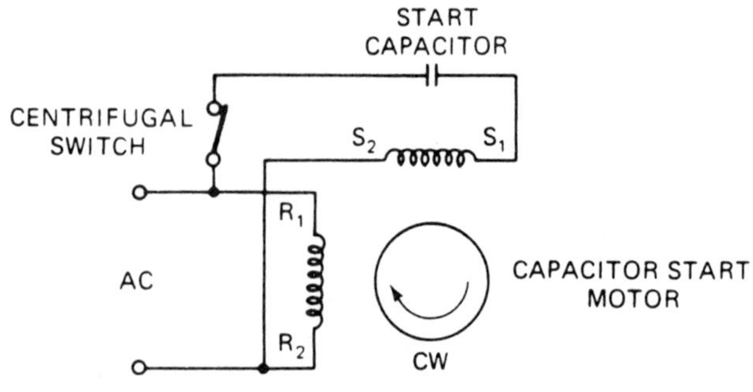


**Figure 14–15**

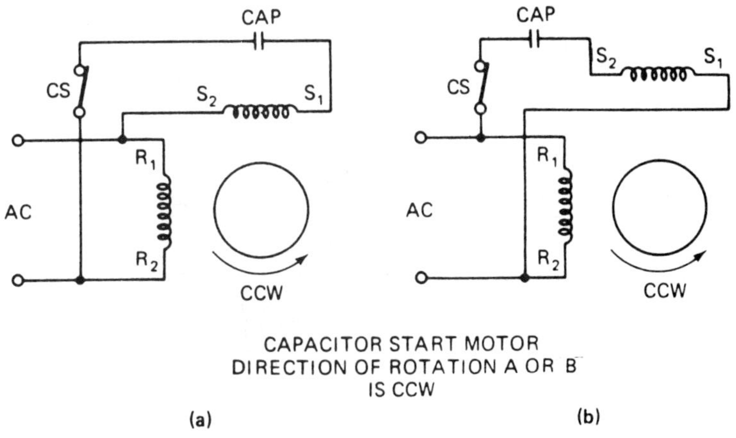


**Figure 14–16**

A diagram of a capacitor-start, capacitor-run motor is shown in Figure 14–17. High-starting torque is obtained through the use of an electrolytic (start) capacitor, $C_2$, in parallel with the permanent oil-filled (run) capacitor, $C_1$. When power is applied to the motor and shaft rotation approaches operating speed, the control switch, $S_1$, opens. The large value capacitor, $C_2$, is removed from the circuit, and total current draw by the motor is reduced. The permanent capacitor, $C_1$, remains in the circuit, providing for constant torque as well as improving the overall power factor of the motor.

The capacitor-start (capacitor-run) motor is probably the most common type of motor used in the air conditioning hermetic compressor system. The run and start windings of the motor are connected together internally, and three connections are brought out to motor terminals. These terminals are marked R for run winding, S for start winding, and C for common connection. This arrangement is shown in Figure 14–18.

There are times when a service technician may have to determine the proper terminal marking for a single phase compressor. An ohmmeter is all that is necessary in making this determination. The following steps should be taken when checking for motor terminal connections:

1. Mark the terminals 1, 2, and 3.
2. Measure the resistance between each set of terminals.
3. The resistance between one set of terminals will be higher than the resistance between the other two sets. The terminal not included in the highest resistance measurement is the common terminal. Mark that terminal C.

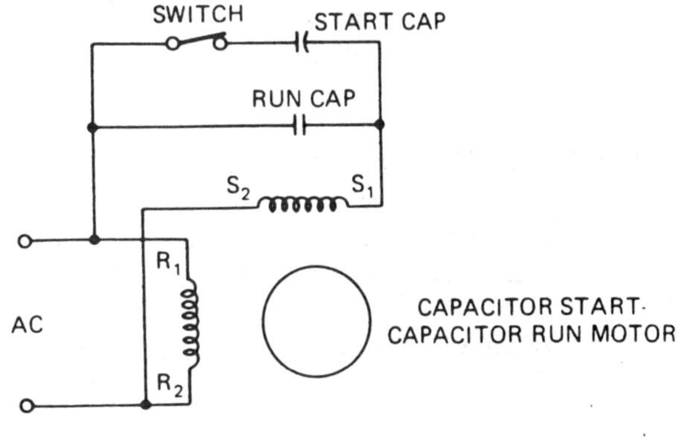


**Figure 14–17**

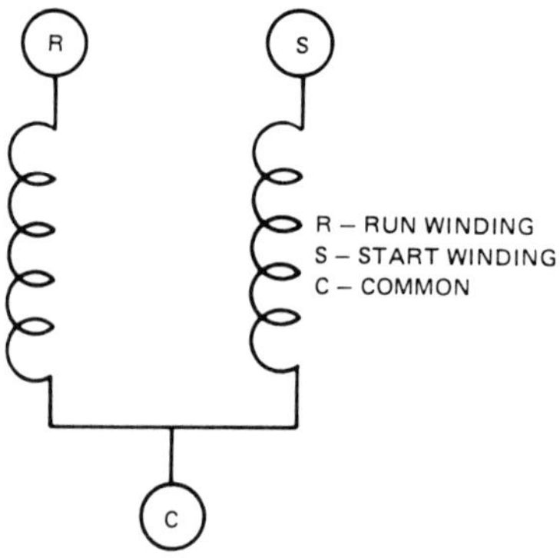


**Figure 14–18**

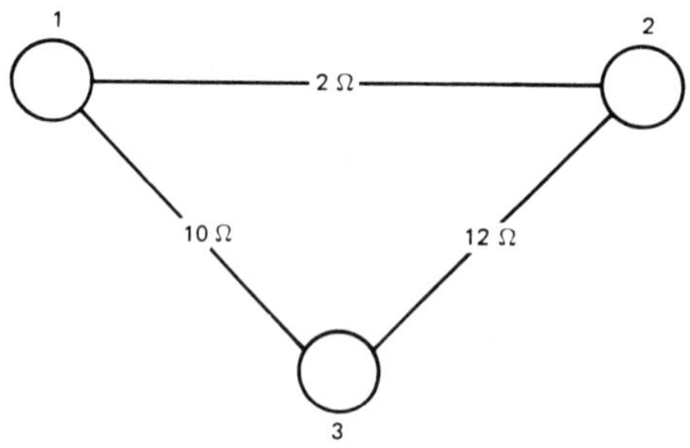


**Figure 14–19**

4. Measure the resistance between the common terminal and the other terminals.
5. The indication of lower resistance from a terminal to common C determines the run (R) winding.
6. The indication of higher resistance from a terminal to common C determines the start (S) winding.

An example of the winding resistance is shown in Figure 14–19.
The ohmmeter measurements are in Figure 14–19.

1. The highest resistance measured, 12 ohms, is measured between terminals 2 and 3. Terminal 1 must be the common terminal and should be marked C.
2. The resistance between terminals 1 and 2 is 2 ohms, and the resistance between terminals 1 and 3 is 10 ohms.
3. The lowest resistance, 2 ohms, indicates that terminal 2 is the run (R) winding.
4. The next higher resistance, 10 ohms, indicates that terminal 3 is the start (S) winding.

## *REPULSION-START INDUCTION-RUN MOTORS*

Repulsion-start induction-run motors are not common in industry today. Older operating units may be found with this type of motor still in use.

The repulsion-start induction-run motor includes coil windings on the rotor (Figure 14–20). The individual ends of the coil winding are connected to a commutator. A centrifugal mechanism controls the start of the motor, where a set of brushes connects the ends of some of the rotor windings together. As the motor approaches

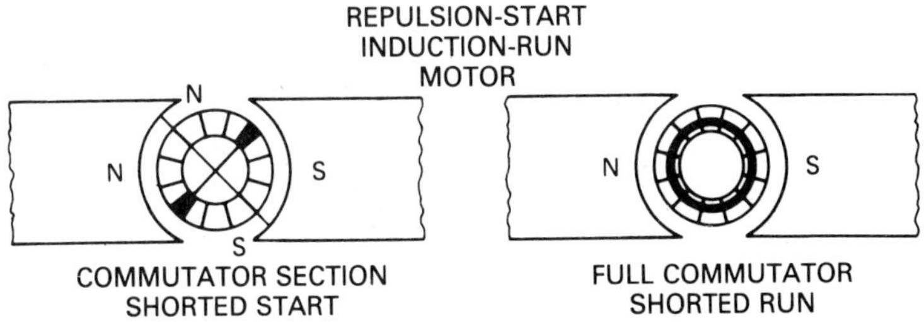


**Figure 14–20**

operating speed, the centrifugal device completely shorts the commutator segments together through the use of a shorting ring. The rotor essentially becomes a squirrel-cage rotor. The high manufacturing cost of this motor, compared to other types of motors, ended its popularity.

## *POLYPHASE MOTOR (THREE-PHASE)*

Polyphase motors are simple in construction. They do not require starting windings, centrifugal switches, or start capacitors.

When a three-phase voltage is applied to the stator of a three-phase motor, a rotating magnetic field is automatically present around the stator (Figure 14–21). A set of magnetic pole pairs is indicated by the numbers 1 ①, 2 ②, and 3 ③. With the application of the three phases of voltage to the windings of the poles, a rotating field around the face of the motor stator results.

First pole 1 is a north pole, then pole 2, and then pole 3. Pole ① then becomes a south pole as the current changes direction. Poles ①, ②, and ③ were south poles in order as 1, 2, and 3 were north poles. The magnetic field rotates around the stator face continuously. A squirrel-cage rotor rotates in a magnetic field of this type. No other starting system is needed.

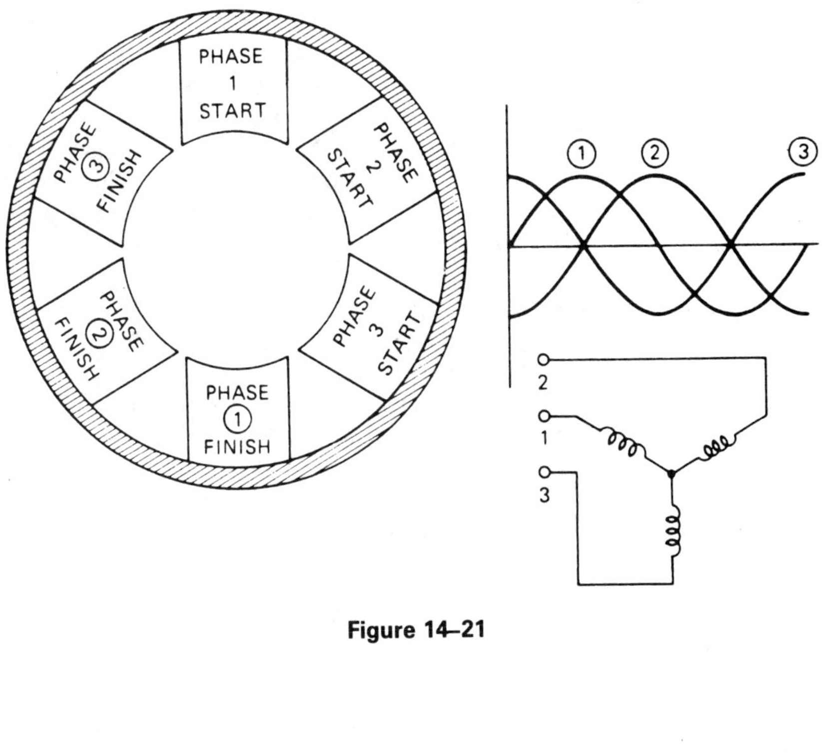


**Figure 14–21**

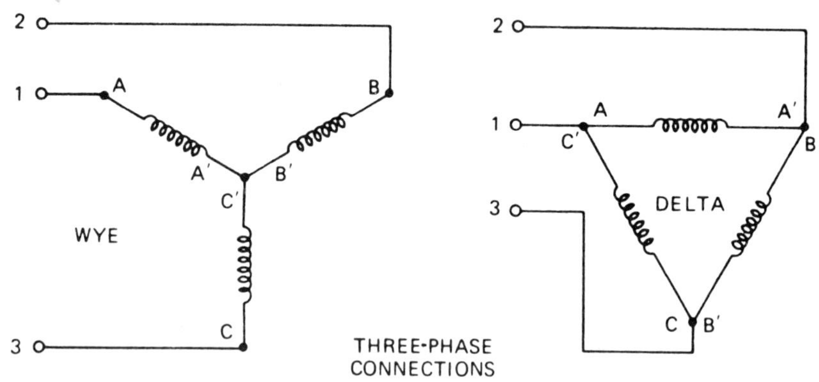


**Figure 14–22**

### Winding Connection

The windings of three-phase motors are connected in either the Wye or delta configuration, depending on voltage and current requirements. As a current-limiting scheme, three-phase motors are sometimes started in the Wye configuration and are switched into the delta configuration after the motor comes up to speed. In this manner, the starting current is limited. Figure 14–22 shows examples of Wye and delta connections.

### Direction Control of Three-Phase Motors

In order to reverse the direction of a three-phase motor, it is only necessary to interchange any two of the three supply voltage lines. In this manner, the direction of rotation of the magnetic field is reversed and, consequently, so is the direction of rotor rotation.

## *MOTOR-STARTING CIRCUITS*

As indicated earlier, in this unit there are different types of ac motors. Many of these motors require some special control in order to start them.

Shaded-pole motors are generally used as a small fan motor or as timing motors. The shaded-pole motor does not require a special starting control. When power is supplied to the motor winding, the motor rotates. When power is removed from the motor winding, the motor stops but is ready to start again with the application of power. The starting feature of the motor, the shaded poles, is always in the circuit, ready to produce the required starting action.

### Permanent Split-Capacitor Motor

The permanent split-capacitor motor, like the shaded-pole motor, does not require special starting considerations. Simple switch devices, either manual or automatic, may be used to start this motor. When power is applied, the motor runs.

### Split-Phase Motor (Current Relay)

The split-phase motor is started by the use of a centrifugal switch. The centrifugal switch removes the start (auxiliary) winding from the circuit when the motor shaft approaches operating speed. This is a practical system for starting the motor when it is used in open compressor-type applications.

**Figure 14–23** Series-start relays.

One of the most common uses of a split-phase motor is to "drive" a compressor in small-capacity refrigerators and freezers. The split-phase motor in this application is hermetically sealed along with the compressor as a combined unit. In a sealed unit, a centrifugal switch starting system would not be practical. (See Figure 14–13). Arcing at the switch contacts would contaminate the refrigerant-oil combination in the sealed system. Another problem is that oil on the switch contacts would increase contact resistance. For these reasons, centrifugal switches are not used in hermetically sealed compressors.

The most common type of starting control used with the small-capacity, hermetically sealed compressors is the *series-starting relay*. This type of relay is known as a *current relay* (Figure 14–23). The coil of the current relay consists of a few turns of relatively large wire connected in series with the run winding of the motor. The contacts of the current relay are normally open and are connected in series with the start winding of the motor. A circuit diagram of a split-phase motor with a current relay is shown in Figure 14–24.

The relay coil is connected in series with the run winding of the motor. The open contacts of the relay are connected in series with the start winding of the motor. When the control switch, $S_1$, is closed, 120 volts are applied to the run winding of

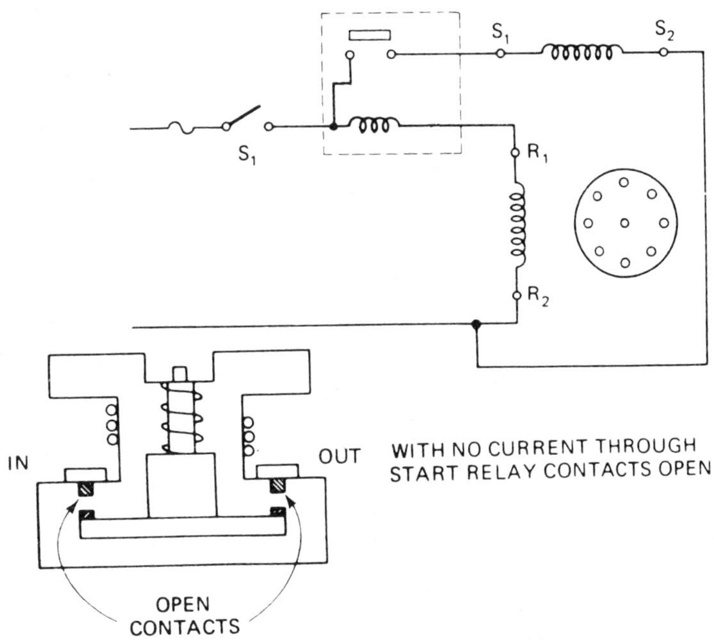


**Figure 14–24**

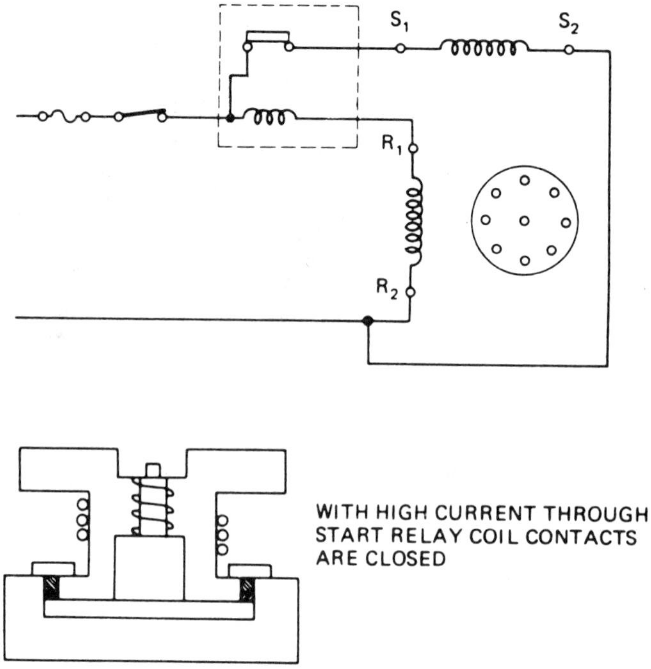


**Figure 14–25**

the motor through the few turns of the relay coil. The start winding is not connected to the power source because the relay contacts are open. A high current flows in the run winding because the motor has not yet started. This same current flows through the relay coil (Figure 14–25).

With a high current flow in the coil, the relay energizes; the normally open (NO) contacts of the relay close. A circuit is completed through the relay contacts, and 120 volts are applied to the start winding of the motor. The motor starts to rotate and gains speed. When the motor approaches operating speed, the current flow in the run winding decreases toward the normal operating current (Figure 14–26). This reduction causes the relay to deenergize; the relay contacts then return to their normally open position, again removing the start winding from the circuit.

Current relays are commonly used with motors up to $\frac{3}{4}$-horsepower capacity. Larger motors commonly use the potential relay for starting control.

Many current-type starting relays operate with the weight of the relay core as a factor in opening the relay contacts. This type of relay must be mounted in the position prescribed in the manufacturer's specifications.

Current relays are designed to pick up (energize) and to drop out (deenergize) at specific current levels. The current relay used in a compressor motor circuit must be matched to the compressor motor. When it is necessary to replace a current relay, an exact replacement must be used.

## Capacitor-Start Motors (Current Relay)

Capacitor-start motors often use the same type of current relay that is used with a split-phase motor.

A diagram of a capacitor-start motor circuit using a current relay is shown in Figure 14–27. When switch $S_1$ is closed, high current flows through the coil of the current relay and the run winding of the compressor motor. The relay contacts close, connecting the start winding of the motor to the supply voltage through the start capacitor.

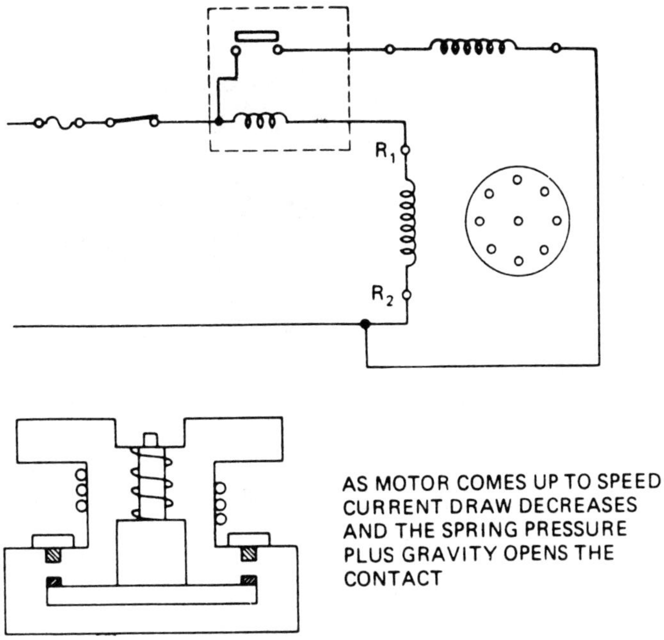


**Figure 14–26**

As the motor approaches operating speed, the run winding current decreases. When the current reaches a predetermined level, the relay deenergizes, removing the capacitor and the start winding from the circuit.

It is important to remember that the electrolytic capacitor used with capacitor-start motors is for intermittent use only. It is only in the circuit for a few seconds during the motor start period.

### Capacitor-Start, Capacitor-Run Motors (Potential Relay)

Capacitor-start, capacitor-run motors are usually provided with a potential relay as a starting control. This type of motor is generally used in motor applications from $\frac{3}{4}$ to 10 horsepower or larger.

The potential relay that is used in motor start control is quite similar in design and appearance to many other types of relays (Figure 14–28). The potential relay consists of a coil and a set of normally closed (NC) contacts. Potential relays are voltage sensitive; they are rated with pick-up (energize) and drop-out (deenergize) voltage. Exact replacement parts should always be used. Although many potential relays have a *sensitivity adjustment* associated with the contacts, adjustments should

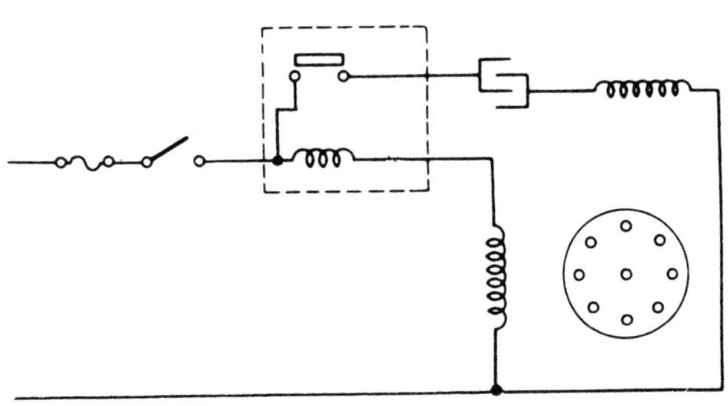

**Figure 14–27**

**Figure 14–28** Potential relay.

not be made "in the field." If a potential relay does not function properly, it should be replaced. It is not a good practice to attempt adjustment or repair of relays.

The circuit of a capacitor-start, capacitor-run motor with potential relay start control is shown in Figure 14–29. The relay contacts are only used to connect and remove the *start* capacitor from the circuit. The run capacitor remains in the circuit at all times.

When switch $S_1$ is closed, 220 volts are applied to the motor circuit. The run winding is connected directly to the 220-volt source. The start winding is connected to the 220-volt source through both the start and run capacitors. Electrically, the two capacitors are connected in parallel through the closed contacts of the potential relay.

The potential relay coil is connected in parallel with the start winding of the motor. The relay energizes when the voltage across the start winding rises to the predetermined level, depending on the "specifications" of the potential relay used. As voltage is applied to the motor, it immediately starts to rotate. High torque is present because of the high current in the start winding. The speed of the motor increases, and the back EMF across the start winding also increases. When the back EMF reaches the predetermined level, the relay energizes, opening the normally closed contacts. The start capacitor is then electrically removed from the circuit. The motor continues to rotate, with the run winding fed directly by the power source. The start winding is fed power with the run capacitor in series with the winding (Figure 14–30).

Potential relays, though manufactured by many companies, are often interchangeable. When replacement of a potential relay is necessary, an exact "specification" replacement must be used. The pick-up and drop-out voltage range of this type of relay is most important.

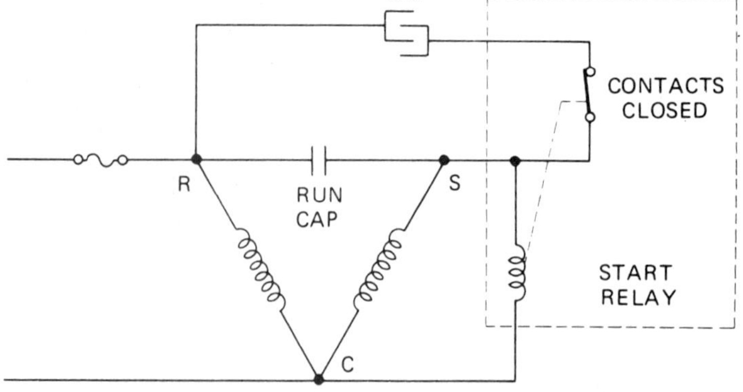

**Figure 14–29**

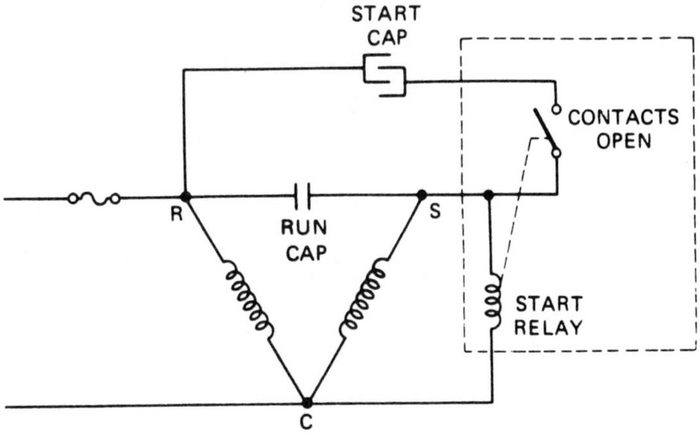


**Figure 14–30**

Contacts of the potential relay may become pitted if high voltage stored in the start capacitor is discharged through the contacts as they close. In order to reduce the pitting problem, it is common practice to connect a 15,000-ohm–18,000-ohm, 2-watt bleed resistor across the terminals of the electrolytic capacitor. The resistor does not affect the operation of the capacitor or motor, but it does provide a discharge path for the capacitor when it is not in use.

### Push-Button On-Off Motor Control

A common control used in the start operation of motors is the push-button magnetic-type starter switch. One excellent feature of this system is that the control may be located away from the motor. A diagram of a pushbutton system is shown in Figure 14–31.

The motor is controlled by a magnetic contactor, $K_1$. A thermal overload protector is included in each line of the motor control.

Should excessive current be drawn in either line, the thermal overload will overheat, causing the associated set of contacts to open, interrupting the circuit.

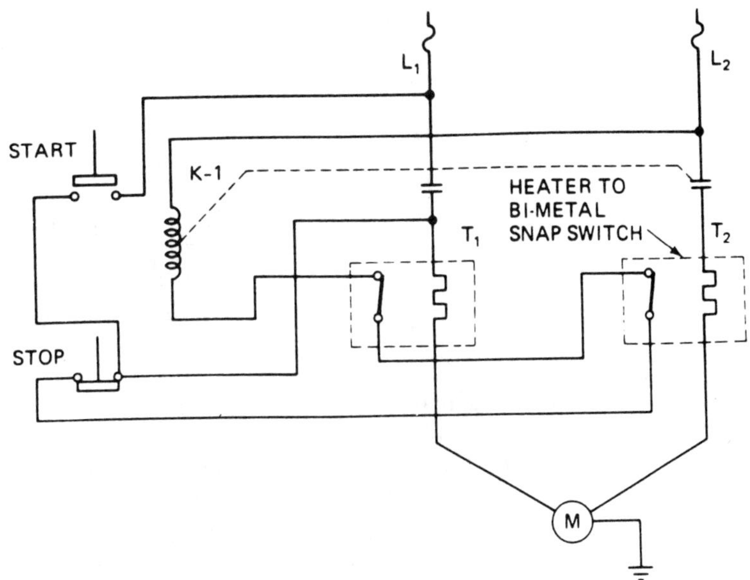


**Figure 14–31**

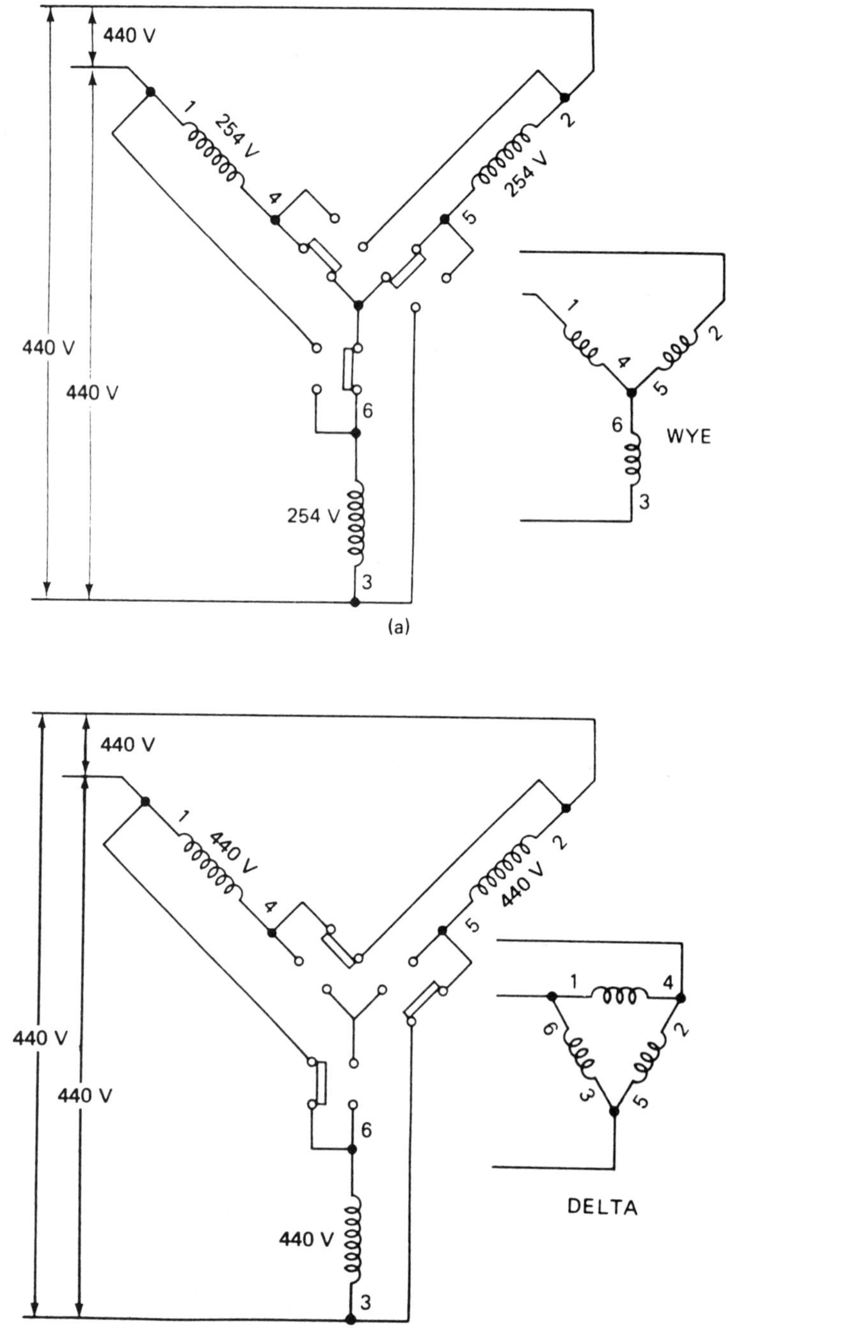


**Figure 14–32**

The normally open start button is pressed in order to start the motor. The current path is from line $L_1$, through the start button, now closed, through the normally closed stop button, through both thermal overload contacts to the bottom of the contactor coil. The top of the contactor coil is connected directly to line $L_2$. The contactor energizes, supplying power to the motor. The contacts between $L_1$ and $T_1$, when closed, complete the circuit, bypassing the start button. The contactor coil remains energized. The contactor coil deenergizes if the contacts of a thermal overload open or if the stop button is pressed.

### Three-Phase Motor

Three-phase motors are not started through the use of phase-shifting devices. A rotating magnetic field is present about the motor stator whenever three-phase voltage is applied to the windings. With larger units, a starting sequence is sometimes used to reduce initial starting current. During starting, the motor is electrically connected in the Wye configuration; it is electrically reconnected in the delta configuration when operating speed is approached. Through the use of such connections, voltage across the motor windings is reduced during the starting sequence but is at the full level during operating speeds.

In Figure 14–32a, the three-phase motor connection for start up is in the Wye configuration. The voltage across each winding is 254 volts. The source voltage is 440 volts. Full coil voltage operation with the delta connection is shown in Figure 14–32b. The operation of the contactor or relay used to switch from Wye to delta operation may be controlled by a time-delay mechanism. When power is first applied to the motor, the relay is deenergized, and the circuit is Wye connected. After a fixed time delay, during which the motor picks up speed, the relay energizes, changing the motor to a delta connection. The voltage across the motor windings is 440 volts.

### Single Phasing of Three-Phase Motors

When a three-phase motor is connected to a three-phase power source, it starts and eventually reaches normal operating speed. If, after the motor is started, one of the three input power connections becomes open, single-phase power is supplied to the motor. The motor continues to operate as a single-phase motor; however, excessively high current is drawn. The motor overheats; if the motor is allowed to continue to operate in a single phase, permanent damage to the motor may result. Thermal protective devices are connected in most three-phase motor circuits to electrically disconnect the motor if it becomes overheated.

## *LOW/HIGH VOLTAGE OPERATION*

Many motors are wound with coils in pairs. With such motors, it is possible to connect them electrically to operate on either high (220) or low (120) voltage. In Figure 14–33a, each coil is designed to function with 120 volts applied. In Figure 14–33b, the coils are connected in parallel for operation from a 120-volt source. So connected, each coil has 120 volts applied to it.

In Figure 14–33c, the coils are connected in series for operation from a 220-volt source. As shown, each coil has 110 volts across it. For a better understanding of this arrangement, review Units 4 and 5.

### Single-Phase Motors–Low and High Voltage

Many dual-voltage, single-phase motors have two coils for the run winding and two coils for the starting winding. Winding connections for low- and high-voltage operation are shown in Figure 14–34.

Other dual-voltage, single-phase motors have two coils for the run winding and a single coil for the start winding. Winding connections for low- and high-voltage operation are shown in Figure 14–35. When connected for low-voltage operation, the two coils of the run windings ($T_1$ $T_2$ and $T_3$ $T_4$) are connected in parallel across the 120-volt power source. The start winding $T_5$ $T_6$ is also connected directly across the 120-volt power source.

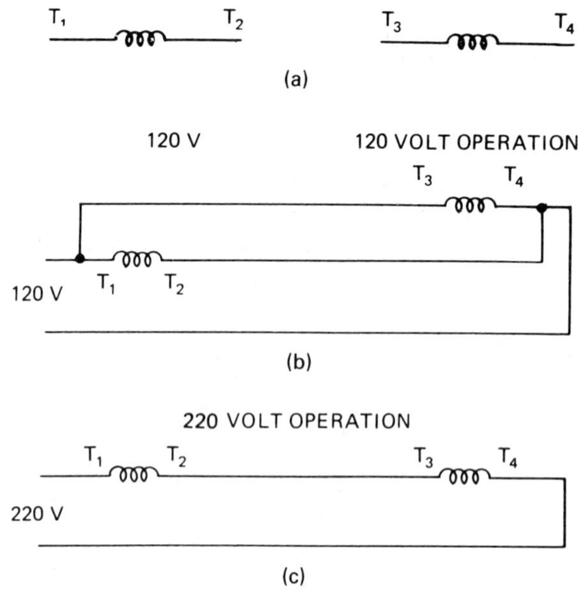


**Figure 14–33**

For operation at higher voltage, the run windings are connected in series. The start winding is connected to the junction of $T_2$ and $T_3$. Slightly less than $\frac{1}{2}$ of the source voltage appears across the start winding during motor start as shown in Figure 14–36a. After the motor comes up to operating speed, the switch device removes the start winding from the circuit. A schematic diagram of the circuit is shown in Figure 14–36b.

### Three-Phase Motors—Low and High Voltage

The connection of windings for operation on high or low voltage with three-phase motors is a series or parallel consideration. In Figure 14–37, the motor is connected

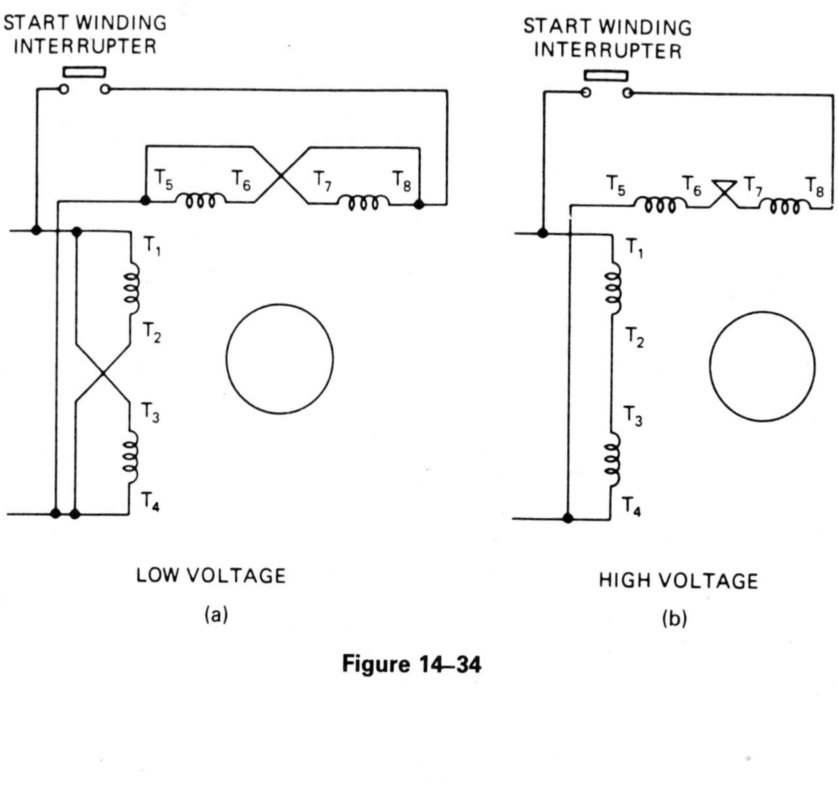


**Figure 14–34**

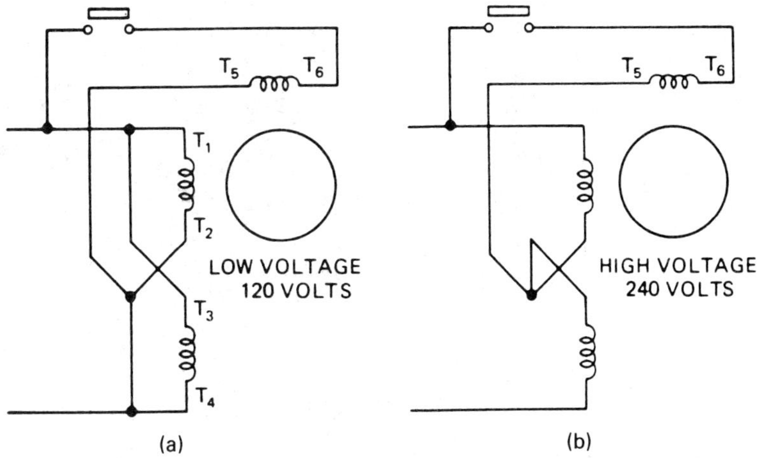


**Figure 14–35**

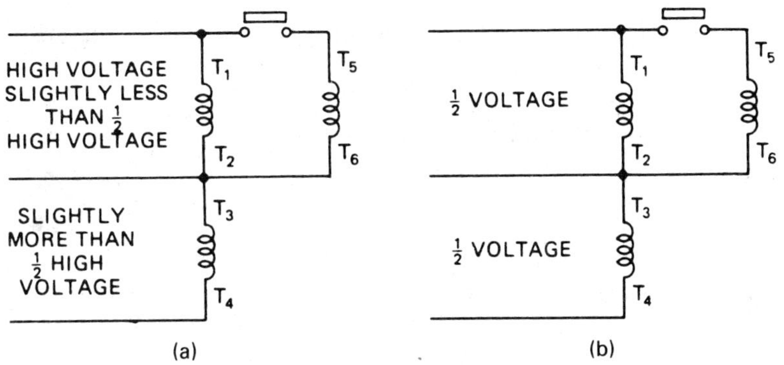


**Figure 14–36**

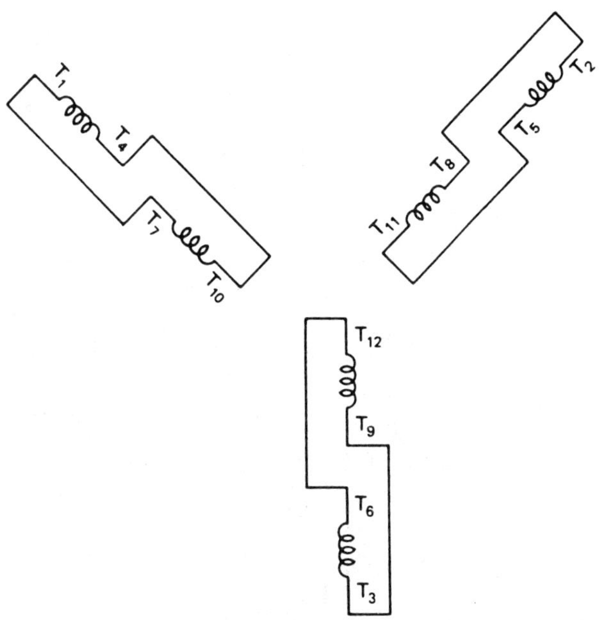


**Figure 14–37**

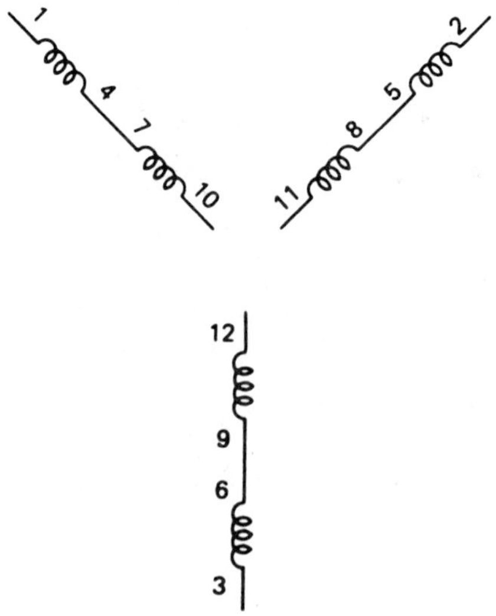


Figure 14–38

for operation at the low voltage. The winding pairs of each phase are connected in parallel.

For high-voltage operation, the winding pairs for each phase are connected in series. This configuration is shown in Figure 14–38.

High- or low-coil voltage on a three-phase motor is a consideration during the start of the motor. As with any motor, starting current is high. Current may be excessive if full voltage is applied to the coil's winding of each phase during the start process.

In order to maintain starting current at a reasonable level, many three-phase motors are started with the motor windings connected in Wye. (As the motor approaches operating speed, a relay switches the winding connection to delta.)

The motor starts with line voltage divided by $\sqrt{3}$ across each winding. As the motor approaches operating speed, the relay energizes, and full line voltage is placed across the windings. A simple delay timing device may be used to control the relay. (See Figure 14–39.)

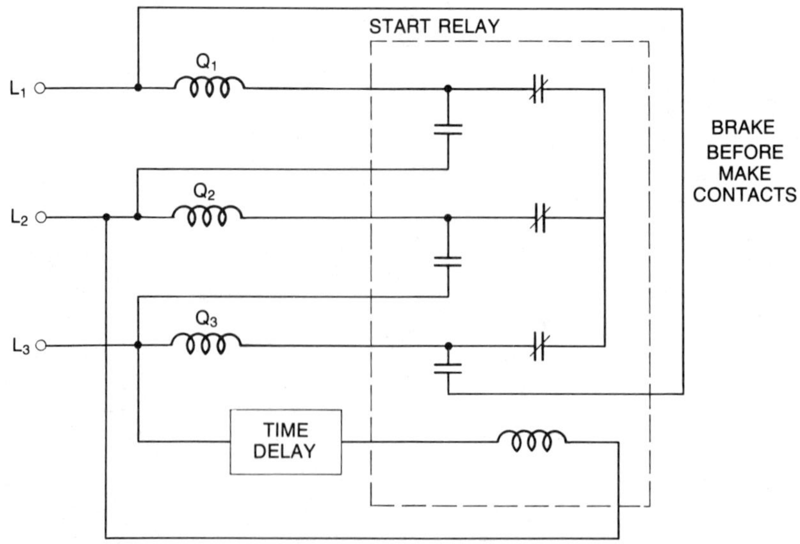


Figure 14–39

A motor commonly used with small electronic equipment is operated with dc power. A few years ago dc motors were rather expensive. This was mainly because the cost of construction included field coils. Advances in the design of magnetic materials have made it possible to provide permanent magnets for use as the field-pole pieces. The cost of dc motors has decreased and their use has increased.

Another motor coming into common use today is the *stepper motor*. This motor will operate from different control voltages. The motor provides for exact positioning, as well as in other operations, controlled speed.

These special motors require controls whose explanation is beyond the scope of this book. Special materials are available covering their operation.

## REVIEW QUESTIONS

T ______ F ______ 1. DC motors are the most common type used to drive high-powered equipment.

T ______ F ______ 2. In a shaded-pole motor, the shaft always rotates in the direction of the shaded pole.

T ______ F ______ 3. The direction of rotation of a split-phase motor may be changed by interchanging the connections to the start winding.

T ______ F ______ 4. Centrifugal switches are common in high-HP motors.

T ______ F ______ 5. Shaded-pole motors are more efficient than PSC motors.

T ______ F ______ 6. Capacitor-start motors have high starting torque.

T ______ F ______ 7. The run capacitor used with capacitor-start, capacitor-run motors is of a larger value than the start capacitor.

T ______ F ______ 8. Three-phase motors use capacitors in their start circuit.

T ______ F ______ 9. In order to reverse the direction of a three-phase motor, all three input wires must be changed.

T ______ F ______ 10. Three-phase motors are sometimes started in the Wye configuration and then switched to the delta configuration after the motor comes up to speed.

T ______ F ______ 11. The shaded-pole motor is used in low-power applications.

T ______ F ______ 12. A permanent split-capacitor motor requires a special starting device, called a centrifugal switch.

T ______ F ______ 13. The starting device used in split-phase motors in hermetically sealed compressors is the centrifugal switch.

T ______ F ______ 14. The potential-relay coil is connected in series with the run winding of the motor.

T ______ F ______ 15. The current relay has contacts that are open when the relay is deenergized.

T ______ F ______ 16. The potential relay has contacts that open when the motor approaches operating speed.

T ______ F ______ 17. It is not good practice to make adjustments or repairs on relays.

T ______ F ______ 18. All current relays may be operated in any position.

T ______ F ______ 19. Many high-horsepower motors use potential relays.

T ______ F ______ 20. Current or series relays are used in high-horsepower applications.

# UNIT FIFTEEN

# *CONTROL DEVICES*

There are many different types of control devices used with electric power circuits. These controls may be classified into four main divisions:

1. Personal protection
2. Circuit protection
3. Operation controls
4. Operation protection

The devices in each classification are covered in this unit.

## *PERSONAL PROTECTION*

Perhaps the most important control associated with electrical devices is the main power interrupter. With domestic refrigerators or through-the-wall air conditioning units, the power interrupter may simply be the wall plug. The technician may ''pull the plug'' and work on the unit knowing that power to the unit has been removed.

In larger systems in which power is supplied from a distant power panel, a power disconnect switch is usually located near the condensing unit. The technician may use this switch to interrupt power for maintenance or inspection. Whenever the power is interrupted by the technician for the purpose of maintenance or inspection, the switch should be *locked* in the open position (Figure 15–1). *Use an actual lock.* This is of particular importance if the switch is not located in the immediate area in which the technician is working.

Many technicians have received serious electrical shock while working on units when power was thought to have been disconnected. Be certain: Use the lock.

A common safety switch includes fuses for overload protection in the circuit. The safety switch may be used to disconnect power from a unit requiring maintenance

**Figure 15–1** Main power switch locked open.

or inspection. It is sometimes necessary to replace the fuses in a safety switch. Whenever a fuse is to be replaced in a system with which you are not familiar, a voltage check should be made at the fuses. Safety switches have been wired in backwards creating a dangerous situation. The proper way to wire in a safety switch is shown in Figure 15-2a. When the safety switch is open, the fuses are isolated from power.

The incorrect connection is shown in Figure 15–2b. With this connection when the switch is open, power is still connected to the fuses. An attempt to replace a fuse might result in a dangerous shock.

Before attempting to replace any fuse, check for voltage between the ends of the fuses as shown in Figure 15–3 and also from each fuse terminal and ground.

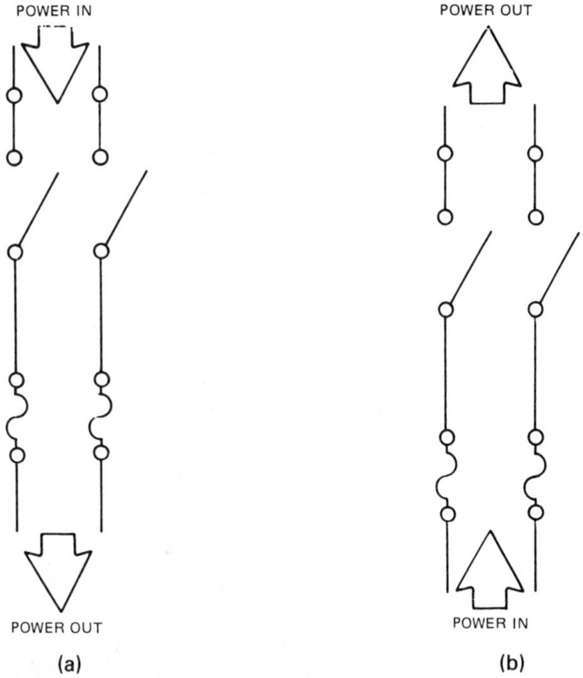


**Figure 15–2** (a) Correct connection of safety switch. (b) Incorrect connection of safety switch.

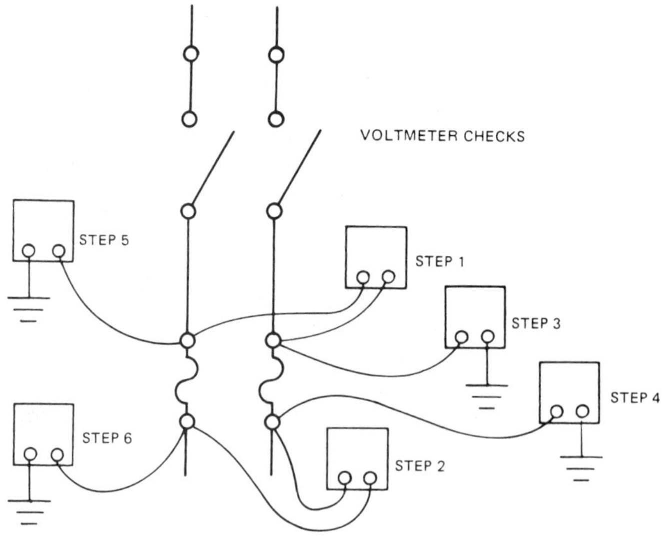


**Figure 15–3**

## *CIRCUIT PROTECTION*

The most common type of electrical circuit protective device is the simple *fuse*. A fuse is a one-time overcurrent protective device. When "blown," it must be replaced. The electrical portions of a fuse are the fuse link and the input and output connection terminals. Whenever current above the rated-value limit of the fuse is drawn (through the fuse), sufficient heat develops to melt the fusable link; this permanently opens the circuit.

Fuses are available in many physical sizes and shapes, as shown in Figure 15–4. Many applications require the use of *time-delay* fuses. A time-delay fuse is one in which current beyond its rated value may be drawn for a limited time without a burn out. This fuse allows for the brief start up of motors (high starting current) but still protects in case of continuous overload or short circuits.

Many cartridge fuses, particularly the larger, amperage-rated, knife-blade type, are constructed in a manner to provide a means of replacing a burned-out element. Fuse elements of the same ampere rating should be used for replacement. Whenever a fuse with a replacement element is to be installed, it should be disassembled first in order to insure that the correct ampere-rated element is in place.

Edison-base fuses are often found in older installations in applications requiring up to 30 amperes. The Edison-base fuse should not be used in newer installations. The S-type fuse with an S-adaptor should be used in new installations. The problem with the Edison-base fuse is that fuses of any value can be installed in any fuse socket. With the S-type base, only the prescribed amperage-value fuse fits in the base. The

**Figure 15–4** Fuses.

**Figure 15–5** Magnetic circuit breaker.

base usually includes a locking device to insure that the base cannot be removed from the socket and replaced with a larger amperance-value S-type base.

### Circuit Breakers

Circuit breakers are available in two general types—magnetic and heat overload. Although the popularity of the magnetic circuit breaker is increasing, the number of circuit breakers (in use) that function in this manner has not yet approached the number that employ a heat-operated overload (heat proportional to current).

Both types function in a similar manner in a portion of their operation. In the "on" (closed) position, the switch mechanism is spring-loaded and locked. The control (release) mechanism functions with the lock portion of the circuit breaker. When the lock is released, the spring mechanism forces the circuit breaker switch to the "off" (open) circuit position. It remains in this position until it is manually reset.

*Magnetic Circuit Breakers.* In the magnetic circuit breaker, the line current passes through an electromagnet (Figure 15–5). The electromagnet controls the lock mechanism, which holds the circuit breaker (switch) in the "on" position. When an overload (high) current passes through the electromagnet coil, the magnetic field is of a magnitude sufficient to release the lock mechanism. The spring-loaded switch snaps to the "off" position.

Magnetic circuit breakers are fast acting. The circuit breaker may be manually reset to the "on" position as soon as the problem causing the overload has been corrected.

*Heat Overload Circuit Breaker.* The most common type of circuit breaker operates on the principle of heat overload reacting on a sensitive bimetal strip. The bimetal strip controls the lock-in mechanism (Figure 15–6). Line current enters the circuit breaker at the top and passes through the switch contacts, the bimetal strip, and the output terminal. The bimetal strip and associated levers and springs mechanically control the lock-in mechanism. In the case of overload current, the bimetal strip heats and physically changes shape.

As the bimetal strip changes shape, it engages the trip mechanism and releases the lock mechanism. The locking arm is released, the compression spring is released, and the expanded spring pulls the contacts open.

Before the circuit breaker can be reset, the overload condition must be removed from the line. There is an extra delay because the bimetal strip must cool down before the lock and trip mechanism may be reengaged.

The lock mechanism may be interconnected to more than one set of control contacts. Figure 15–7 is a photograph of a 40-ampere circuit breaker of the "heater" type; if an overload occurs in either line, the interlocking mechanism opens both sets of contacts. The same type of mechanism may open three sets of main contacts for three-phase applications.

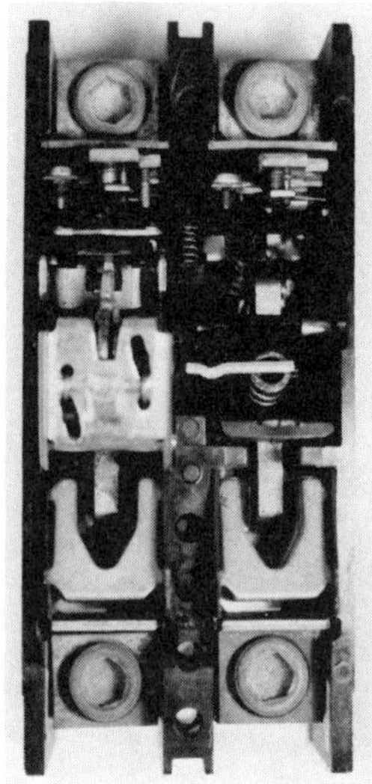

**Figure 15–6**
Heat overload circuit breaker.

**Figure 15–7**
Interlocked circuit breakers.

## *OPERATION CONTROLS*

The controls associated with motors are, for the most part, standardized. As discussed in Unit 14, for low-horsepower motor applications, simple switches are used to run them "off" and "on."

As the horsepower rating is increased and/or the type of motor is changed, the number and types of control devices associated with the compressor motor increases.

### Primary Controls

The primary controls associated with compressor motors are those that have a direct effect on the operation of the motor. In most applications, the controls consist of

1. Main contactor or power relay
2. Start relay
3. Overload (thermal)

*Main Contactor.* The main contactor (power relay) is used to control the application of high- (line-) voltage electric power to the motor circuits. The contactor is usually controlled by a low-voltage coil. When the coil is energized, contacts are mated, and high voltage becomes available to the circuits as required. In Figure 15–8, a photograph of a power relay is shown. The circuit diagrams of the device are also given.

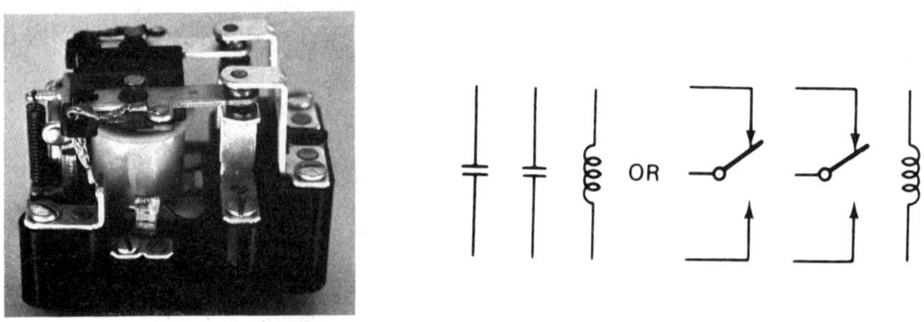


**Figure 15–8** Power relay.

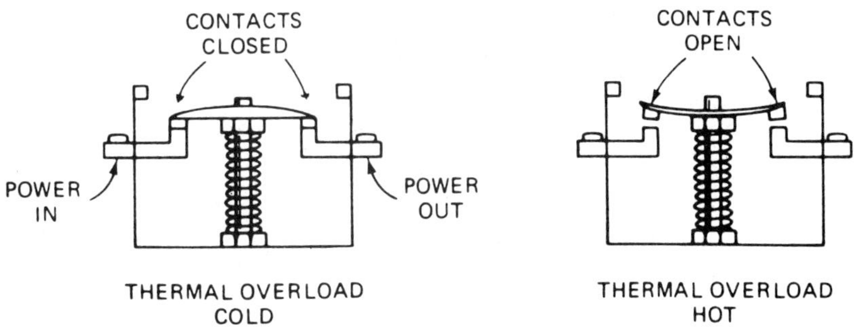


**Figure 15–9**

*Motor Start Relays.* The single-phase motor start relays used in standard circuits are of the current type or the potential type. Both relays were covered in Unit 14.

Most of the secondary controls that affect the operation of large motors are connected so as to control the application of low voltage to the coil of the main motor contactor.

*Thermal Overload.* A thermal overload is used with most motors in present-day operations. The overload opens in order to disconnect power to the motor whenever excess heat causes the internal bimetal disc to snap open.

The thermal overload is placed in physical contact with the motor case to "sense" heat from within the motor (Figure 15–9). If the motor becomes too hot for safe operation, the bimetal disc bends and snaps open, removing electric power from the motor common terminal, C.

In some older systems, a three-terminal thermal overload is used (Figure 15–10). With this type of overload, the temperature of the motor, as well as excessive starting current, causes the overload to open. The start winding is connected in series with the overload heater. Excessive current or start current drawn for too long of a time period heats up the overload, causing the bimetal disc to snap open. The two- and three-terminal overloads are connected in different electrical positions (Figure 15–11).

## Multiple-Contact Relays

Multicontrol (contact) relays are used for a special application. The purpose of a multiple-contact relay is to provide for an automatic multiswitching sequence. In Figure 15–12, a photograph of a multiple-contact relay is shown. The schematic diagram of the relay is also shown.

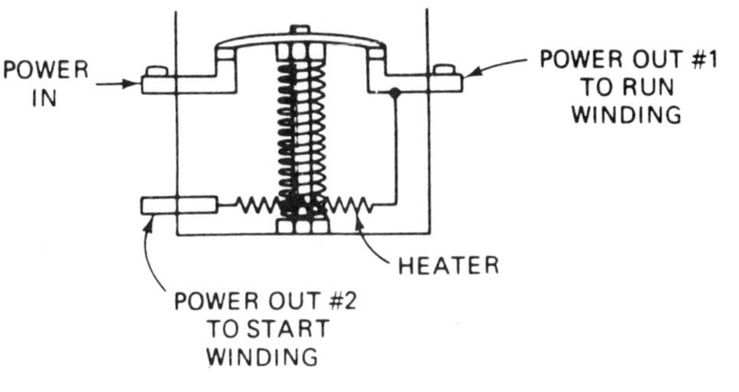


**Figure 15–10**

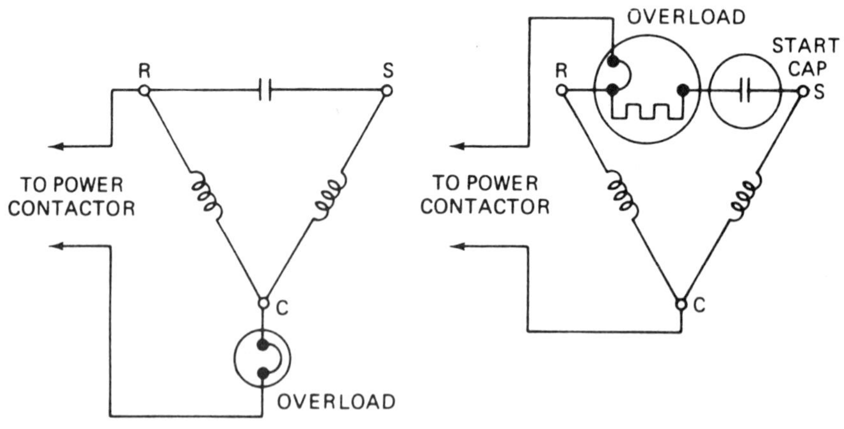


Figure 15–11

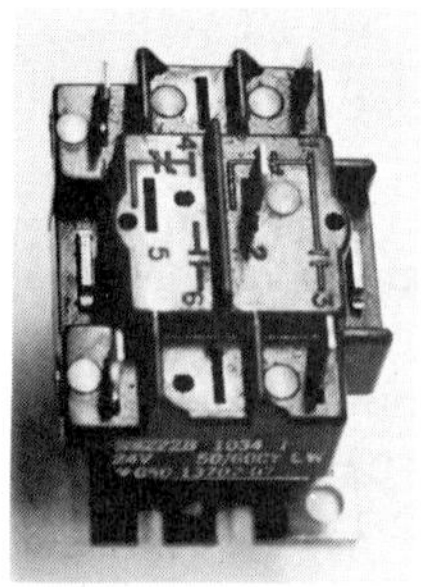

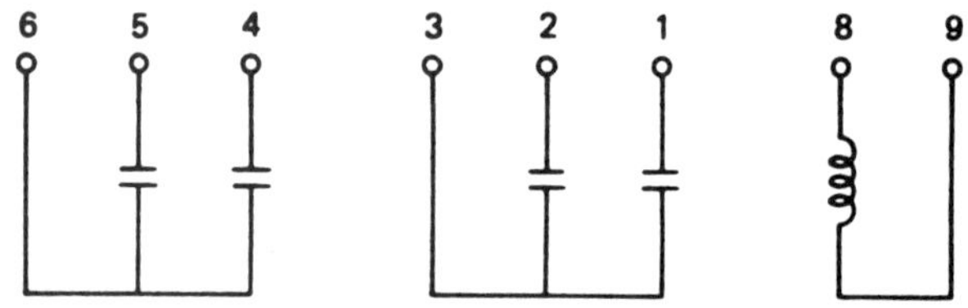


Figure 15–12

### Thermal Relay (Time-Delay)

The thermal time-delay has been used in system control for a number of years. The relay consists of a heater, a bimetal strip, and a set of contacts (Figure 15–13). When power is applied to the control terminals 1 and 2, the internal heater warms up. The bimetal strip bends, completing the circuit between terminals 4 and 3. Thermal time-delay relays are available with variable delays between control power application and contact closure. A photograph of a time-delay relay is shown in Figure 15–14.

### Solenoids

Solenoids are used in industry to provide a means of electrical control of mechanical devices. A sketch of a solenoid valve without power supplies is shown in Figure 15–15. The spring-loaded valve arm holds the valve needle tight against the valve seat, obstructing flow from the input connection to the output connection. When power is applied to the solenoid coil, the magnetic attraction of the coil overcomes the spring

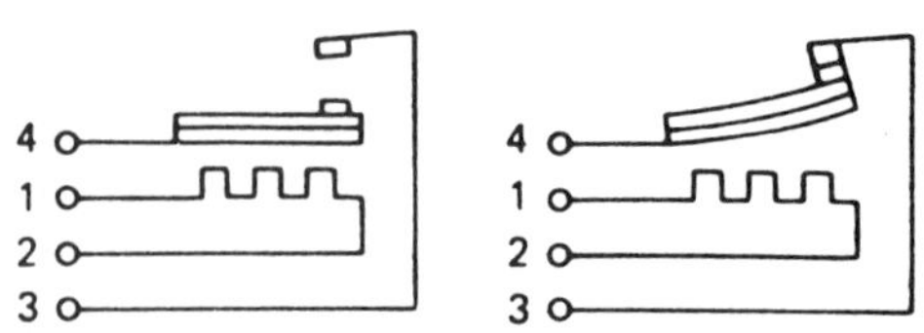


Figure 15–13

Figure 15–14

and pulls the solenoid armature with the valve needle up into the solenoid core. The valve needle is upseated from the valve seat, and a flow path is opened from the input connection to the output connection.

Caution must always be taken in the connection of input and output lines. When the lines are properly connected, the pressure from the input line tends to keep the valve closed (Figure 15–16). If the lines are connected in reverse, the pressure from the input line pushes against the valve-head surface, tending to open the valve.

Two-way solenoid valves are also available and are used in some systems (Figure 15–17). When the solenoid coil is deenergized, the spring-loaded valve head is held down against the output 2 valve surface. A flow path is open from the input past the valve arm to output 1.

When the solenoid is energized, the valve arm is pulled up. The valve head is positioned against the top valve surface, closing off output 1. As the valve moves up, the output 2 path for flow is opened.

### Pressure Control Switches

Pressure control switches are used as safety and/or temperature-sensing devices. The heat-limit switch shown in Figure 15–18 is a pressure switch that is sensitive to temperature change.

All pressure switches regardless of type have a common element, the bellows. The bellows is an expanding-contracting element that operates the switch or switches. The limit switch shown in Figure 15–18 is an example of a bellows-operated switch.

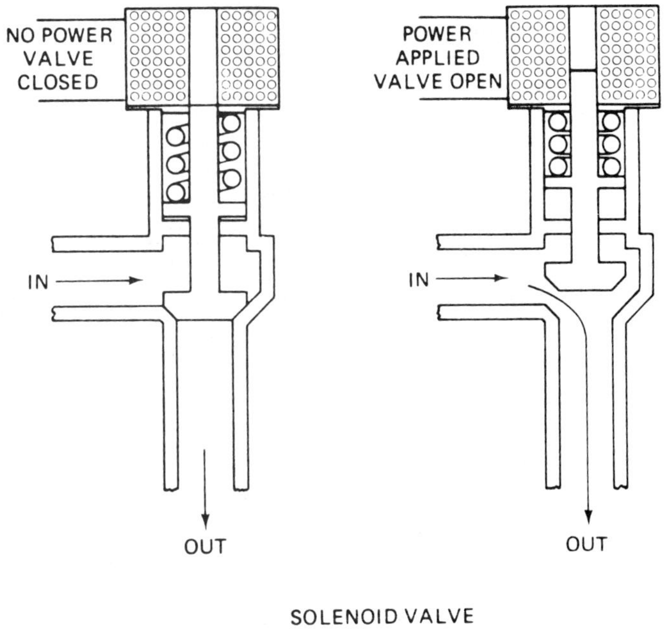


Figure 15–15

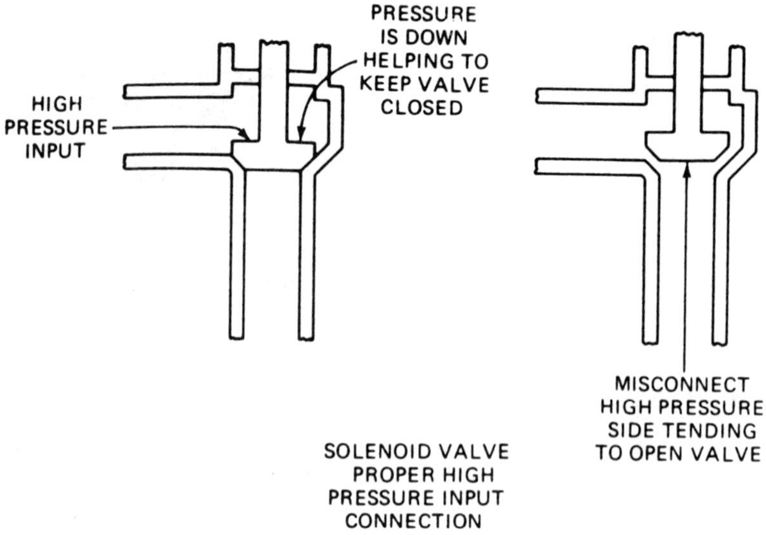


**Figure 15–16**

The switches associated with the pressure mechanism are both normally open. They could, however, have both been normally closed, or one with the other. Switch functions NO or NC or any combination are found as required by the application of the device.

*Temperature-Sensitive Pressure Switch.* When the pressure input (capillary) tube to a pressure switch is closed at the end and is filled with an inert gas or liquid, it becomes sensitive to temperature (Figure 15–19). Quite often, a remote bulb is soldered to the input tube to hold a larger supply of inert material in the miniature closed "system." As the temperature rises, the inert material expands, causing the bellows to move. As the temperature falls, the inert pressure decreases,

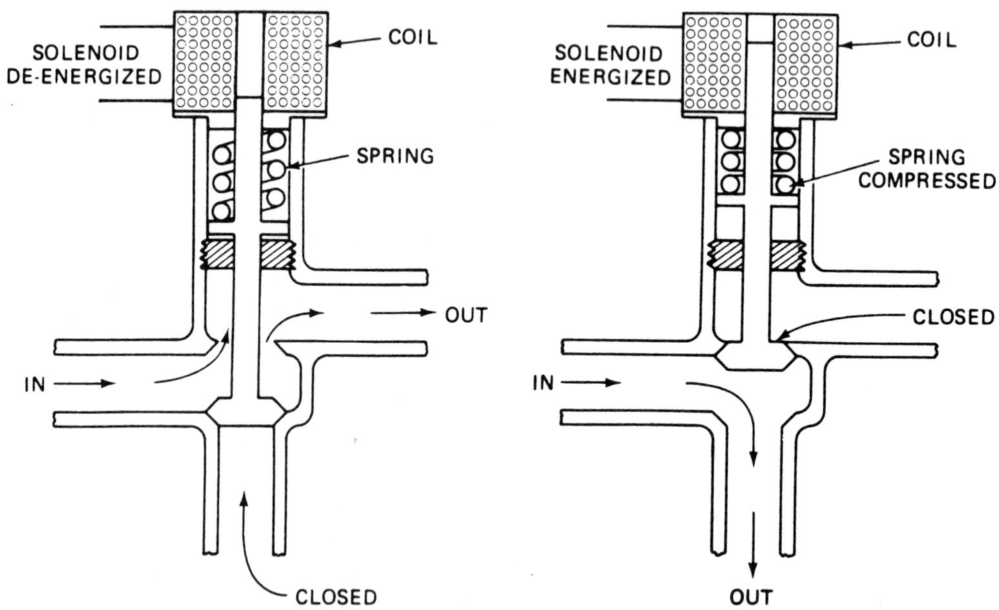


**Figure 15–17**

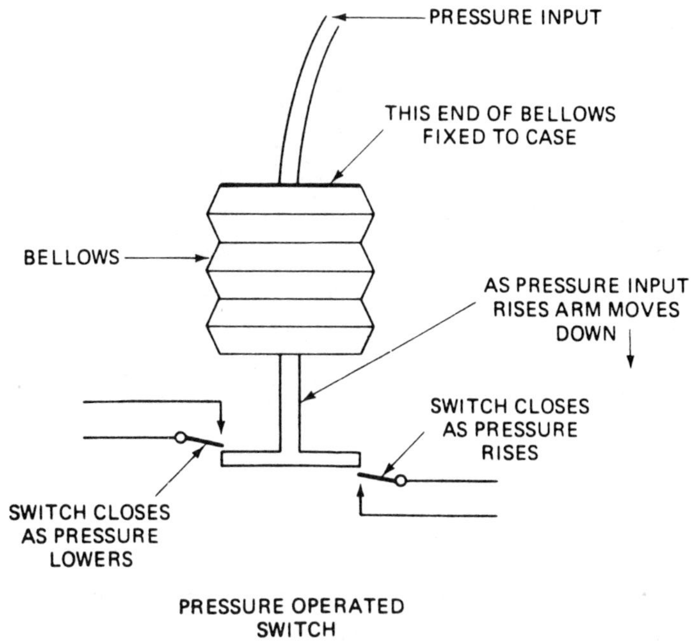


**Figure 15–18**

and the bellows move in the opposite direction. Switching, as required by the design, is controlled by the moving end of the bellows.

*Limit-Pressure Controls.* Limit-pressure controls are used to connect or disconnect electrical power from compressor components when the pressure being monitored is above or below the preset limit value. The pressure switch used may be a high-pressure, a low-pressure, a dual-pressure, or a pressure differential switch. In Figure 15–20, examples of the pressure switches are shown.

High- and low-pressure switches are similar in appearance. The pressure range at which the switch bellows mechanism operates determines whether it is a high- or

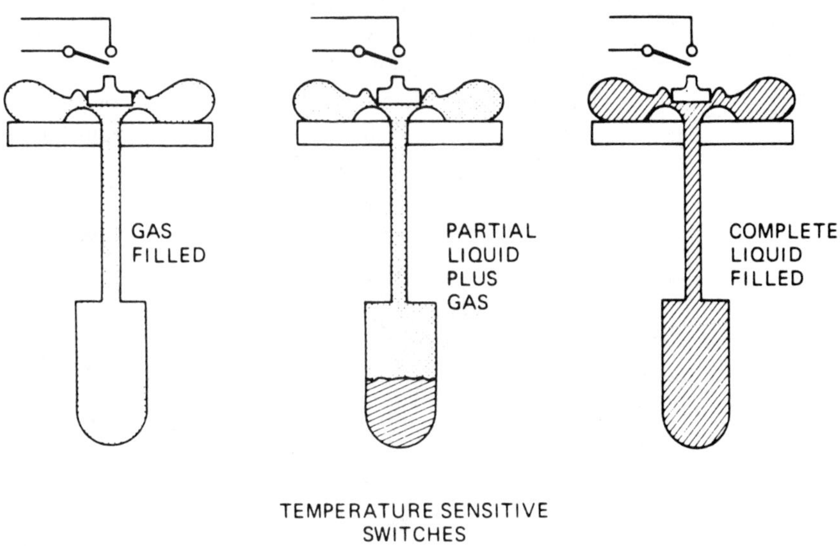


**Figure 15–19**

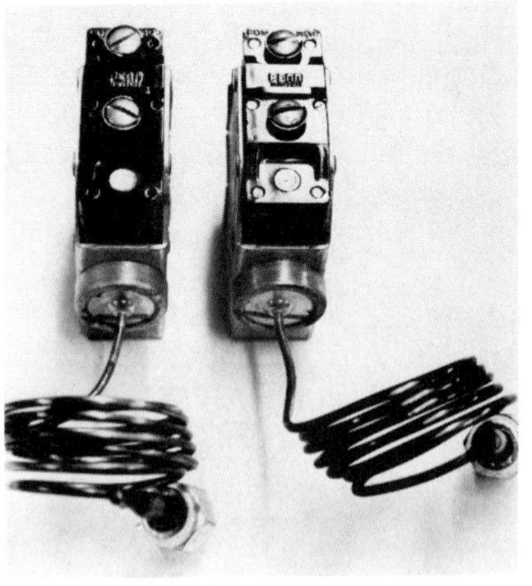

**Figure 15–20** Pressure switches.

a low-pressure control. Quite often, the pressure at which the switches actuate is adjustable.

There is a difference in the pressure range at which a system should be operational, depending on the altitude at which the system is located. The desired temperature range is another reason why the pressure switches are adjustable.

## *OPERATION PROTECTION*

### Pressure Controls

Pressure controls are used for safety protection with compressor systems. A high-pressure control actuates in the system if high-side system pressure exceeds the preset level. When the switch in the pressure control opens, the compressor is stopped. This type of safety control is required to prevent the compressor motor from overloading and to prevent the system from rupture because of excessive internal pressure.

In Figure 15–21, the electrical connections of both a high- and a low-pressure control are shown. The switch systems associated with pressure controls are available in many combinations. Pressure controls are available with manual or automatic reset features. Automatic controls reset when the pressure stabilizes and is back within the operating range. If the problem causing the abnormal pressure condition is not corrected, the control again opens.

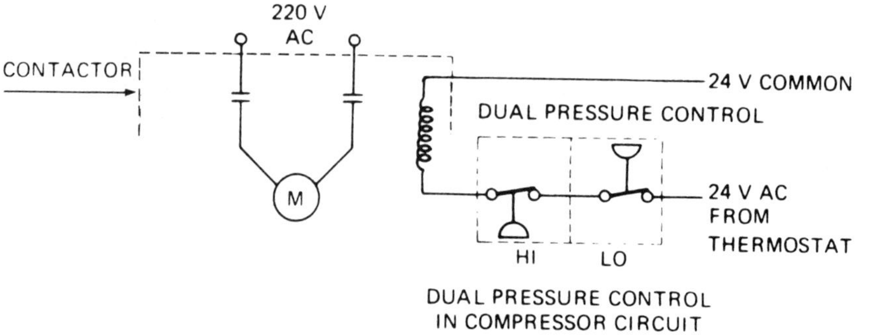


**Figure 15–21**

## REVIEW QUESTIONS

T ______ F ______ 1. Edison-base fuses are replacing S-type fuses in industry.

T ______ F ______ 2. Thermal circuit breakers may be reset immediately.

T ______ F ______ 3. Current relays are connected in parallel with motor windings.

T ______ F ______ 4. Thermal time relays operate on dc power only.

T ______ F ______ 5. Magnetic starters are used to control large motors.

T ______ F ______ 6. Magnetic starters with pushbutton controls have thermal overload contacts in series with the stop-push button.

T ______ F ______ 7. Some thermal snap overloads contain an electric heater.

T ______ F ______ 8. Thermal snap overloads open if the motor-housing temperature rises above the safe limit.

T ______ F ______ 9. Solenoid-controlled valves may be used in water lines to control water flow.

T ______ F ______ 10. Connection of input and output lines to a solenoid valve may be interchanged.

T ______ F ______ 11. Twenty-four volts ac is a common control signal voltage.

T ______ F ______ 12. In a temperature-sensitive bellows system, as temperature decreases, pressure increases.

T ______ F ______ 13. Fuses should be placed in the power-in side of the disconnect switch.

T ______ F ______ 14. Magnetic circuit breakers are slower-acting than heat-overload circuit breakers.

T ______ F ______ 15. The heater of the three terminal thermal overload is connected in the run-winding circuit.

T ______ F ______ 16. The two-terminal thermal overload is normally connected in the common line.

T ______ F ______ 17. Pressure control switches may provide both NO and NC contacts.

T ______ F ______ 18. Sensitive switches are normally used to control relay coils rather than power circuits.

# UNIT SIXTEEN

# *TRANSFORMERS*

Transformers are nonmechanical electric devices used in many applications for the control of electrical power. High voltages used on transmission lines for electrical power transfers are dropped (reduced) with the use of transformers. Voltages, at low levels, used in electrical control systems are obtained through the use of transformers.

For example, the transmission cable may be 2,400 volts while utility service requirements are 240 volts. A transformer is used to reduce the 2,400-volt transmission to a usable 240 volts. This 240 volts supplies power to the high-voltage electrical components. Another transformer within the electrical system steps this 240 volts down to 24 volts for the low-voltage control system.

## *VOLTS PER TURN*

In Unit 12, the back EMF developed in a coil of wire carrying an ac current was discussed. It was shown that the inductance of the coil has the major effect in the control of current. The varying magnetic field around a coil carrying an ac current cuts through the coil and produces a back EMF that limits the total current. The small current that flows through the coil is called the magnetizing current.

In Figure 16–1, a coil of wire wound on an iron core is shown. The coil of wire has 600 turns in it, connected in series. The back EMF developed in the coil is very close to 100 volts. Another way of stating the situation is that for every six turns of wire on the coil, approximately 1 volt is developed.

A second coil of wire is wound on the same iron core (Figure 16–2). The second coil is called the *secondary winding*; the first coil is called the *primary winding*. The same magnetic field that cut through the primary winding, developing back EMF, cuts through the secondary winding. A voltage is developed in the secondary winding at 1 volt per 6 turns. Because there are 60 turns in the secondary, a total of 10 volts is developed across the secondary winding ($60 \div 6 = 10$).

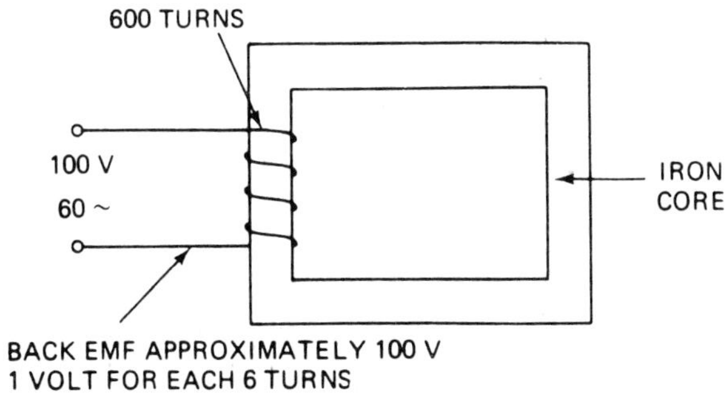


**Figure 16–1**

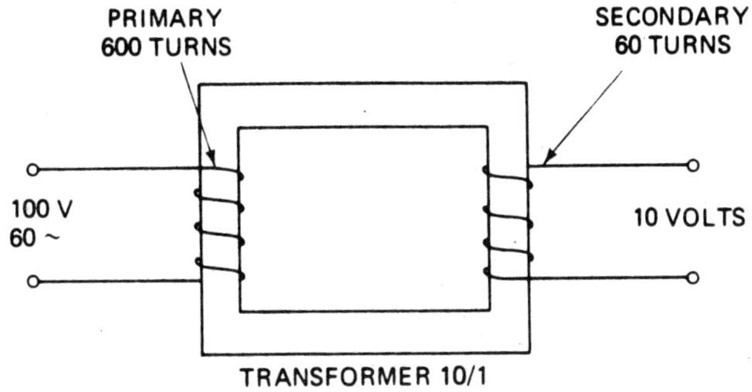


**Figure 16–2**

There is voltage relationship between the primary and secondary windings of a transformer. Simply stated, the turns ratio is equal to the voltage ratio. The transformer shown in Figure 16–2 has a turns ratio of 600 to 60 or 10 to 1; the voltage ratio is also 10 to 1.

Figure 16–3 shows the same transformer with a load resistance of 1 ohm connected across the secondary. The 10-volt output across the secondary winding will be impressed across the 1-ohm resistor after the switch is closed. According to Ohm's Law:

$$I = \frac{E}{R}$$

$$I = \frac{10}{1}$$

$$I = 10 \text{ amperes}$$

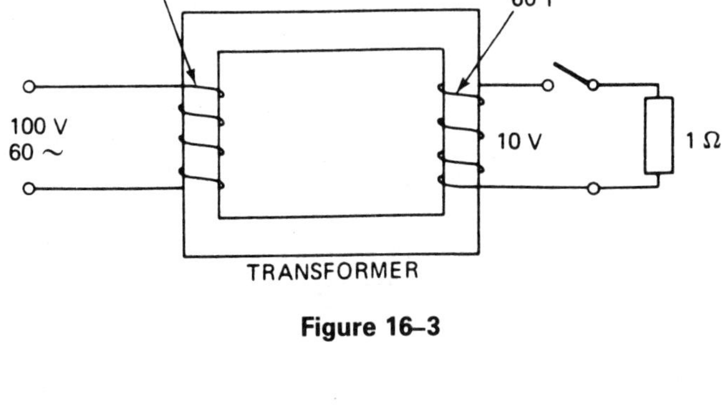


**Figure 16–3**

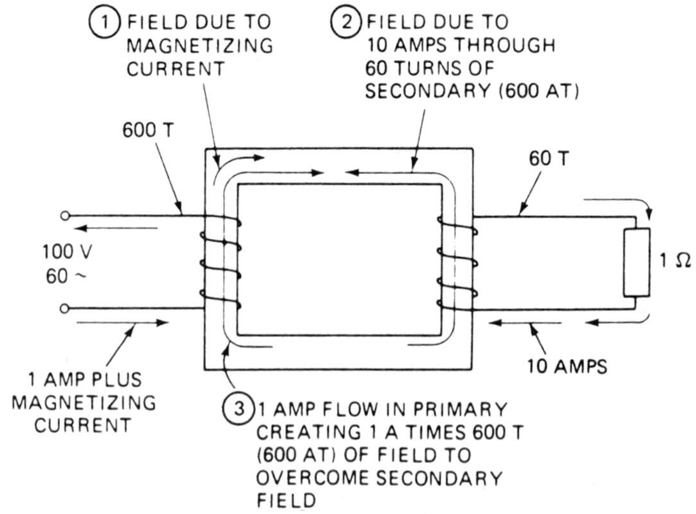


**Figure 16–4**

The 10 amperes of current flow through the 1-ohm resistor. Electrical power is supplied by the secondary of the transformer. This power must be coming through the primary of the transformer from the 100-volt, 60-hertz power source. The power developed in the secondary circuit is:

$$W = I \times E$$

$$W = 10 \times 10$$

$$W = 100 \text{ watts}$$

When the primary of a transformer is connected to a source voltage, a small current, called the *magnetizing current*, flows through the primary. The magnetic field set up by this magnetizing current is indicated in Figure 16–4 as 1.

When a load is connected across the secondary, current flows in the secondary. This current produces a magnetic field in the core of the transformer in direct opposition to the field set up by the magnetizing current. In Figure 16–4, this magnetic field is indicated as 2. The field strength is based on 600 ampere-turns and tends to reduce the back EMF of the primary. The decreased back EMF allows for an increase in current in the primary.

A current flow producing a magnetic field based on 600 ampere-turns is required in the primary in order to overcome the 600 ampere-turns in the secondary. With 600 turns in the primary, 1 ampere of current is needed to produce this field. The field is shown as 3 in Figure 16–4.

The relationship between the currents in the primary and the secondary can now be determined. The currents are inversely proportional to the turns ratio. For example, in Figure 16–3:

1. The turns ratio is 600/60 or 10/1.
2. The voltage ratio is 600/60 or 10/1.
3. The current ratio is 60/600 or 1/10.

The power developed in the secondary has been shown to be 100 watts. The power developed in the primary is:

$$W = I \times E$$
$$W = 1 \times 100$$
$$W = 100 \text{ watts}$$

Actually, the current flow in the primary is slightly more than 1 ampere since the magnetizing current also flows in the primary. The transformer is not 100 percent efficient: There is some power loss.

## POWER LOSS IN TRANSFORMERS

There are three main reasons for power loss in a transformer:

1. Resistance of the windings,
2. Hysteresis, and
3. Eddy currents.

*Resistance of the Windings.* The power loss in the resistance of the windings needs no further explanation. Whenever current flows through resistance, heat is developed and power is lost. This power loss is known as the $I^2R$ loss.

*Hysteresis.* When alternating current flows in a coil, the magnetic field about the coil changes direction with the current. If the coil has an iron core, as in a transformer, the direction of the magnetism of the iron core must also change with the current: Each magnetic molecule of the iron core must change direction with each alternation of the current. This continuous motion of the magnetic molecules produces heat due to friction, which is a power loss in the transformer.

*Eddy Currents.* Eddy currents are currents set up in the iron core of a transformer. The core of the transformer is a metal, and it acts the same as a secondary winding. A voltage is induced in the core and current flows in the core. Eddy currents flow throughout the core. If solid iron cores were used in transformers, the losses due to eddy currents would be quite high.

In order to overcome the high eddy current losses, the cores of transformers and most other magnetic devices using alternating current are laminated. Laminations are thin, sheet metal parts that may be mechanically fastened together to produce the required core. By the use of thin sheets of metal, the eddy current is confined to the individual lamination. Because the resistance path for eddy currents is increased, the total eddy current draw is decreased. Laminated iron cores are found in most ac devices, such as transformers, motors, solenoids, relays, and contactors.

## POWER TRANSFORMERS

Power transformers are used throughout the world to step up voltages for transmission and distribution. They are also used to step down voltages for use at the destination point.

For example, a power company may generate electric power at a voltage level of 13,800 volts at the generating station. The main factor in determining the voltage level at the generator is the insulation problems related to operating generators at higher voltages. Before the electrical power is distributed on transmission lines across country to the individual localities, the voltage level is stepped up, using transformers.

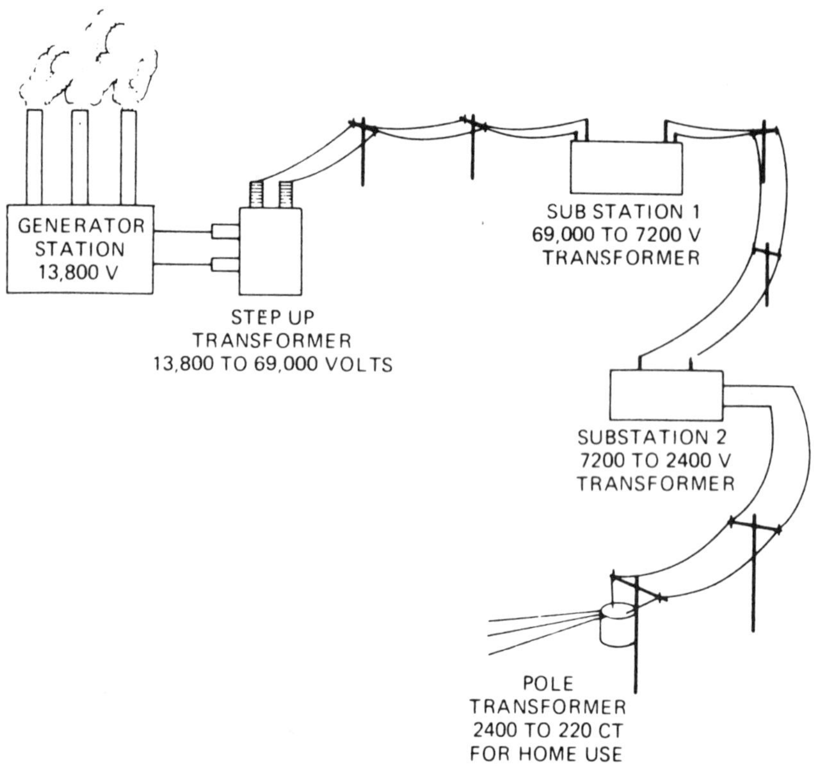


**Figure 16–5**

A standard transmission line voltage level is 69,000 volts. Substations (transformers) are used to reduce the voltage to safer levels (about 2,400 volts) for distribution around populated areas. Finally, at the end point of use, the voltage is further reduced by transformer to the level required by the consumer, usually 220/110 volts. Figure 16–5 shows an example of a full system from generation to final use of electrical power by the residential customer. Voltage levels shown are representative of industry standards.

Power transformers at substations are usually large, bulky items (Figure 16–6). High-voltage insulators of ceramic material are usually visible at the terminals of

**Figure 16–6** Transformer substation.

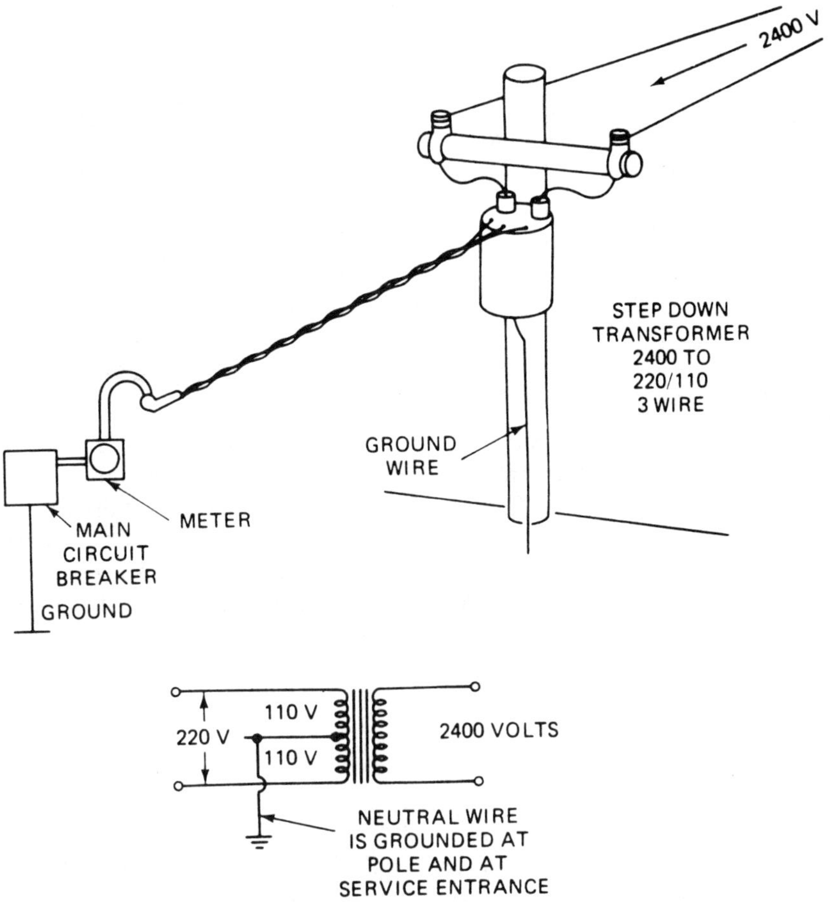


**Figure 16–7**

the transformers. Power transformers used to step voltage down from a final substation (2,400 volts) to the voltage needed in residential use (220/110 volts) are located on power-line poles (Figure 16–7). A single transformer will usually supply several homes. The final step-down transformer is center tapped and therefore can provide the required 220 volts for power circuits and also the 110 volts required for lighting and small appliances.

## Transmission Losses

Voltages are stepped up whenever power is to be distributed over long distances in order to decrease power losses. For example, consider a manufacturing plant located approximately 10 miles from a power-generating station (Figure 16–8). The manufacturing plant requires 100,000 watts of electrical power. The generating station is

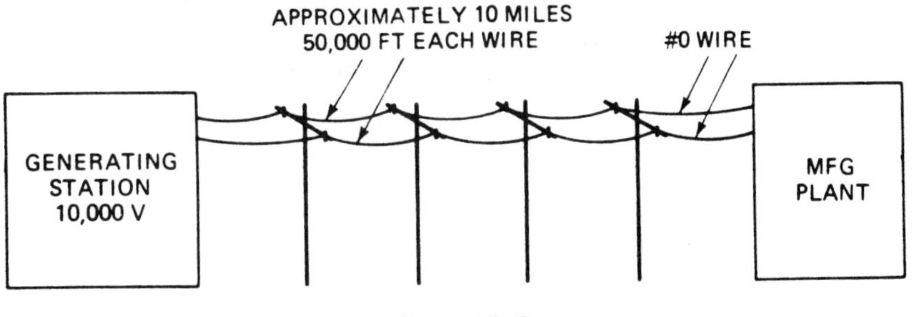


**Figure 16–8**

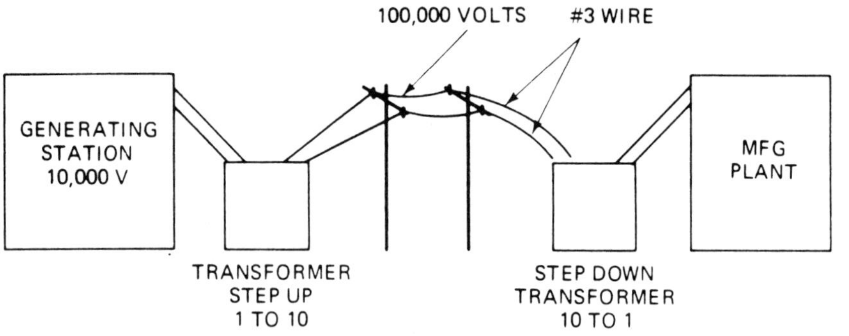


**Figure 16–9**

producing power at a 10,000-volt level. A current of 10 amperes is needed at the 10,000-volt level in order to provide the 100,000 watts.

Two wires are required in order to get the electrical power to the manufacturing plant; each wire is 50,000 feet long. If size 0 wire were used, the total resistance of the two wires would be about 10 ohms because size 0 wire has 0.1 ohm per 1,000 feet. According to Ohm's Law, the voltage developed on the resistance of the transmission line would be 100 volts.

$$E = I \times R$$
$$E = 10 \times 10$$
$$E = 100 \text{ volts}$$

This would leave only 9,900 volts to be delivered to the manufacturing plant. The power company would continually waste the power developed in the resistance of the transmission lines.

$$W = I \times E$$
$$W = 10 \times 100$$
$$W = 1{,}000 \text{ watts lost}$$

The losses can be reduced and additional savings can be realized by employing transformers, thus making feasible the use of smaller wire in the transmission lines. In Figure 16–9, the power station and the manufacturing plant are the same. A step-up transformer has been added at the generator station, and a step-down transformer has been added at the manufacturing plant. The transmission lines have been changed to size 3 wire, which has about 0.2 ohms per 1,000 feet. The resistance of the total transmission line, 2 lengths of 50,000 feet each, is about 20 ohms. With the step up of the voltage at the power plant from 10,000 volts to 100,000 volts, the current is stepped down from 10 amperes to 1 ampere. The power loss in the lines has been reduced considerably.

The voltage developed in the resistance of the transmission lines is

$$E = I \times R$$
$$E = 1 \times 20$$
$$E = 20 \text{ volts}$$

The power loss is

$$W = 1 \times E$$
$$W = 1 \times 20$$
$$W = 20 \text{ watts}$$

The voltage at the primary of the step-down transformer is

$$100{,}000 - 20 = 99{,}980 \text{ volts}$$

At the secondary, the voltage is 9,998 volts, which is much closer to the 10,000 volts being delivered by the generating station.

In actual practice, the voltage required at a manufacturing plant is seldom that generated at a power station. For example, the principles of using transformers to step up and to step down voltages as well as to save power are similar to those in actual situations.

## *SIGNAL TRANSFORMERS*

Transformers are used in almost all industrial situations to provide safe, low-level ac voltage for control purposes. The input voltages for systems may be 110, 220, or 440 volts; 220 volts is the most common input for residential control systems. A step-down transformer within the system provides 24 volts ac for use in the control circuit. The control circuit is used to energize high-voltage circuits for the control of motors and other devices.

An example of a low-voltage control circuit of an air conditioning system using a 24-volt step-down transformer is shown in Figure 16–10. A voltage of 220 volts

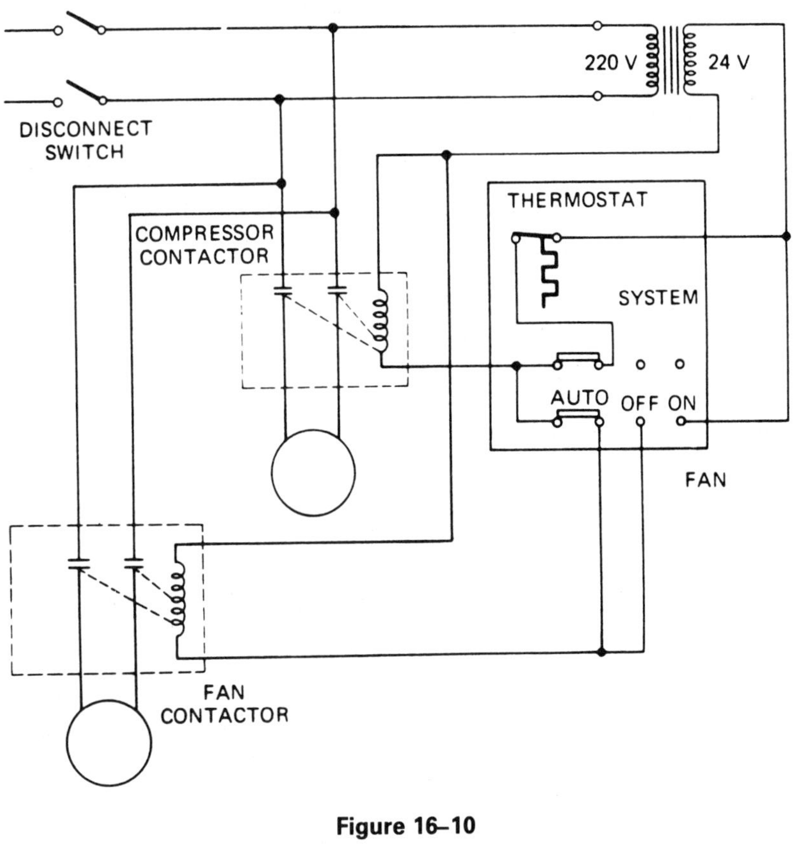


**Figure 16–10**

is fed through the disconnect switch to the primary of the low-voltage transformer and to the top of the two power contactors. With the thermostat switches (covered in Unit 15) as shown, the 24 volts from the secondary of the low-voltage transformer are fed to the coil of the compressor contactor and to the coil of the inside fan motor; both contactors energize. The 220 volts are then fed through both contactors to the motors.

After the room cools, the thermostat contacts open at a preselected temperature. This removes the 24 volts from the contactor coils; the contactors deenergize. The contacts then open and 220-volt power is removed from the motors. The motors will not run until the room temperature rises and the switches (contacts) in the thermostat again close.

Through the use of a low-voltage transformer, a safe system of signal transfer is accomplished. The individual making adjustments at the thermostat is not in close proximity to dangerous high-voltage power circuits.

Another important factor in the use of low voltage in air conditioning control circuits is the cost of installation of control-circuit wiring. If 110 volts were used in the control circuit, wiring between the thermostat and the evaporator/condenser would have to be treated the same as the power circuits for lighting or outlets. If the local building code called for the use of conduits in all line-voltage circuits, then a conduit would have to be installed between the evaporator/condenser unit and the thermostat. Installation costs could vary greatly, depending on the distance between the thermostat and the condenser.

The same installation using a low-voltage control system would be less expensive. Control circuit wiring between the evaporator/condenser and thermostat would consist of a standard signal cable that met building code requirements. If the conduit is not needed, the cost decreases dramatically.

## *TRANSFORMER RATINGS*

Transformers are usually rated with the primary and secondary voltage and the volt-ampere design of the unit. An example of this is a control transformer. This transformer is designed to operate with 220 volts on the primary, and it will produce 24 volts at the secondary terminals. The volt-ampere rating of the transformer is 48VA. This means that a maximum of 2 amperes can be drawn from the transformer secondary (2 amperes $\times$ 24 volts $=$ 48 VA).

## *IMPEDANCE-MATCHING TRANSFORMERS*

Transformers may also be used as impedance-matching devices. Impedance matching is necessary whenever the transfer of maximum ac power from source to load is required.

In Unit 7, in the sections dealing with batteries and internal resistance, it was shown that maximum power is available from a source when the load resistance is equal to the source resistance. This concept is as true in ac as it was in dc.

Many times, the load resistance does not match the source resistance, and loss of power cannot be tolerated. An impedance-matching transformer is needed. Consider Example 1.

**EXAMPLE 1**

Power is required to drive a 75-ohm load. The source resistance is 300 ohms. What is the turns-ratio of the required impedance-matching transformer? The circuit is shown in Figure 16–11.

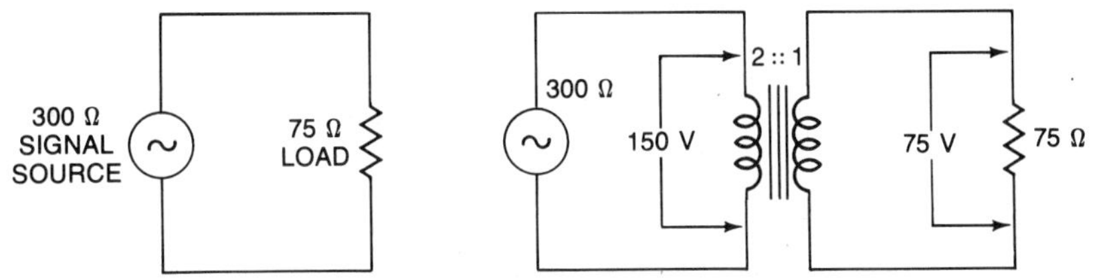


**Figure 16–11** Impedance-matching transformer.

***Solution*:** Assume 75 volts at the secondary. Secondary power equals:

$$W = V^2/R$$
$$W = 75^2/75$$
$$W = 75 \text{ watts}$$

The power supplied by the primary must be 75 watts. (Ignore transformer losses.)

The primary impedance is 300 ohms. The 75 watts must be developed in the source impedance for maximum power transfer. (Maximum power is transferred when the power developed in the load is equal to the power developed in the source.) What is the voltage necessary for the development of 75 watts in the primary circuit?

$$W = V^2/R$$
$$V^2 = W \times R$$
$$V^2 = 300 \times 75$$
$$V^2 = 22{,}500$$
$$V = \sqrt{22{,}500}$$
$$V = 150 \text{ volts}$$

If the primary voltage must be 150 volts while the secondary voltage is 75 volts, then the turns-ratio must be 2 to 1. The turns-ratio is equal to the voltage ratio.

$$\frac{N_P}{N_S} = \frac{V_P}{V_S} = \frac{2}{1}$$

The impedance ratio of the components in the original problem was 300 to 75, or 4 to 1. The relationship between turns-ratio and impedance ratio is:

$$\frac{Z_P}{Z_S} = \left(\frac{N_P}{N_S}\right)^2$$

The impedance ratio is equal to the turns-ratio squared.

Consider the use of impedance-matching transformers in a television antenna system. Figure 16–12 shows a 300-ohm antenna connected to a matching transformer,

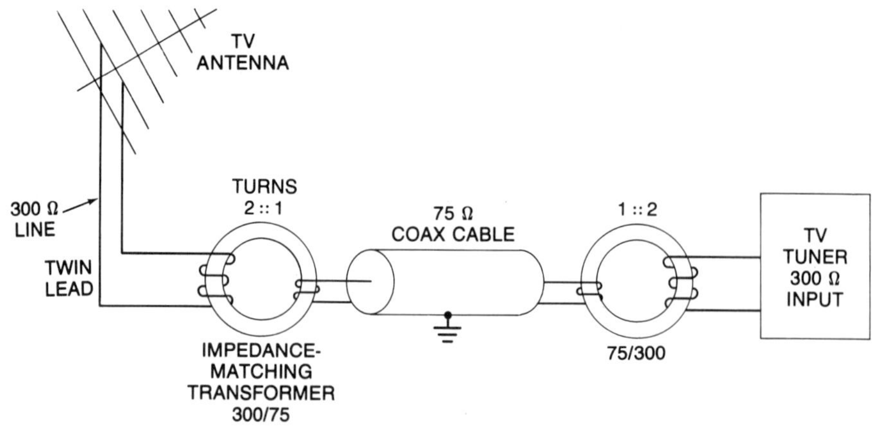


**Figure 16–12** Matching antenna to coax to TV.

300 ohms to 75 ohms. Coaxial cable of 75-ohm characteristic impedance is used to carry the signal to the TV set. The input impedance of the TV tuner is 300 ohms. Another matching transformer is required at the tuner to match the 75-ohm coax cable to the 300-ohm input.

## REVIEW QUESTIONS

1. The voltage ratio of a transformer is ______ proportional to the turns-ratio.
2. The current ratio of a transformer is ______ proportional to the turns-ratio.
3. The power supplied to the primary is approximately ______ to the power taken from the secondary.
4. The amount of power loss in transmission lines will be (higher, lower) ______ if the power is transmitted at higher voltages.
5. Why are low-voltage transformers used in air conditioning systems?
6. A transformer has 800 turns on its primary and 400 turns on its secondary. The primary voltage is 200 volts. What is the secondary voltage?
7. In the transformer in Question 6, if 4 amperes are drawn in the secondary circuit, how much current flows in the primary?
8. What are the three main losses in a transformer?
9. Which transformer loss is related to the motion of molecules?
10. What ratings are usually provided on control transformers?
11. A transformer is providing 25 watts of power to its secondary circuit. If the current in the secondary is 1 ampere while the current in the primary is 5 amperes, what is the primary voltage?
12. An impedance-matching transformer is used to match a 500-ohm load to a 3,000-ohm source. The transformer has 84 turns on the output side. How many turns are there on the 3,000-ohm input side?

# UNIT SEVENTEEN

# SEMICONDUCTOR DEVICES

Electronic control of electrical devices is increasing at a rapid rate. Timing circuits that used to include clock motor devices now use electronic timers. Power-level control devices that used to include variacs or rheostats now operate using silicon-control rectifiers or triacs. The new electrical technician must be familiar with the electronic devices presently being used. These devices include most of the semiconductors used as discrete components as well as many of the integrated circuits.

## DEFINITION

A semiconductor is a material that falls somewhere between the good conductors—most metals—and the poor conductors or insulators—such as glass or wax. Semiconductors function in electronic circuits because of their crystalline structure and because they are doped with materials called *impurities*. The impurities provide characteristics to the semiconductor material that make the material useful in rectifiers and amplifiers.

The basic semiconductor materials are germanium and silicon. Silicon is the most popular and is covered here. When the pure silicon material is doped with an impurity, it becomes either P-type silicon or N-type silicon, depending on the material added as the impurity. The P stands for positive, and the N for negative. The impurity combines with the silicon crystals and modifies the crystalline structure. If P-type impurity is added, the crystals of silicon form improperly, leaving a space for an electron in the structure; this space is called a *hole*, and it is movable in the silicon. If N-type impurity is added, the crystals of silicon form with an extra electron available; this extra electron is movable.

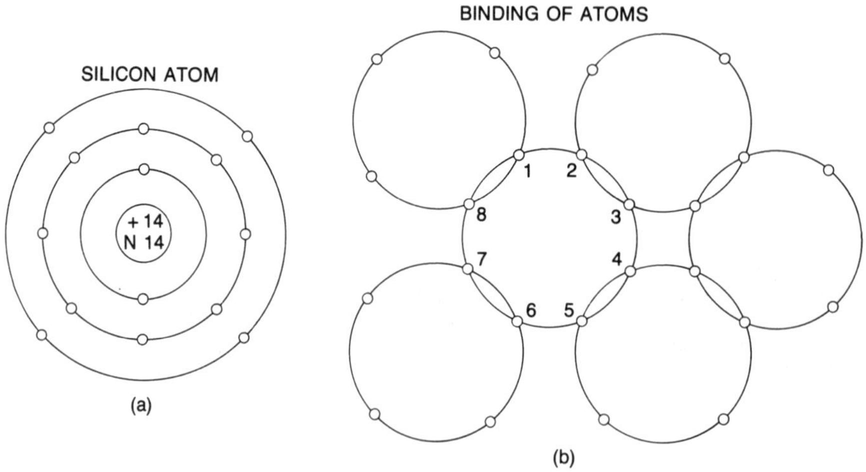


**Figure 17–1** Silicon.

### *SILICON*

Silicon has an atomic number of 14. (See Appendix B.) In its pure form, the material is a fairly good insulator, having a resistance approximately $1.34 \times 10^{11}$ that of copper. In its pure form, silicon forms in a crystal. If the individual atoms could be observed, a general pattern would be seen.

Silicon has 14 electrons in orbit about the nucleus. Formed in the standard pattern, there are two electrons in the first orbit, eight electrons in the second orbit, and four in the third and last orbit. See Figure 17–1a.

In the formation of silicon crystals, four atoms share an electron of their outer ring with the original atom of silicon, as seen in Figure 17–1b. This is a crystalline material where each atom forming the crystal has the equivalent of eight electrons in its outer orbit.

As already indicated, pure silicon has a rather high resistance and will not work as the material for electronic components.

### *DOPING OF PURE SILICON*

In order to use silicon in semiconductor-conductor components, the characteristics of the silicon must be changed. This is accomplished by adding some impurity to the pure silicon. This process is called *doping*.

Adding a small amount of phosphorus to the pure silicon produces some dramatic changes. When phosphorus is added to silicon, the material still forms into crystals. Phosphorus has an atomic number of 15. This means 15 electrons orbit the nucleus. The standard electron orbits will prevail with two electrons in the first orbit, eight electrons in the second orbit, and five electrons in the third orbit.

When the four atoms of silicon form in a crystalline structure with the fifth atom of phosphorus, there will be an excess of one electron. (See Figure 17–2.) This new material is still electrically neutral because electrons have not been moved. But there is an excess electron in the crystalline structure; the material is called N-type silicon.

In the doping process, if aluminum were added to pure silicon, a P-type material would be formed. Aluminum has an atomic number of 13. In the electron orbits about the nucleus, there are two, eight, and three electrons in the three orbits. (See Figure 17–3.)

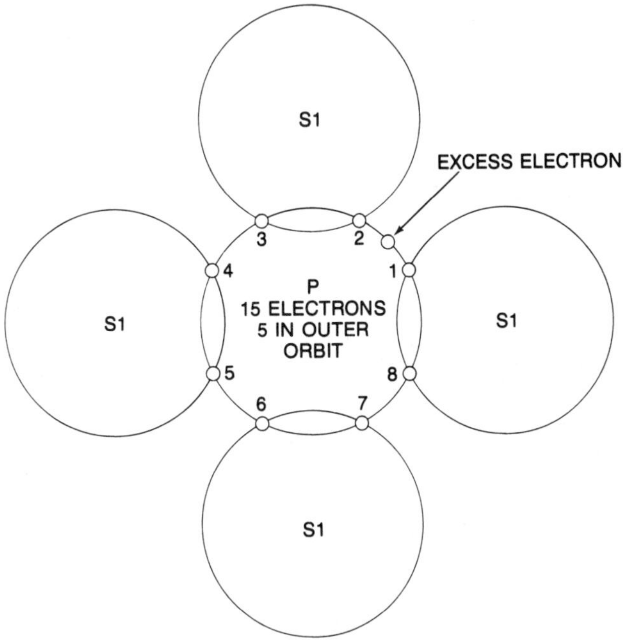


**Figure 17–2** N-type silicon.

Many crystals would form of the pure silicon. Where an aluminum atom exists, the crystal would form about the aluminum atom with only seven electrons in the outer orbit. A hole would exist in the crystalline structure. A hole is a space where an electron should exist. Because it is a lack of an electron, a hole is considered positive.

Remember that the N-type silicon and P-type silicon are electrically neutral. There is no electric charge, simply because the silicon has been doped. One important change that takes place when the pure silicon is doped is the great reduction in resistance.

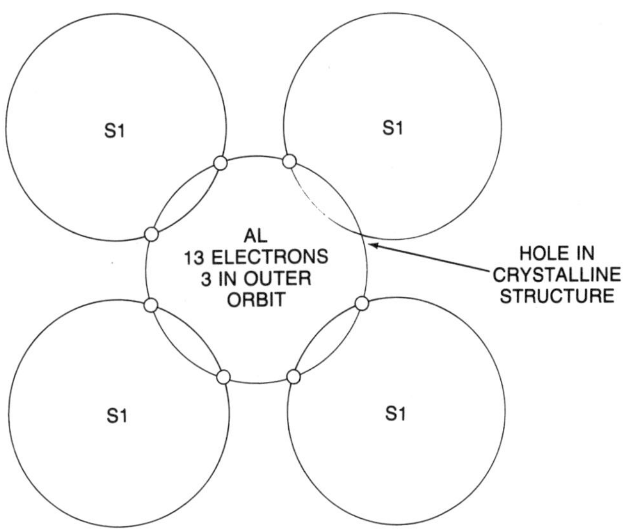


**Figure 17–3** P-type silicon.

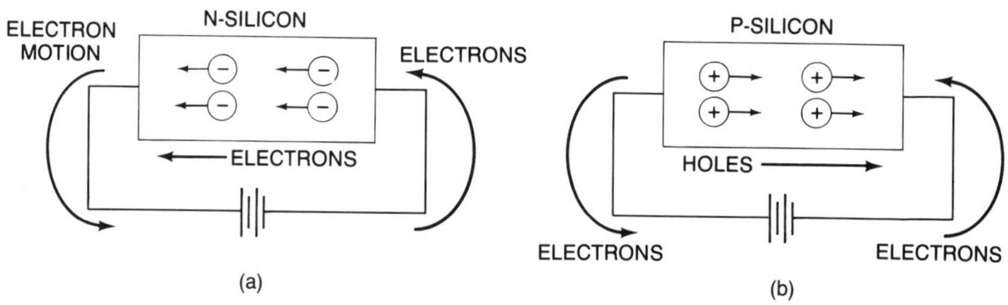


**Figure 17–4**

## *CONDUCTION IN DOPED SILICON*

The resistance of pure silicon is very high. The resistance is high because the crystals form with a bond between the electrons in the outer orbit of each atom. Each silicon atom is bound with four other silicon atoms. This provides for eight electrons in the outer orbit of each atom, a very stable condition. None of the electrons are free to move.

In pure silicon, heat causes agitation of the outer-orbit electrons, and some electrical conduction can take place. The higher the temperature, the greater the conduction, but still the conduction is very low.

In N-type silicon, there are excess electrons in the crystalline structure. These excess electrons are free to move. When a voltage is applied to N-type silicon, an electron current flows. See Figure 17–4a.

In P-type silicon, there are holes in the crystalline structure. An electron may move into a hole, leaving a hole where the electron came from. The hole is the thing that moves in P-type silicon. If a voltage is applied to P-type silicon, there will be hole movement in the silicon toward the negative terminal. There is electron motion in the external circuit. Observe Figure 17–4b. The positive potential draws an electron from the P material at the left side, creating a hole. The hole is free to move through the P-type silicon to the negative terminal where it combines with an electron provided from the battery negative terminal. Another hole is created at the left terminal as an electron is drawn off to the battery positive terminal. In the external circuit electron, current flows from negative to positive. Internally, in the P-type silicon, holes are moving from positive to negative.

## *DIODES*

A *diode* is a device that allows electrons to flow through it in only one direction. Diodes are useful in converting alternating current into direct current.

In the construction of a diode, P-type silicon and N-type silicon are joined together (Figure 17–5). Each is electrically neutral before they are joined together. The excess electron in the N-type material and the hole (the place for an electron) in the P-type material relate only to the crystalline structure. When the two materials are joined together, some of the excess electrons in the N-type migrate over and fill the holes in the P-material (Figure 17–6a).

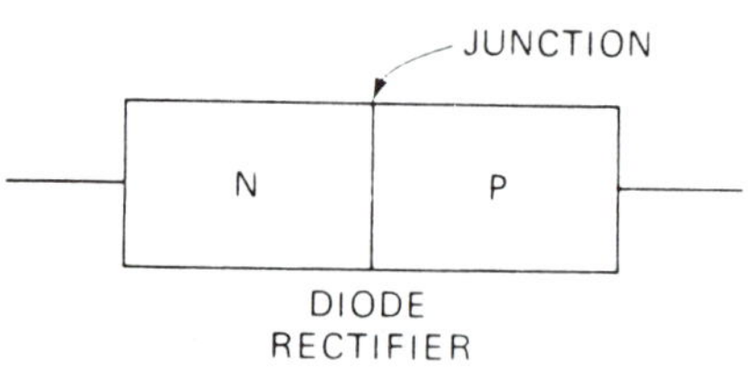


**Figure 17–5**

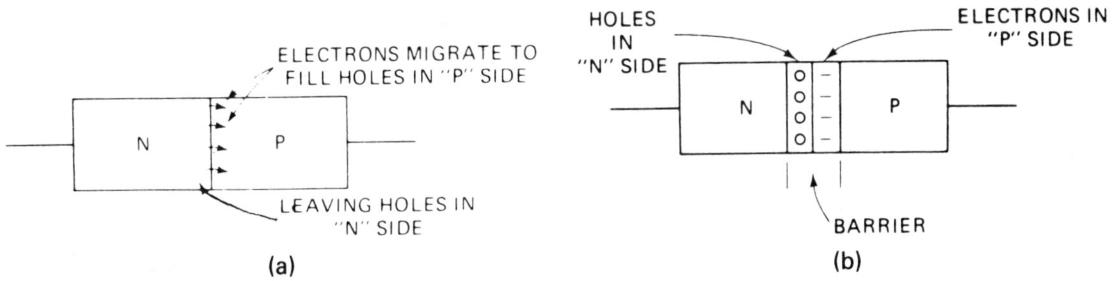


**Figure 17–6**

When the electrons from the N-side move over to the P-side, a potential is developed across the junction. The potential is called a *barrier*. The N-material loses some electrons and is positively charged; the P-material gains electrons and is negatively charged (Figure 17–6b). No further action takes place unless an outside potential is applied to the diode.

If a voltage is applied to the diode, as shown in Figure 17–7, electron current does not flow. Electrons cannot flow because the application of the external potential increases the width of the barrier. The positive terminal of the battery pulls a few more electrons from the N-side, increasing the number of holes and widening the barrier. The negative terminal of the battery adds few electrons to the P-side, widening the barrier here. The application of the battery potential to the diode increases the resistance of the diode.

If the battery is reversed, a completely different situation exists (Figure 17–8). The application of the positive potential to the P-material allows for the movement of an electron from the barrier area toward the positive potential. An electron from the N-side may move across the junction to fill the hole left by the moved electron in the P-side. An electron may travel through the battery and out the negative terminal to the N-side of the diode to replace the electron that moved across the junction.

The procedure is continuous, and electron current flows through the diode (Figure 17–9). The resistance of the diode is reduced to a low level. The symbol for a diode is shown in Figure 17–9. Electrons flow against the arrowhead.

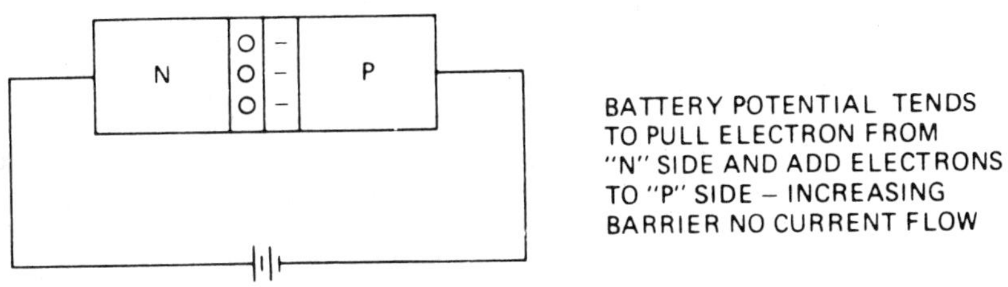


**Figure 17–7**

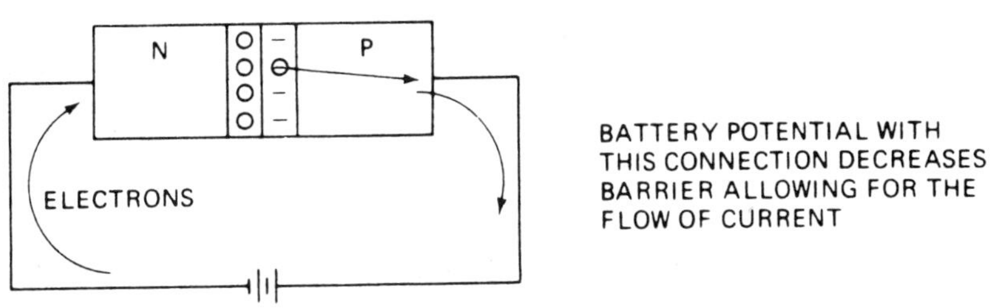


**Figure 17–8**

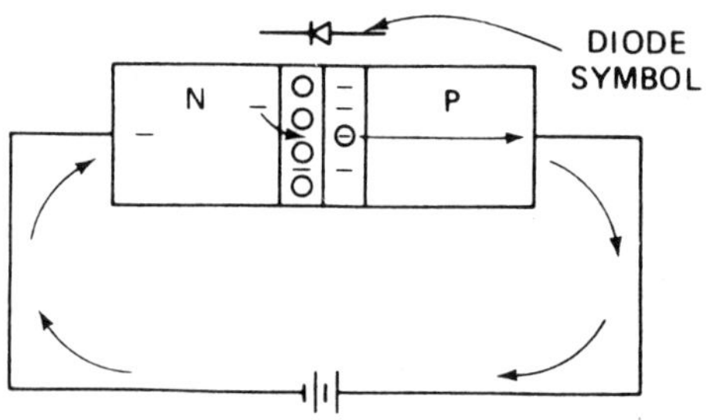


Figure 17–9

### Forward Characteristics of Silicon Diodes

As the voltage is increased across a diode in the forward direction, current will start to flow. At first, the current is very low. With silicon diodes, when the forward voltage reaches about 0.6 volt, the resistance of the diode drops off significantly. (See Figure 17–10.) The diode becomes a voltage-variable resistor with small changes in voltage providing for large changes in resistance.

Figure 17–10 shows a forward voltage of 0.7 volt providing current flow of 300 ma. At a forward voltage of 0.75 volt, the current flow is 500 ma. The diode resistor is 2.33 ohms when the voltage is 0.7 volt and 1.5 ohms when the voltage is 0.75 volt.

As the voltage across the diode increases slightly, the resistance decreases. In a circuit containing other resistive components, the voltage across the diode will be relatively constant. Diodes are usually rated in maximum-forward current and maximum-reverse voltage. They are available with voltage ratings from low values up to thousands of volts. Current ratings start at low values, and diodes are available that can carry hundreds of amperes.

### Germanium Diodes

The forward characteristics of germanium diodes are quite similar to those of silicon diodes. The knee of the curve will be between 0.2 and 0.3 volts. Germanium has a much lower forward-voltage drop. The variable resistance characteritic is quite similar to the silicon diode. Germanium diodes are used where the low forward-voltage drop is of significance. For example, in a radio, high-frequency signals are amplified before being detected (rectified) to obtain the audio frequences. If germanium diodes are used to detect, only 0.3 volt will be lost. If silicon is used, the first 0.6 volt are lost. Smaller signals may be detected and used with germanium diodes. Just about all other characteristics of germanium make it inferior to silicon for diode use. Most diodes manufactured and used today are silicon diodes.

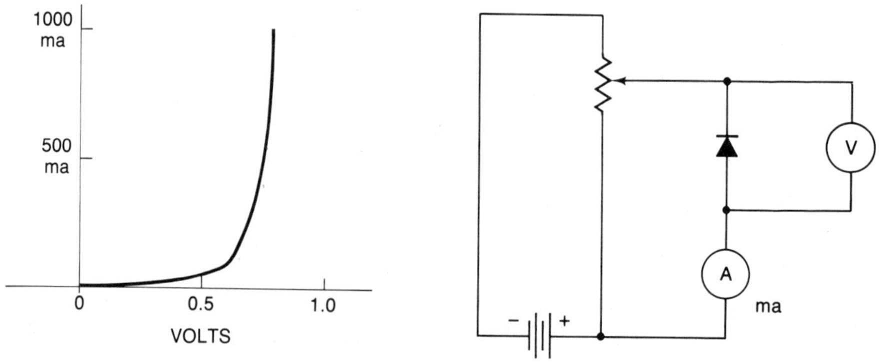


**Figure 17–10** Forward characteristics.

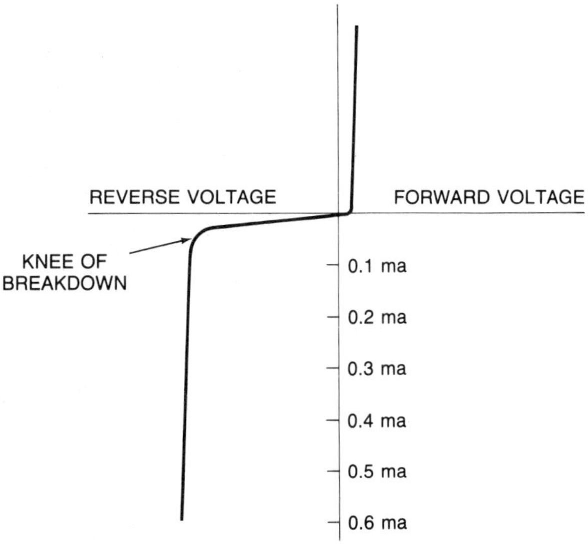


**Figure 17–11** Diode characteristics.

## Reverse Bias Characteristics

If the reverse voltage across a diode is increased, the resistance of the diode remains extremely high until the breakdown voltage is reached. At breakdown, the diode becomes a voltage-variable resistance. Small changes in voltage provide large changes in resistance. (See Figure 17–11.)

Conduction in the reverse direction before breakdown approaches 100 microamperes for germanium diodes and something in the low nanoamperes for silicon. Silicon has the better reverse-current characteristics.

## Breakdown Voltage

The *breakdown voltage* of a diode is based on the doping ratios of the P and N materials that comprise the diode. Different doping ratios provide different breakdown voltages. Diodes are presently available with breakdown levels rated from relatively low values (50 volts) up to thousands of volts.

One way to look at the breakdown level is to consider the reverse biased diode. Without bias, the diode develops its own potential hill. For silicon, this level is about 0.6 volt. The potential hill develops as electrons from the N material move into the P material, leaving holes in the N material. This is shown in Figure 17–12a. As reverse bias is applied to the diode, the potential hill widens (Figure 17–12b). If the

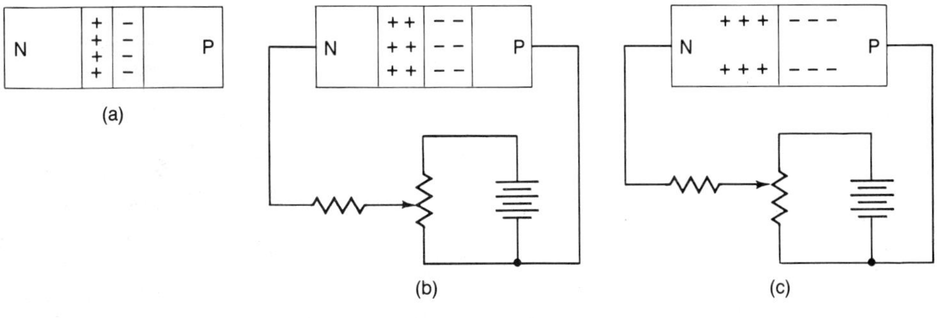


**Figure 17–12** Diode breakdown.

reverse bias is further increased, the width of the hill takes up the complete diode (Figure 17–12c) and the diode breaks down. Further attempts to increase the voltage across the diode result in sharp increases in current and little change in voltage.

Reaching the breakdown voltage does not hurt the diode as long as the current is not excessive. Actually, it is heat that will hurt the diode. The diode will be all right as long as the power developed in the diode does not reach the specified limit.

### *DIODE RECTIFIER*

If an alternating voltage is applied to the diode, the diode conducts every other half cycle. In the circuit in Figure 17–13, current flows through the diode and $R_1$ every other half cycle. The input voltage to the resistor diode combination is represented as $E_{in}$, the output voltage across $R_1$ is represented as $E_{out}$. The circuit is called a *half-wave rectifier*. A capacitor may be added to the circuit to provide filtering. In Figure 17–14, the connection of the capacitor and the effects on the output waveform are shown.

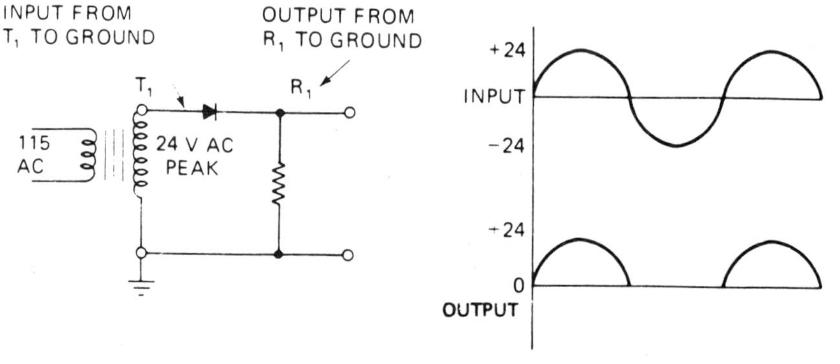


**Figure 17–13** Half-wave rectifier.

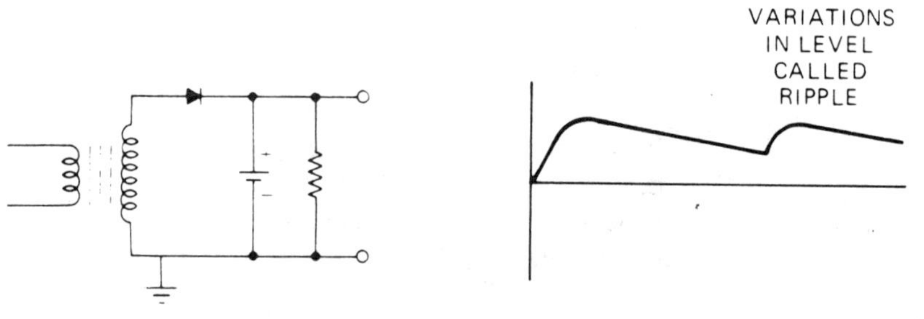


**Figure 17–14** Filter action.

### *FULL-WAVE BRIDGES*

The *full-wave bridge* circuit is used to provide an output voltage with less ripple. The circuit is shown in Figure 17–15. In Figure 17–15a, the path for electron flow during the first half cycle is shown. Figure 17–15b shows the path for electron flow during the second half cycle. Note that the ripple amplitude has decreased in the filtered output.

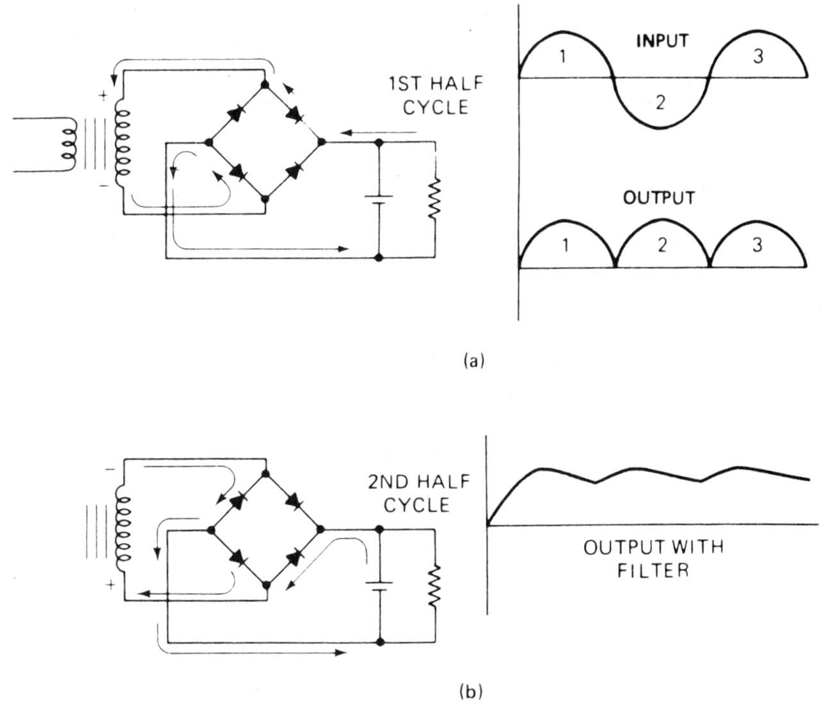


**Figure 17–15** (a) Full-wave bridge. (b) Filtering.

## ZENER DIODES

The breakdown voltage across a diode remains fairly constant and is set by the doping ratios. This characteristic provides a means of developing voltage regulators. Voltage-regulating diodes, or *zener diodes*, are diodes developed to operate in circuits where the applied voltage exceeds the diode's breakdown voltage. The diode operates in the breakdown region and has a relatively constant voltage across it. The symbol for a zener diode is shown in Figure 17–16.

Consider the circuit in Figure 17–16 where a 25-volt dc power supply is connected through $R_1$, a 100-ohm resistor, to $Z_1$, a 15-volt, 2-watt zener diode. The switches $S_1$ and $S_2$ control the connection of load resistors $R_1$, 500 ohms, and $R_2$, 500 ohms.

With switches $S_1$ and $S_2$ open, the +25 volts from the supply exceeds the breakdown voltage of the zener diode. Current will flow up through the zener and the resistor $R_1$ to the source voltage. The current will rise to 0.1 (100 ma) ampere where 10 volts will appear across the resistor $R_1$, and 15 volts will be developed across the zener diode. The zener diode has 15 volts across it and 100 ma through it. By Ohm's Law, its resistance is 150 ohms.

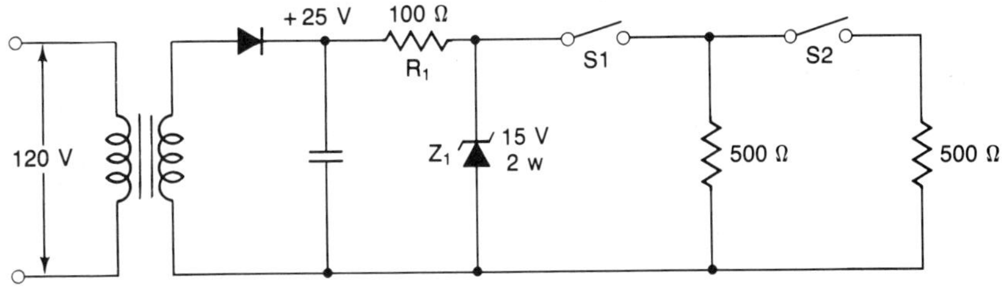


**Figure 17–16** Zener regulator.

When switch $S_1$ is closed, current will flow through resistor $R_2$. This would tend to increase the voltage across $R_1$ and thereby decrease the voltage across the zener. When the voltage across the zener decreases, the zener resistance increases. The current through the zener decreases from 100 ma to 70 ma. With 70 ma through the zener diode and 30 ma through the load resistor $R_2$, the current through $R_1$ remains at 100 ma. The 100 ma through $R_1$ continues to provide a 10-volt drop across $R_1$, and the output voltage remains 15 volts.

Closing switch $S_2$ increases the load current by another 30 ma to a total of 60 ma. The current through the zener drops to 40 ma. The current through $R_1$ remains at 100 ma, and the output voltage remains at 15 volts.

Actually, the voltage across the zener will decrease slightly as the current drawn by the load is increased. It is the slight decrease in voltage across the zener that provides for the change in zener resistance.

### Zener Power Rating

The zener diode used in Figure 17–16 was a 2-watt diode. The maximum power is developed in the zener when the load current is minimum; in this case, zero. With 100 ma of zener current at 15 volts, the power developed in the zener is 1.5 watts.

This level is approaching the 2-watt limit on the diode. If the zener were to dissipate a higher amount of power, a higher-power-rated zener would be required.

## *TEMPERATURE EFFECTS*

As it does with most materials, heat causes a change in the conduction of semiconductors. In the junction diode, the forward current will be higher at high temperature for a given forward voltage. When biased in the reverse direction, the reverse current will be higher at the higher temperature below breakdown voltage. At high temperature, breakdown occurs at a higher voltage. The curves of Figure 17–17 show the relationship. The temperature effect on reverse current flow in diodes is of major importance in transistor circuit design.

It is important to note in Figure 17–17 that the reverse current at high temperature is greater than at low temperature for low reverse voltages. The fact that reverse current increases with temperature makes the stabilization of some transistor circuits

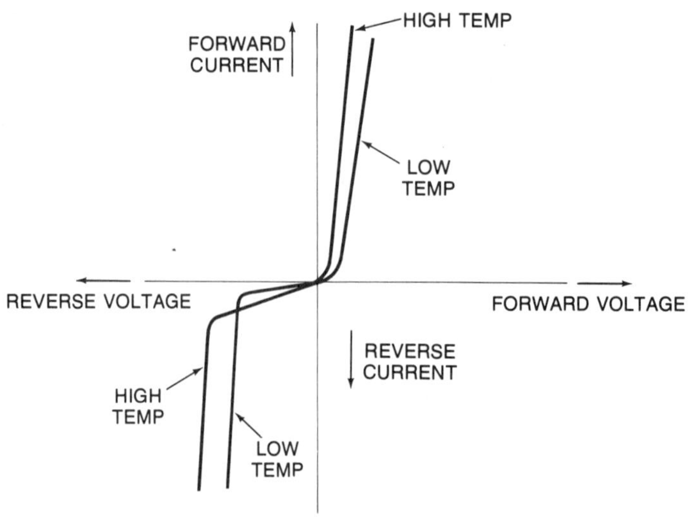


**Figure 17–17** Temperature effects.

necessary. Stabilization prevents thermal runaway. *Thermal runaway* is a situation where an increase in temperature causes an increase in current. The increase in current causes an increase in temperature, and so on.

## BIPOLAR TRANSISTORS

Bipolar transistors are three-terminal semiconductor devices that can be used as amplifiers or switches. They are called bipolar because they are made up from a PN and an NP junction. Two types of bipolar transistors are available. They are the PNP and the NPN transistor. (See Figure 17–18.)

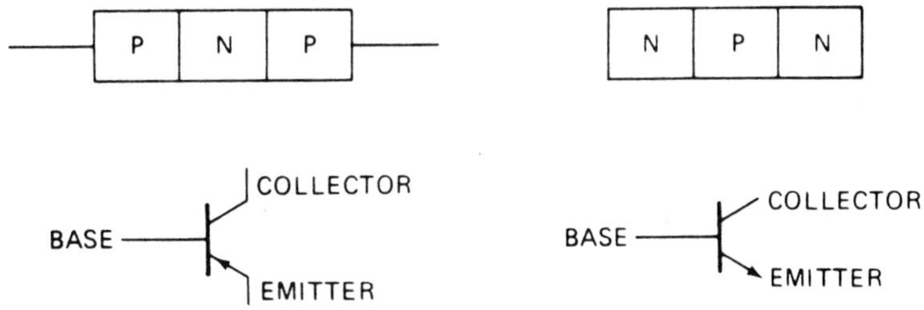


**Figure 17–18** Transistors.

The main characteristic of transistors is that a small current flow in the emitter base circuit controls a larger current flow in the emitter collector circuit. Transistors, like diodes, require the proper polarity of applied voltage for proper operation. The emitter base junction must have voltage applied in the forward direction for conduction. The applied voltage must oppose conduction in the base collector junction. An example of transistor amplifiers is shown in Figure 17–19. Note the reverse polarity of applied voltages on the NPN and PNP circuits.

**Figure 17–19** Transistor bias.

## TRANSISTOR TERMINALS

The three terminals of a transistor are called the *emitter*, *base*, and *collector*. In comparison, the emitter and collector are large, whereas the base is rather thin and is located between the two. The emitter and collector are heavily doped, while the base is lightly doped. The transistor can operate as an amplifier because a small forward-bias current flowing in the emitter-base circuit can overcome the reverse bias of the collector-base junction and allow for current passage from emitter through the base to the collector.

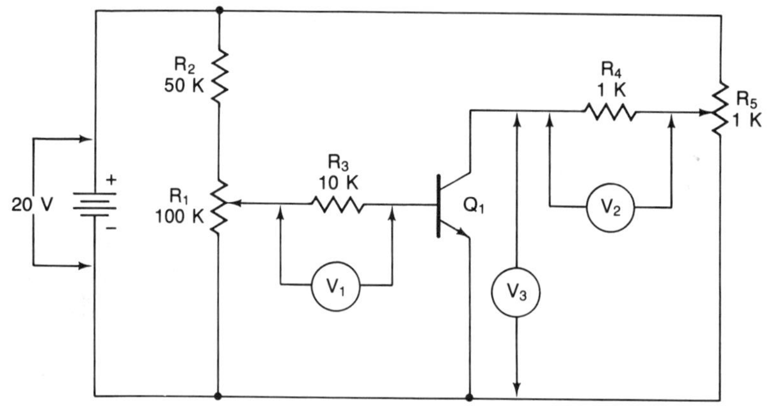


Figure 17–20

## NPN TRANSISTOR OPERATION

Because a small current in the emitter-base circuit controls a larger current in the emitter-collector circuit, the transistor is said to have *current gain*. If an experimental circuit were set up, the relation of these currents could be shown. In Figure 17–20, a small signal transistor test circuit is shown. The circuit tests the transistor in the common-emitter connection.

Potentiometer $R_1$ in Figure 17–20 is used to adjust the base current. Base current is adjusted in 25-microampere steps. The current is determined by measuring the voltage across $R_3$, 10 k. A voltage of 0.25 volt indicates 25 microamperes of base current. Voltmeter $V_2$ is used as a collector-current indicator. A currrent of 1 ma is indicated by a voltage reading of 1 volt on $V_2$. The relation is direct between voltage and current, 2 volts = 2 ma, and so on. The data taken on a transistor in a circuit such as that shown in Figure 17–20 could be used to indicate the characteristics of the transistor.

Note the polarity of the voltages applied to the transistor. To function as an amplifier, the emitter-base junction of the NPN transistor must be forward biased. The base is positive with respect to the emitter. Electron current will flow against the arrowhead indicator of the emitter. The collector-base junction must be reverse biased with the collector of the NPN transistor positive with respect to the base. Potentiometer $R_5$ provides for this capability.

### Characteristic Curves $I_c/I_b$

In order to develop the characteristic curves of the transistor, the base current is to be adjusted using $R_1$ in 10-microampere steps while monitoring the collector current through the full range of voltage, 0 to 20 volts. The curves indicate two main transistor characteristics—current gain and independence of collector current to collector voltage. (See Figure 17–21.)

### Current Gain: $h_{FE}$ (dc) or $h_{fe}$ (ac)

The current gain of the transistor may be determined using the information contained in Figure 17–21. The curves indicate that with a base current of 10 microamperes (μa), the collector current is about 1 ma. The term used for current gain is AI and the symbol used is $h_{FE}$. By formula:

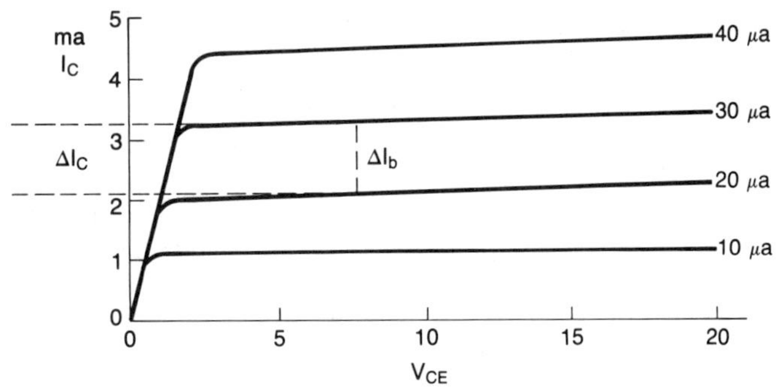


**Figure 17–21**

$$h_{FE} = I_c/I_b$$

$$h_{FE} = 1\ \text{ma}/10\ \mu\text{a}$$

$$h_{FE} = 100$$

The current gain of the transistor varies slightly with the level of operation. At a base current of 40 μa, the collector current is 4.3 ma.

$$h_{FE} = I_c/I_b$$

$$h_{FE} = 4.3\ \text{ma}/40\ \mu\text{a}$$

$$h_{FE} = 107.5$$

A small increase in current gain at higher collector currents is common for most transistors.

### AC Current Gain

Note the vertical line drawn in Figure 17–21 between the curve for 20 μa and 30 μa of base current. This line indicates a change (Δ) in base current of 10 μa. A corresponding change in collector current is shown to the left where the collector current changed from 2 ma to 3.25 ma. This provides a $\Delta I_c$ of 1.25 ma. The ac current gain of a transistor is given by the formula:

$$h_{fe} = I_c/I_b$$

$$h_{fe} = 1.25\ \text{ma}/10\ \mu\text{a}$$

$$h_{fe} = 125$$

The ac current gain is normally slightly higher than the dc current gain.

*Beta* is another term sometimes used to indicate current gain. When beta is used, the suffix dc or ac will be added to show the relation to the factor being considered.

### Collector Voltage

It is worthwhile noting the relation of collector voltage to collector current. At any base-current setting, the collector current rises rapidly as collector voltage is increased above the base voltage. The collector current then remains almost constant as the voltage is increased to the maximum of 20 volts available in this circuit. The change in collector voltage has little or no effect on collector current.

## *MAXIMUM RATINGS AND ELECTRICAL CHARACTERISTICS*

The manufacturer's data on transistors provide important information on transistor operation. A partial data sheet is shown below. It provides the type of information that is considered when a transistor is to be used in a circuit. Make a special note of the range of current gain on the single transistor type. It is common for transistors of the same type to have a range of current gain.

**TYPE 2N3904 SILICON**

Maximum Ratings

| *Rating* | *Symbol* | *Value* | *Unit* |
|---|---|---|---|
| Collector-emitter voltage | $V_{ceo}$ | 40 | Vdc |
| Collector-base voltage | $V_{cb}$ | 60 | Vdc |
| Emitter-base voltage | $V_{eb}$ | 6.0 | Vdc |
| Collector current | $I_c$ | 200 | ma |

Electrical Characteristics on Condition

| | *Symbol* | *Min* | *Max* |
|---|---|---|---|
| Current gain dc | $h_{FE}$ | 100 | 300 |
| $I_c$ = 10 ma $V_{ce}$ = 1.0 Vdc | | | |
| Current gain ac | $h_{fe}$ | 100 | 400 |

## REVIEW QUESTIONS

T ______ F ______ 1. Each atom of silicon has 24 electrons in orbit about its nucleus.

T ______ F ______ 2. The resistance of pure silicon is a little over 1.34 times that of copper.

T ______ F ______ 3. Silicon crystals form by sharing outer-ring electrons with four other silicon atoms.

T ______ F ______ 4. Doping of pure silicon with a material having three electrons in its outer orbit provides P-type silicon.

T ______ F ______ 5. Doping of pure silicon with a material having five electrons in its outer orbit provides N-type silicon.

T ______ F ______ 6. The two types of material used in semiconductors are cobalt and silicon.

T ______ F ______ 7. Diodes are used to change dc to ac.

T ______ F ______ 8. Diodes normally conduct in both directions.

T ______ F ______ 9. The output of a full-wave rectifier is easier to filter than the output of a half-wave rectifier.

T ______ F ______ 10. Transistors are two-terminal semiconductors.

T ______ F ______ 11. In a transistor, a small signal current in the collector controls a large signal current in the base.

T ______ F ______ 12. Transistors may be used to make amplifiers.

T ______ F ______ 13. Complete amplifiers may be formed as miniature integrated circuits.

T ______ F ______ 14. Silicon control rectifiers allow for current flow in both directions.

T ______ F ______ 15. Triacs require a negative gate for turn on.

T ______ F ______ 16. The breakdown voltage of a diode is dependent on the size of the diode.

T ______ F ______ 17. One of the breakdown voltages across a diode is exceeded. The diode is destroyed.

T ______ F ______ 18. One of the main uses of a zener diode is as a voltage regulator.

T ______ F ______ 19. Making the base of an NPN transistor more positive will increase the collector current.

T ______ F ______ 20. When the base current of a transistor changes from 12 μa to 24 μa, and the collector current changes from 1.4 ma to 2.9 ma, the transistor has an $h_{fe}$ of 125.

# UNIT EIGHTEEN

# *SEMICONDUCTOR CIRCUITS*

The purpose of this unit is to provide basic information on different semiconductor circuits. Information will be presented on bipolar transistor circuits, field-effect transistors, MOSFETs, and thyristor circuits.

The information presented is intended to provide the reader with fundamental concepts of how semiconductor circuits work. This is general information, not specific design material.

Present-day technicians in the electrical field must have a basic grasp of electronic circuits if they are to succeed in the industry.

## *TRANSISTOR AMPLIFIER CIRCUITS*

There are three circuits used for transistor amplifiers. They are

1. Common-emitter circuits.
2. Common-collector circuits.
3. Common-base circuits.

The term *common* refers to the transistor terminal that is related to both the input and output circuits. The material covered in the discussion of voltage gain and reflected impedance of amplifiers is only to be considered as an approximation. The actual formulas for voltage gain and reflected impedance contain many terms, and they are beyond the scope of this book.

The circuits of the common-emitter, common-collector (emitter follower), and common-base configuration are shown in Figure 18–1. Note the location of the input and output ports. The characteristics of the circuits are as follows:

Common-Emitter Circuit

1. current gain—$h_{fe}$
2. voltage gain—high
3. power gain—high
4. input impedance—low
5. output impedance—medium

Common-Collector (emitter follower) Circuit

1. current gain—$h_{fe}$
2. voltage gain—less than one
3. power gain—medium
4. input impedance—high
5. output impedance—low

Common-Base Circuit

1. current gain—less than one
2. voltage gain—high
3. power gain—medium
4. input impedance—very low
5. output impedance—high

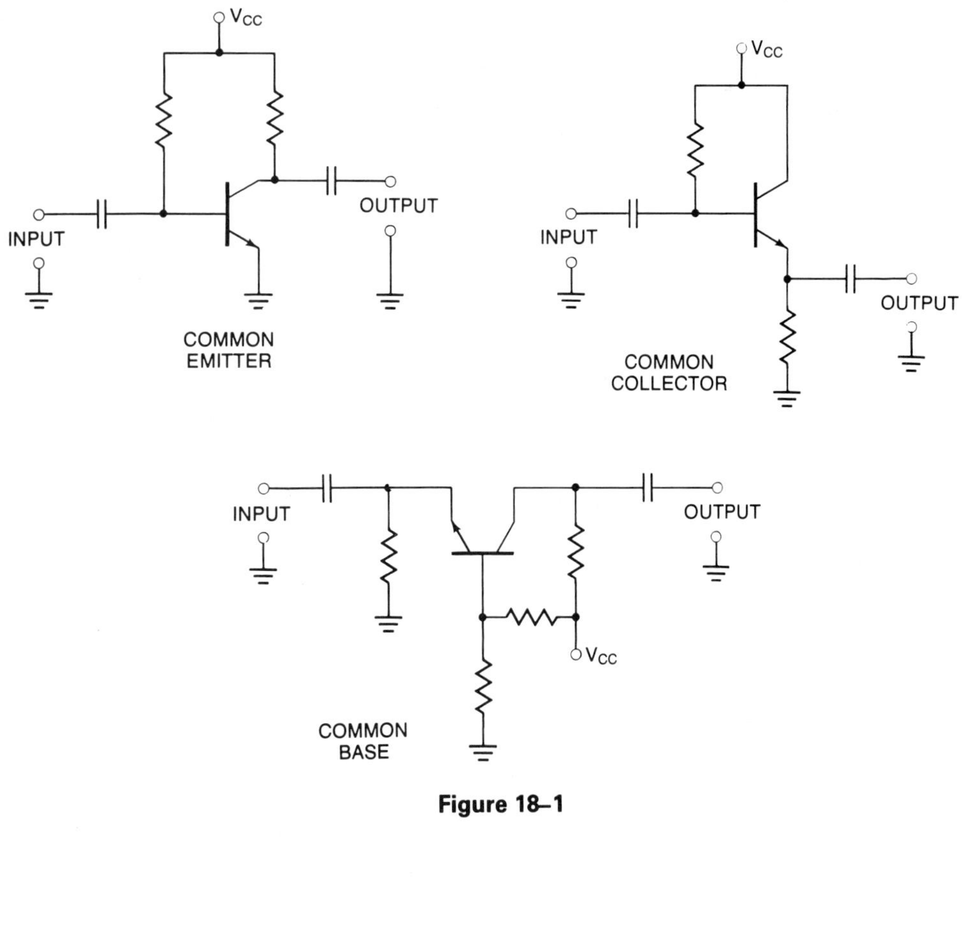


**Figure 18–1**

## *COMMON-EMITTER CIRCUIT*

The *common-emitter circuit* is the circuit most often used in transistor amplifiers. This circuit provides both current gain and voltage gain. Accordingly, the circuit provides high-power gain. Figure 18–2a shows the circuit of a simple common-emitter amplifier.

The voltage gain of this amplifier is dependent on the current gain of the transistor. The circuit is $h_{FE}$, or beta dependent.

In order to simplify circuit consideration, the $h_{FE}$ or beta of all transistors considered from here on, unless otherwise noted, is given as 100.

The resistor $R_1$ in Figure 18–2a is connected to provide forward bias to the emitter-base junction. Current flow is from ground through the emitter to the base, through $R_1$, 1.94 megohms, to +20 volts collector supply called ($V_{cc}$.) Because the emitter-base junction is a diode and is forward biased, the diode will conduct.

The voltage across a conducting silicon diode is about 0.6 volt. The voltage across the emitter-base junction is about 0.6 volt. The emitter-base junction and $R_1$, 1.94 megohms, form a series circuit to $V_{cc}$ of 20 volts. With 0.6 volt across the emitter base, there must be 19.4 volts across the resistor $R_1$.

The 19.4 volts across $R_1$—a 1.94-megohm resistor—provide a current of 10

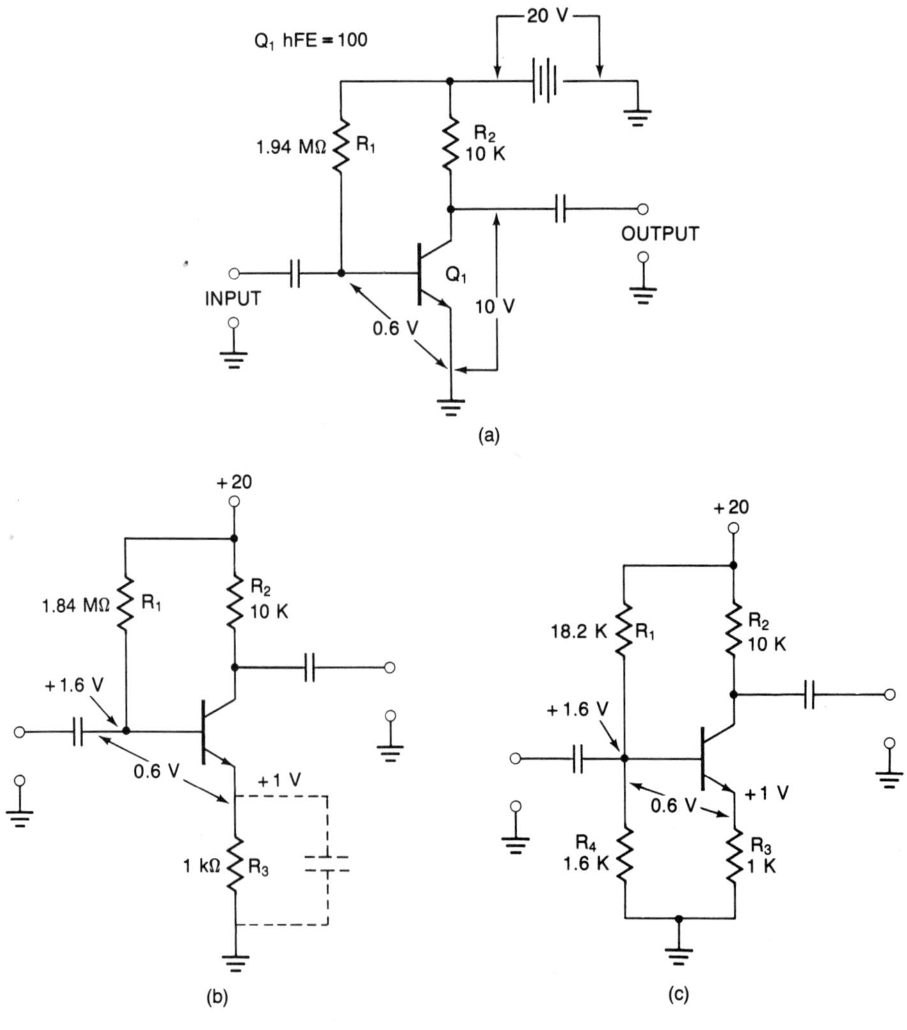


**Figure 18–2**

microamperes. This is the base current of the transistor. Because the $h_{FE}$ or current gain of the transistor is given at 100, the collector current can be calculated.

$$I_c = I_b \times h_{FE}$$

$$I_c = 10\ \mu a \times 100$$

$$I_c = 1\ ma$$

The collector current (1 ma) flowing through $R_2$ (10 kilohms) will provide for 10 volts across the resistor.

$$V = I \times R$$

$$V = 1\ ma \times 10\ k$$

$$V = 10\ volts$$

Subtracting $V_{R2}$ (10 V dc) from $V_{cc}$ (20 V ac) leaves a collector to emitter voltage, $V_{CE}$, of (10 V dc). This completes the calculation for the dc operating conditions of the transistor.

## *THERMAL EFFECTS*

The circuit shown in Figure 18–2a will work reasonably well at lower temperatures. If the transistor should become warm, the effects on the reverse-biased collector-base current become important. As shown in Unit 17, in the section titled "Temperature Effects," and in Figure 17–17, as the temperature of a diode increases, the reverse-bias current increases. The collector-base junction of a transistor is a reverse-biased diode. As the temperature increases, the leakage current from base to collector increases. Follow the effects in steps:

1. The temperature of the transistor increases.
2. The reverse-biased collector base current increases.
3. The extra current flows through emitter base.
4. The increased $I_b$ is amplified by $h_{FE}$, causing increased $I_c$.
5. The increased $I_c$ causes transistor to heat further.
6. Return to Step 1 and repeat.

It is obvious that the circuit in Figure 18–2a is unstable.

## *STABILITY*

One of the most often-used methods of providing stability to transistor amplifiers is to add a resistor in the emitter circuit. This circuit is shown as Figure 18–2b. If temperature rise should tend to cause an increase in base-collector current, the collector current will start to increase as in the previous case. The increase cannot be as great though because the increase in collector current flows through the emitter resistor. The voltage developed across $R_3$ makes the emitter more positive. Making the emitter more positive decreases the forward emitter base bias, which tells the collector current to decrease. The collector current will therefore have a tendency to remain constant with temperature increase. A capacitor is commonly connected across the emitter resistor to provide low impedance to ac signals. This eliminates any negative feedback for ac signals.

A second method of increasing the dc stability of the amplifier is to provide voltage-divider base bias as shown in Figure 18–2c. In this circuit, the bias on the transistor is controlled by the ratio of $R_1$ to $R_4$ by the formula

$$V_b = (R_4/R_1 + R_4) \times V_{cc}$$

The current through the emitter base can be ignored for the most part as far as determining bias as the divider current is normally much higher than the base current.

The base-voltage divider network adds stability since a second path is offered to base-collector reverse-bias current. The current can flow through $R_4$ rather than the emitter base circuit. As long as the increase in collector-base current does not flow through the emitter base junction, it cannot cause an increase in collector current.

## AC POWER AMPLIFIERS

The purpose of most amplifier circuits is to take in a small signal power and amplify it to a large signal power to drive some power-consuming device such as a loudspeaker.

### Emitter Followers

As covered in battery circuits, maximum power is transferred from source to load when source impedance matches load impedance. A common circuit used to match impedance with transistors is the *emitter follower* or *common-collector* amplifier. The circuit of an emitter follower ac amplifier is shown as Figure 18–3.

The circuit is called an emitter follower because the output is taken from the emitter, and the output is of the same phase as the input. In most cases, the output is about the same amplitude as the input. The dc characteristics of the circuit in Figure 18–3 are based on a selection of 2 ma for emitter current. The 100-ohm emitter resistor would have +0.2 volt across it. With the standard of 0.6 volt between emitter and base, the base-to-ground voltage (0.2 + 0.6) is 0.8 volt. There is therefore 19.2 volts left across $R_1$. The base current is 20 μa because $I_c = 2$ ma and $I_b = I_c/h_{FE}$. In order to drop 19.2 volts from $V_{cc}$ of 20 volts, the base resistor would have to be 960 kilohms.

When an ac signal of 100 millivolts peak-to-peak is fed to the base, it will cause an increase in emitter current when the signal goes positive and a decrease in emitter current when the signal goes negative.

Remember, it takes but a small change in voltage across a forward-biased diode to provide a large change in current. The emitter-base junction of a transistor is a

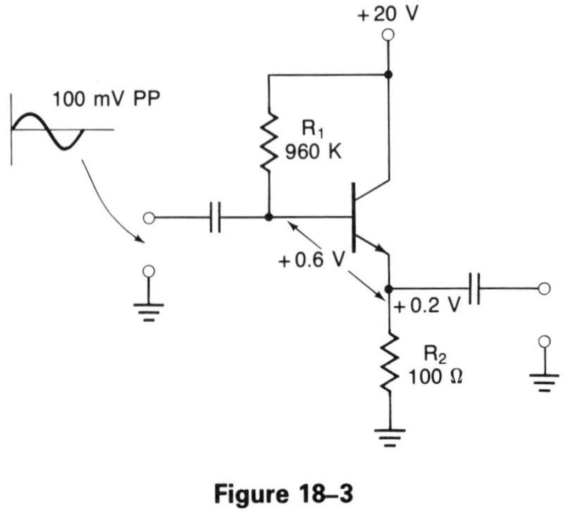


Figure 18–3

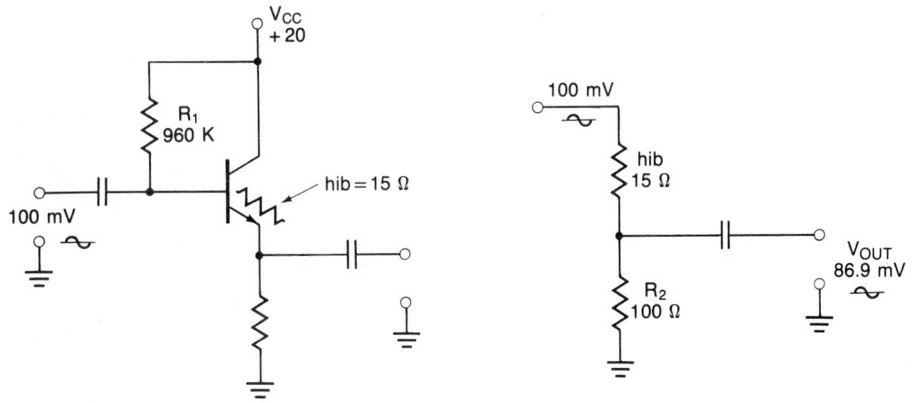


**Figure 18–4**

diode. Also, keep in mind that the emitter current, which is slightly larger than the collector current, is $h_{FE}$ times the base current.

The amplitude of the output signal at the emitter is controlled by the relation between an internal resistance of the transistor and the emitter resistance. The internal resistance is called hib. The resistance value of hib changes with the value of the emitter current. The value of hib is equal to 30 divided by the magnitude of the emitter current in milliamps (ma). By formula, hib = $30/(I_E$ in ma).

In the circuit of Figure 18–3, the emitter current is 2 ma. The value of hib is therefore 30/2, or 15. The input voltage is impressed across hib in series with the emitter resistor as shown in Figure 18–4.

The output voltage may be found using the voltage-divider formula:

$$V_{out} = V_{in} \times \frac{R_{out}}{hib + R_{out}}$$

$$V_{out} = 100 \text{ millivolts} \times \frac{100}{100 + 15}$$

$$V_{out} = 86.9 \text{ millivolts (mV)}$$

Because gain is equal to output divided by input,

$$\text{Gain} = \frac{V_{out}}{V_{in}}$$

$$\text{Gain} = \frac{86.9 \text{ mV}}{100 \text{ mV}}$$

$$\text{Gain} = 0.869$$

As with all emitter followers, the voltage gain is less than one. There is power gain as there is a current gain of 100.

The voltage gain of this amplifier may be improved by increasing the emitter current. The emitter current may be increased by increasing the base current. For example, if $R_1$, the base resistor, were changed from 960 k to 184 k, the base current would increase to 0.1 ma and the emitter current would increase to 10 ma. The value of hib would decrease from 10 ohms to 3 ohms.

$$hib = \frac{30}{10} = 3$$

The output voltage becomes:

$$V_{out} = V_{in} \times \frac{R_{out}}{hib + R_{out}}$$

$$V_{out} = 100\text{ mV} \times \frac{100}{3 + 100}$$

$$V_{out} = 97.1\text{ mV}$$

$$\text{Gain} = \frac{V_{out}}{V_{in}}$$

$$\text{Gain} = \frac{97.1}{100}$$

$$\text{Gain} = 0.971$$

The gain of the emitter follower has been increased from 0.869 to 0.971.

## *COMMON-EMITTER VOLTAGE AMPLIFIER*

The standard transistor voltage amplifier is the common-emitter circuit shown in Figure 18–5a. As shown earlier, the circuit will have a collector current of 1 ma. Solve the formula for hib:

$$hib = \frac{30}{(I_c\text{ ma})}$$

$$hib = \frac{30}{1}$$

$$hib = 30$$

An equivalent-resistance circuit is shown in Figure 18–5b. The three resistors make up the transistor equivalent.

When an ac signal is impressed on the base of the transistor, it will appear across the internal emitter base resistor hib. This is emitter-follower action. Because there is no emitter resistor, all of the signal must appear across hib.

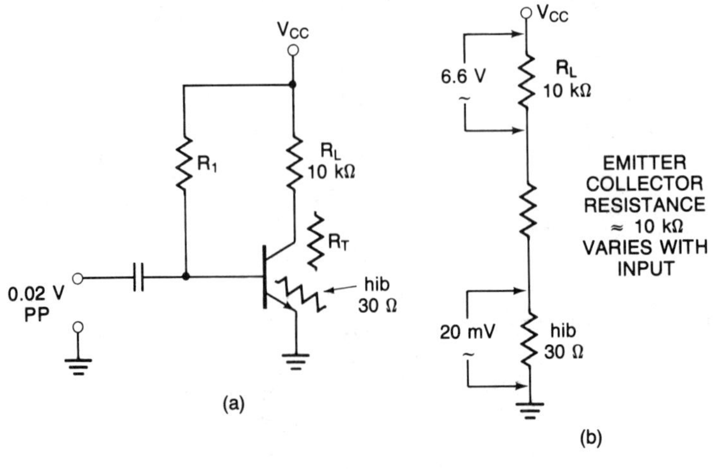


**Figure 18–5**

In the emitter-collector circuit, the internal resistance hib is in series with the emitter-collector resistance of the transistor and the 10-kΩ load resistor as shown in Figure 18–5b. Any series resistive network may be considered as a voltage-divider network. In this case, the 20 mV input signal appears across hib of 30 ohms. The ac emitter and collector currents are about the same value, $I_c = I_e - I_b$. Remember, $I_b$ is $1/100 \times I_e$ where $h_{fe} = 100$. The load resistor (10 kilohms) is 333 times larger than hib; therefore, the ac voltage at the load resistor is almost 333 times the voltage across hib; in this circuit, 6.6 volts. The voltage gain of the amplifier is about 333. A simple formula used for common-emitter amplifiers is:

$$Av = \frac{0.9\,R_L}{hib + Re}$$

The factor $0.9\,R_L$ is used to compensate for terms left out of this formula, and Re is used to take care of situations where an unbypassed emitter resistor might be used. For example: What is the approximate output voltage from the amplifier of Figure 18–6?

In the circuit, a collector current of 1 ma is indicated by the 0.12-volt emitter voltage. The 0.12 volt across 120 ohms of emitter resistance indicates an emitter current of 1 ma. (The 1 ma emitter current is a practical standard for small-signal amplifier design.) The 90-ohm resistor is bypassed by a capacitor that eliminates its effect as far as ac is concerned. Its purpose is to provide stability. The 0.12 volt at the emitter and 0.6 volt between emitter base provide for 0.72 volt base to ground. This means 19.28 volts across $R_1$ and a base current of 10 μa through the 1.98-megohm resistor.

The collector current of 1 ma will develop 8 volts across the 8-kilohm load resistor leaving a collector voltage of +12 volts.

The gain of the unloaded stage is:

$$Av = \frac{0.9\,R}{hib + Re}$$

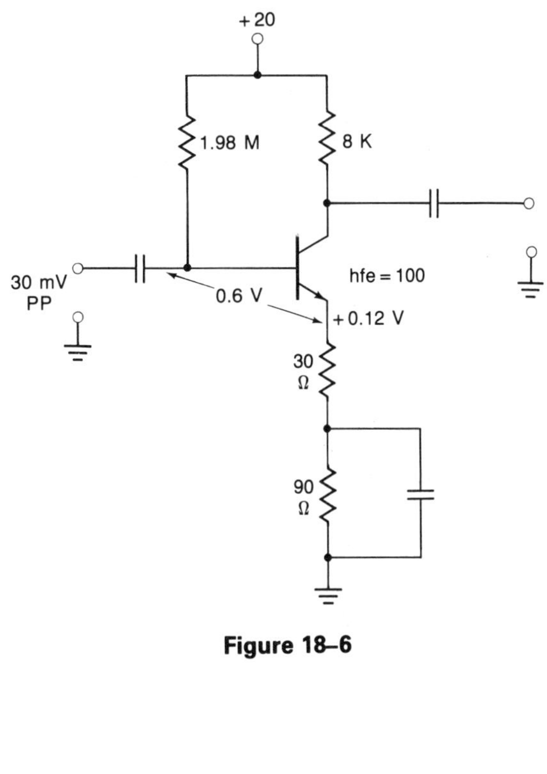


**Figure 18–6**

Observe the small e in the term Re. It stands for the ac emitter resistance. Because the 90-ohm resistor is bypassed, it does not count in the ac circuit. Therefore:

$$Av = \frac{0.9 \times 8{,}000}{30 + 30}$$

$$Av = \frac{7{,}200}{60}$$

$$Av = 120$$

With 30 mV peak-to-peak input, the output will be 3.6 volts peak to peak.

$$V_{out} = V_{in} \times Av$$

$$V_{out} = 30 \text{ mV} \times 120$$

$$V_{out} = 3.6 \text{ volts pp}$$

The voltage gain of a common-base circuit is calculated in a manner similar to that of a common-emitter amplifier. It is the ratio of the emitter impedance to the collector impedance that provides the main factors in computing voltage gain. Again, the common-base circuit is not used very often.

## *AMPLIFIER CIRCUITS*

The formulas developed for ac voltage gain of a transistor provide a means for solving for the voltage gain of a series of voltage amplifiers. Besides the voltage-gain formula, a means must be developed for finding the input and output impedance of individual amplifier stages. Consider the circuit of Figure 18–7, a two-stage amplifier.

To determine the overall gain of the system, the gain of the individual stages must be determined. In order to determine the gain of each stage, the load resistance ($R_L$) must be determined:

$$Av = \frac{0.9 \, R_L}{hib + Re}$$

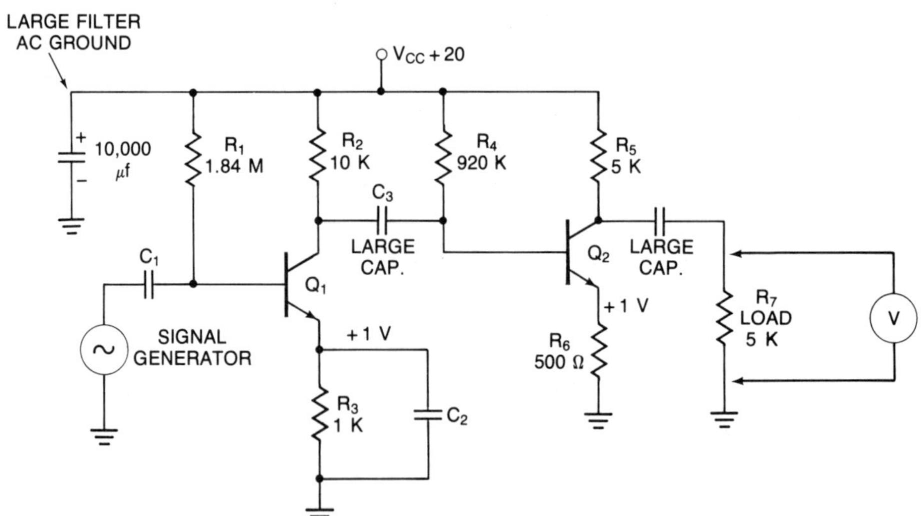


**Figure 18–7**

For transistor $Q_2$, the collector load resistance is 5,000 ohms, but the load resistor of 5,000 ohms is connected in parallel with it through the large coupling capacitor. Remember, $V_{cc}$ is at ac ground potential because there is a large filter capacitor in the power supply.

The ac load resistor for transistor $Q_2$ is therefore $R_5$ and $R_7$ in parallel. That is, 5,000 ohms and 5,000 ohms parallel for an equivalent resistance of 2,500 ohms. The emitter voltage of +1 volts across the 500-ohm emitter resistor indicates a current of 2 ma.

Two ma of emitter current determines the value of hib at 15. The ac voltage gain of transistor $Q_2$ is:

$$Av = \frac{0.9\ R_L}{hib + Re}$$

$$Av = \frac{0.9 \times 2{,}500}{15 + 500}$$

$$Av = \frac{2{,}250}{515}$$

$$Av = 4.4$$

## TRANSISTOR INPUT IMPEDANCE

In normal operation, the emitter-base junction of a transistor is forward biased. This means the emitter circuit is connected through an "ON" diode to the base. Whatever is in the emitter will affect the input impedance at the base. The input impedance of a transistor such as $Q_2$ of Figure 18–7 affects the ac load resistance of transistor $Q_1$. Effectively, the value of $R_L$ for transistor $Q_1$ is $R_2$ in parallel with $R_4$ in parallel with the input impedance of $Q_2$.

A simplified formula for the input impedance of a transistor is

$$Z_{in} = 0.8\ h_{fe} \times (hib + Re)$$

For transistor $Q_2$:

$$Z_{in} = 0.8 \times 100 \times (15 + 100)$$

$$Z_{in} = 80 \times 515$$

$$Z_{in} = 4{,}220$$

### Voltage Gain $Q_2$

The ac load resistance for $Q_1$ consists of $R_2$, $R_4$, and $Z_{in}$ ($Q_2$) in parallel.

$$R_L = 10\text{ k} \parallel 920\text{ k} \parallel 4.22\text{ k}$$

$$R_L = 2.96\text{ k}$$

Transistor $Q_1$ is drawing 1 ma as indicated by +1 volt at the emitter across the 1-kilohm resistor. The value of hib is 30. The voltage gain (Av) of the amplifier is:

$$Av = \frac{0.9R_L}{hib + Re}$$

$$Av = \frac{0.9 \times 2.96\text{ k}}{30} = \frac{2{,}664}{30}$$

$$Av = 88.8$$

Because the emitter resistor is bypassed by capacitor $C_4$, the 1,000-ohm resistor does not appear in the ac circuit.

The overall gain of the two stages is approximately 88.8 × 4.4 = 391. A signal of 10 mV at the input capacitor $C_1$ will produce 3.91 volts across the load resistor $R_7$.

## MULTIVIBRATOR

A *multivibrator* is a free-running system providing square waves at its output. The circuit consists of two transistors. In Figure 18–8, the base of $Q_1$ is controlled by the voltage at the collector of $Q_2$, and $Q_2$ base is controlled by the voltage at the collector of $Q_1$.

If $Q_1$ is conducting, $Q_2$ will be cut off. Capacitor $C_1$ will be charging to $V_{cc}$ through $R_1$. When the voltage at the base of $Q_2$ reaches 0.6 volts, $Q_2$ will start to conduct. The $Q_2$ collector goes to close to zero. This negative-going front is coupled to the base of $Q_1$ cutting it off. Capacitor $C_2$ starts to charge to $V_{cc}$ through $R_2$. When the base of $Q_1$ reaches 0.6 volt, $Q_1$ turns on, and the cycle repeats itself. The wave shapes that appear at the base and collector of $Q_1$ and $Q_2$ are shown in Figure 18–9.

The true duration of the waveform is controlled by the length of time each transistor is held in cutoff. The waveform will be symmetrical if $R_1 C_1$ equals $R_2C_2$. The frequency of the output is given by $f = \frac{1}{1.4RC}$. This formula holds true as long as the negative-going signal at the base does not exceed the zener breakdown of the emitter-base diode. If this occurs, the frequency will be higher than given by the formula.

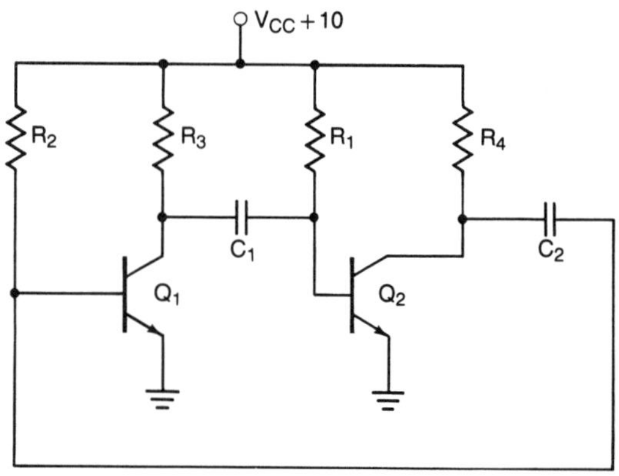


**Figure 18–8**

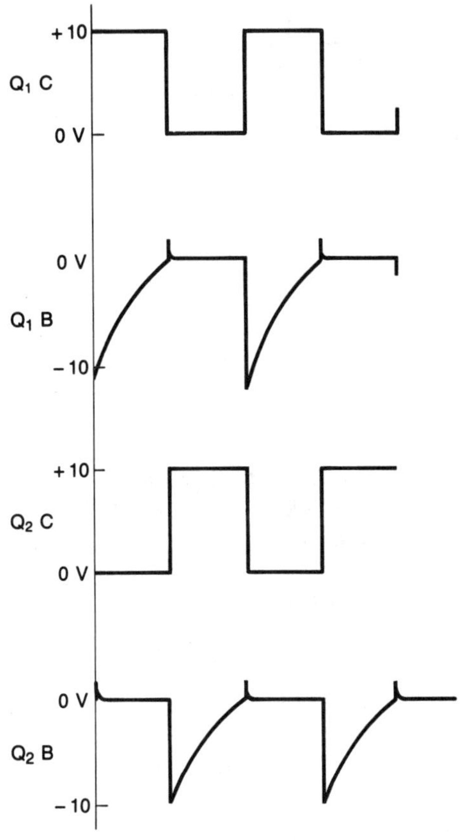


Figure 18–9

## ONE-SHOT MULTIVIBRATORS

*One-shot multivibrators*, often simply called a *one-shot*, are circuits that produce a square output pulse of fixed time duration for each trigger input. A sample circuit is shown in Figure 18–10.

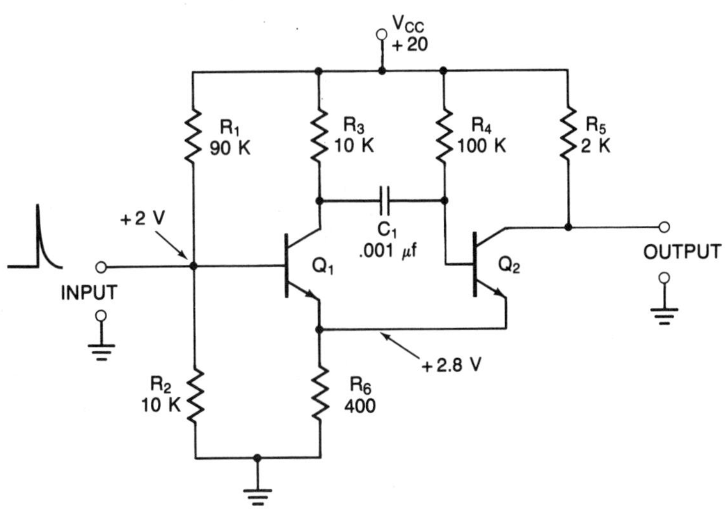


Figure 18–10

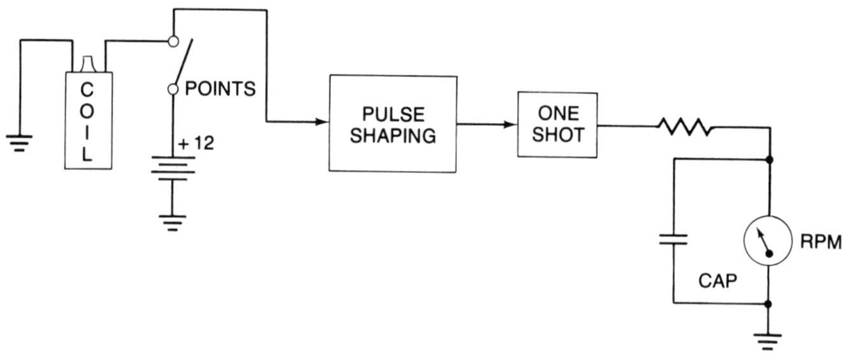


**Figure 18–11**

Transistor $Q_2$, with its base connected to $V_{cc}$ through $R_1$, is drawing approximately 7 ma of collector current. This current coming through the common-emitter resistor $R_6$ produces +2.8 volts at the emitter. The voltage divider $R_1$, $R_2$ produces +2 volts at the base of $Q_1$ keeping $Q_1$ cut off.

When a positive pulse is fed to the base of $Q_1$, it starts to conduct. Capacitor $C_1$ feeds the negative-going $Q_1$ collector voltage to the base of $Q_2$ cutting off at approximately $0.7\,R_4C_1$ or 140 microseconds. As with the multivibrator, if the emitter-base breakdown of $Q_2$ is exceeded, the time duration of the positive output pulse will be shortened.

A practical application of a one-shot is in an automobile tachometer. See Figure 18–11. Every time the points close, a positive voltage is fed into the pulse shaper to produce a sharp trigger pulse at the output. The trigger pulse activates the one-shot, which produces a square positive output pulse. The RPM indicator reacts to the average voltage from the one-shot. The higher the engine RPM, the greater the number of times the points must close and the greater the number of output pulses from the one-shot.

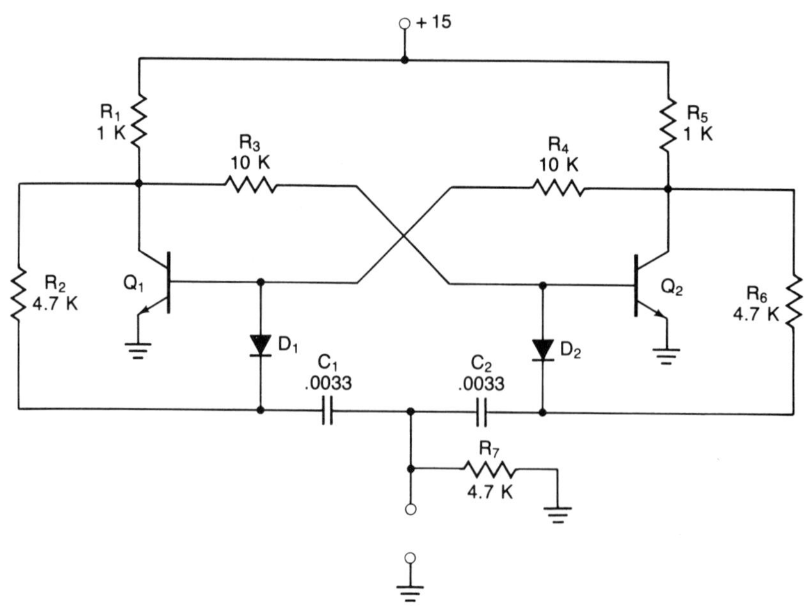


**Figure 18–12**

## FLIP-FLOP CIRCUIT

The *flip-flop circuit* is a common device used extensively in digital computers. It is the circuit that is used to count. A simple flip-flop circuit is shown in Figure 18–12.

This circuit is bi-stable, that is, whichever transistor is conducting tends to remain conducting. The other transistor conducts slightly.

When a positive trigger is applied to the trigger input, the "off" transistor is triggered fully on and the "on" transistor goes to "off."

Assume $Q_1$ in Figure 18–12 is conducting heavily; $Q_2$ is conducting lightly. A positive trigger is fed to both bases, but only the transistor that is "off" is affected by the positive pulse. Thus, $Q_2$ turns on fully, and its collector moves in a negative direction. The negative wave front is connected to the base of $Q_1$ through $R_4$. The negative signal turns the conduction of $Q_1$ down. The circuit remains stable, $Q_2$ conducting, $Q_1$ "off" until the next positive pulse arrives at the input.

## FIELD-EFFECT TRANSISTORS

In the 1950s, the effect of an electric field on current flow in a semiconductor was discovered. Field-effect transistors (FET) thus came into being. The transistor works because an electric field can cause the cross-sectional area of a conduction path to change. A change in cross-section area of conduction in a transistor changes the resistance of the device. A field-effect transistor is a resistive device where the resistance changes according to the application of voltage to the terminal, called a *gate*. It is, in other words, a voltage-variable resistor. In application with supply voltages and other resistors, the FET may be used as an amplifier or a switch.

## MOSFETs

Metal-oxide semiconductor field-effect transistors, or MOSFETs, are the present leaders of FETs in use. They have great application in digital computers. Their use is also increasing as power amplifiers and as switches. Recent developments have provided MOSFETs with an on-resistance of less than 1 ohm. It is obvious that such a device would be useful as a switch. Voltage levels of these devices range from low-voltage devices of 50 volts, up to values above 1,000 volts.

### MOSFET Construction

The most popular MOSFET transistors manufactured today are the *enhancement type*. This simply means a transistor that is normally off. It will turn on with the application of proper voltage. *Depletion-mode* transistors are available but are not as popular. The two enhancement-mode transistors are the N channel and P channel. The letter indicates the conduction path main material, either P- or N-type silicon. In general, the construction is as shown in Figure 18–13.

*Enhancement-mode* means the transistor must be turned on. The construction shows open areas between the source N material and the drain N material. To turn the device on, the gate must be made positive with respect to the source. A positive gate attracts electrons into the area between the N blocks, effectively turning the whole area into N material. This completes the path from source to drain through N material. The magnitude of the positive voltage on the gate determines the number of electrons attracted and therefore the amount of change in resistance between source and drain. In power FETs, this resistance can approach zero ohms.

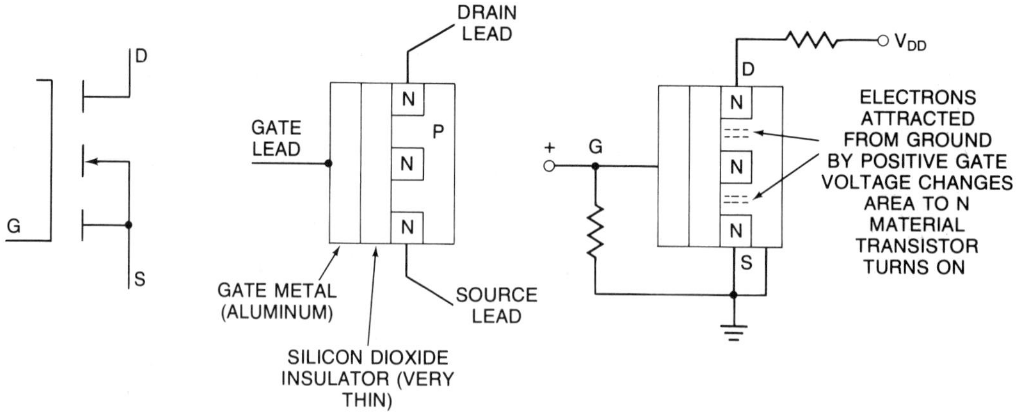


**Figure 18–13** N-type enhancement.

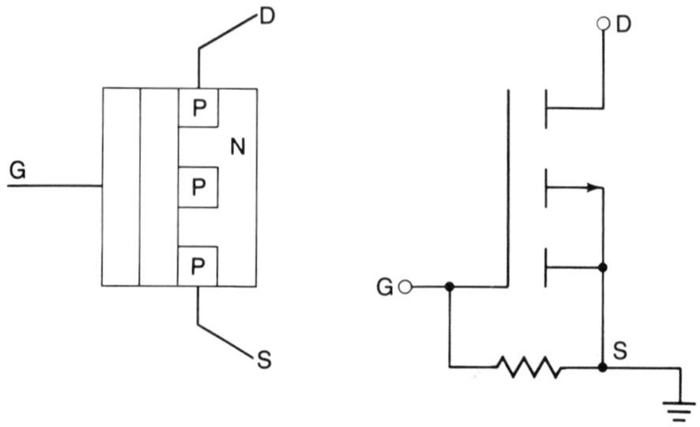


**Figure 18–14** P-type enhancement.

*P-Channel.* A *P-channel* device is shown in Figure 18–14. In order to turn the transistors on, the gate must be made negative with respect to the source. A negative gate will repel electrons from the area between the P blocks, effectively changing this area to P material.

## *THYRISTORS*

### Silicon-Control Rectifier (SCR)

The silicon-controlled rectifier (SCR) is a four-layer, three-terminal semiconductor device used in the control of power. It is essentially a rectifier. The third terminal provides control over when the rectifier may turn on. Other rectifier characteristics—anode positive with respect to the cathode for conduction—still apply.

### SCR Construction and Turn-On

The SCR is a four-layer device as shown in Figure 18–15a. To understand its operation, the four layers should be divided as shown in Figure 18–15b and the effect as shown in Figure 18–15c.

For normal operation, conduction requires the anode be positive with respect

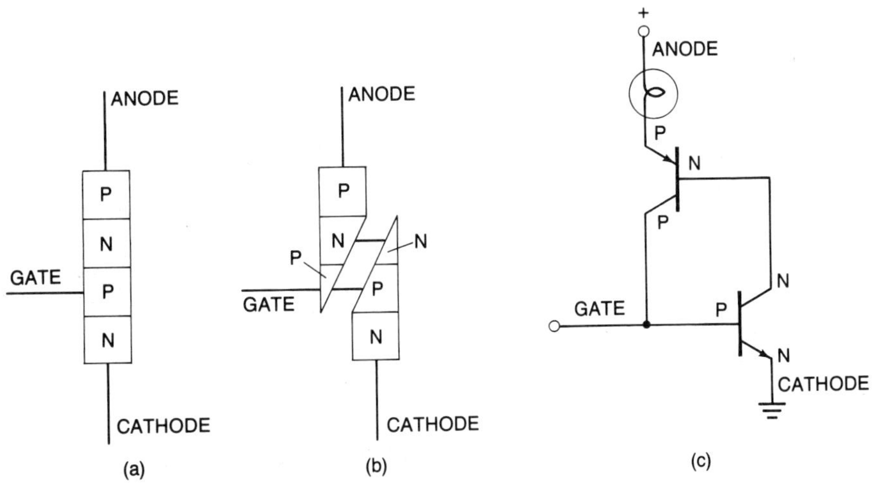


Figure 18–15

to the cathode. Turn-on occurs when the gate is made positive with respect to the cathode.

Observe Figure 18–15c. Without the positive gate, the NPN transistor is OFF. No conduction can take place. If the gate were made positive, the NPN transistor will turn on and conduct to the positive voltage through the base-emitter diode of the PNP transistor. This, in turn, turns on the PNP transistor, providing positive voltage through the PNP transistor to the base of the NPN transistor. Both transistors will remain on until the anode-to-cathode voltage is reduced to a low value.

## Half-Wave Phase Control

Half-wave phase control may be obtained by using a silicon control rectifier as in the circuit shown in Figure 18–16.

Control of the on-time of the SCR and therefore the lamp is adjusted by potentiometer $R_1$. If $R_1$ is fully clockwise (CW) the input voltage is fed directly to the gate of the SCR through diode $D_2$ and appears across $C_1$. Whenever the anode is positive, the gate will be positive, and the SCR will conduct the full half cycle.

If $R_1$ is turned counterclockwise (CCW), the SCR will conduct less than one-half cycle. Capacitor $C_1$ will charge through $D_1$ to the negative peak during each negative half cycle. Capacitor $C_1$ must discharge through $R_1$ before the voltage can move to a positive level allowing the SCR to turn on. If the discharge time of $C_1$ is 120th of a second, the SCR will not be turned on. The range of control would be from zero time on to on for one-half cycle.

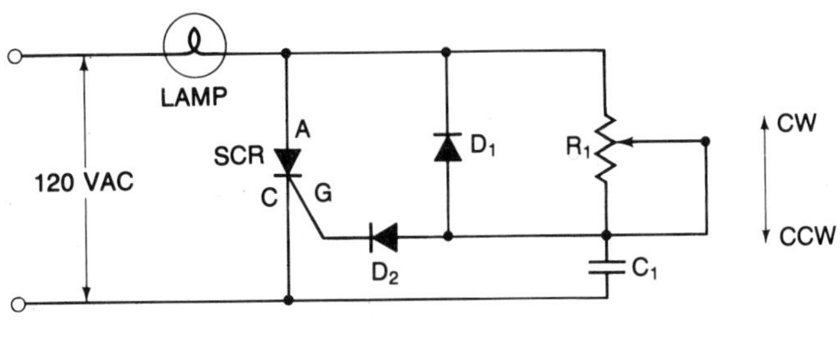


Figure 18–16

## TRIAC

A *triac* is a bidirectional thyristor used in full-wave control of ac power. A triac is similar to two SCR connected in parallel with the anode of one connected to the cathode of the second and vice versa. This would provide a path for current flow in both directions.

One main difference between two SCRs connected back-to-back and a triac is the triggering of a triac. A triac may be triggered by either a positive or negative gate signal. This characteristic makes the triac extremely useful in the control of ac power.

### Lamp Dimmer

A standard circuit use of triacs is in the lamp-dimmer control. The circuit is a two-terminal device that usually fits in a wall outlet box. An example of a lamp dimmer is shown in Figure 18–17.

In the circuit of Figure 18–17, $Y_2$ is the triac. The device $Y_1$ is called a *diac* and is essentially two low-voltage zener diodes connected back-to-back in series. When the input control voltage rises above the zener breakdown, the diac provides a sharp leading-edge trigger voltage to the triac.

The combination $C_1$ and $L_1$ in Figure 18–17 provides for the Federal Communications Commission required repression of radio interference that may be generated by sharp turn-on of the triac.

Phase shift of the input voltage is provided by $R_1$, $R_2$, and $C_2$. The potentiometer $R_2$ provides for various amounts of phase shift and various turn-on time for the triac.

Triacs and SCRs are used in simple dimmer-control circuits and motor-speed control devices such as hand drills. They are also used in high-power, high-horsepower controlled devices. They are available with voltage ratings to thousands of volts and current ratings to hundreds of amperes.

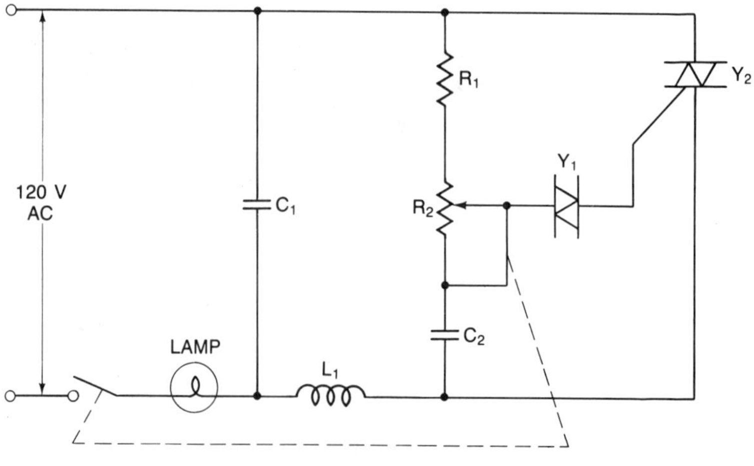


Figure 18–17

## REVIEW QUESTIONS

1. A bipolar transistor will conduct when its emitter-base diode is (forward, reverse) _______ biased.
2. One method of telling whether a transistor is turned on is to measure the voltage across the emitter-base junction. The voltage should be about _______ volts.
3. An NPN transistor will conduct when its base is (positive, negative) _______ with respect to the emitter.
4. The term $h_{FE}$ is equal to _______ divided by _______.
5. The input impedance to a common-emitter amplifier is (low, high)_______.
6. The input impedance to a common-collector amplifier is (low, high) _______.
7. Which of the three bipolar amplifier circuits is most commonly used?
8. If the bipolar transistor amplifier circuit does not contain an emitter resistor, the circuit (is, is not) _______ prone to thermal runaway.
9. The voltage gain of an emitter follower is (greater, less) _______ than one.
10. In order to increase the gain of an emitter follower (increase, decrease) _______ the forward bias on the transistor.
11. The formula for approximate voltage gain of the common-collector amplifier is _______.
12. Increasing the value of an unbypassed emitter resistor will (increase, decrease) _______ the voltage gain of a common-emitter amplifier.
13. The formula for the input impedance of a bipolar transistor amplifier is _______.
14. A multivibrator produces _______ waves in the output.
15. A (positive, negative) _______ pulse is required to trigger the one-shot multivibrator.
16. A flip-flop circuit is (mono-stable, bi-stable) _______.
17. FET stands for _______ _______ _______.
18. MOS in MOSFET stands for _______ _______ _______.
19. An enhancement mode MOSFET must be turned (On, Off) _______.
20. An N-channel enhancement-mode MOSFET requires a (positive, negative) _______ gate voltage to turn on.
21. Silicon control rectifiers conduct in (one, both) _______ direction(s).
22. Triacs conduct in (one, both) _______ direction(s).
23. An SCR will turn on when its anode is (positive, negative) _______ and its gate is made (positive, negative) _______ with respect to its cathode.
24. Once an SCR is turned on, it will remain conducting until its (gate, anode) _______ voltage is reduced to a very low value.
25. Triacs may be triggered by (positive, negative, both polarities of) _______ voltages.

# UNIT NINETEEN

# *INTEGRATED CIRCUITS: OPERATIONAL AMPLIFIERS AND SPECIAL ANALOG CIRCUITS*

## *INTEGRATED CIRCUITS*

Manufacturing processes improved after the development of the transistor. Procedures were discovered for providing in a small package all the components necessary for a complete amplifier. The integrated circuit became a standard in the industry.

A dual-operational amplifier, type 4558, fits in the 8-pin package shown in Figure 19–1. Contained in this package are 30 transistors, 4 diodes, 18 resistors, and 4 capacitors. The dimensions are approximately 0.35 inches long and 0.2 inches wide. This size would provide for a much larger number of components. The problem with the package is the limited space for reasonable-size connecting leads.

The 8-pin package shown in Figure 19–1 is equivalent to a dual 741 operational amplifier. This is one of the standard units used in industry.

**Figure 19–1**

## DIFFERENTIAL AMPLIFIER

A *differential amplifier* is used as the input circuit of most operational amplifiers. One of the main characteristics of a differential amplifier is its ability to convert 180° out-of-phase signals to the same polarity with equal gain. An example of a differential amplifier is shown in Figure 19–2.

Transistor $Q_3$ in Figure 19–2 is a constant-current generator. Zener diode $D_1$ provides 2.6 volts of emitter-base voltage drop. The emitter must be 2 volts positive with respect to $-V$. The 2 volts across $R_1$ (1 k) indicate a current of 2 ma.

Assume both the inverting and noninverting inputs referenced to ground. Each of the two transistors, $Q_1$ and $Q_2$, will conduct 1 ma.

The resistor $R_4$ in the collector circuit of $Q_1$ will have 10 volts across it, leaving $+5$ volts at the collector of $Q_1$.

If a positive signal is applied to the inverting input while the noninverting input remains at zero volts, conduction through $Q_1$ will increase. Because only 2 ma is provided by the constant current source $Q_3$, the increased $Q_1$ current must come from $Q_2$. Assume $Q_1$ current increases to 1.2 ma. The voltage across $R_4$ increases to 12 volts and the voltage out decreases to $+3$ volts. A positive signal at the inverting input causes the output to move in the negative direction from $+5$ volts to 3 volts.

Return the inverting input to zero volts and apply the same positive signal to the noninverting input. When the base of $Q_2$ becomes positive, the current through $Q_2$ increases. Because $Q_3$ is still a constant current source, the increased $Q_2$ current must come from $Q_1$. The changing of currents, in this case from $Q_1$ to $Q_2$, will be equal to the changing of currents from $Q_2$ to $Q_1$ in the previous case.

If the current in $Q_1$ is decreased, the voltage across $R_4$ will decrease. Because the current decrease is the same, the voltage across $R_4$ will decrease by 2 volts. The output rises from $+5$ volts to $+7$ volts. A positive signal applied to the noninverting input raises the output by $+2$ volts. The same positive signal applied to the inverting input lowers the output by $-2$ volts.

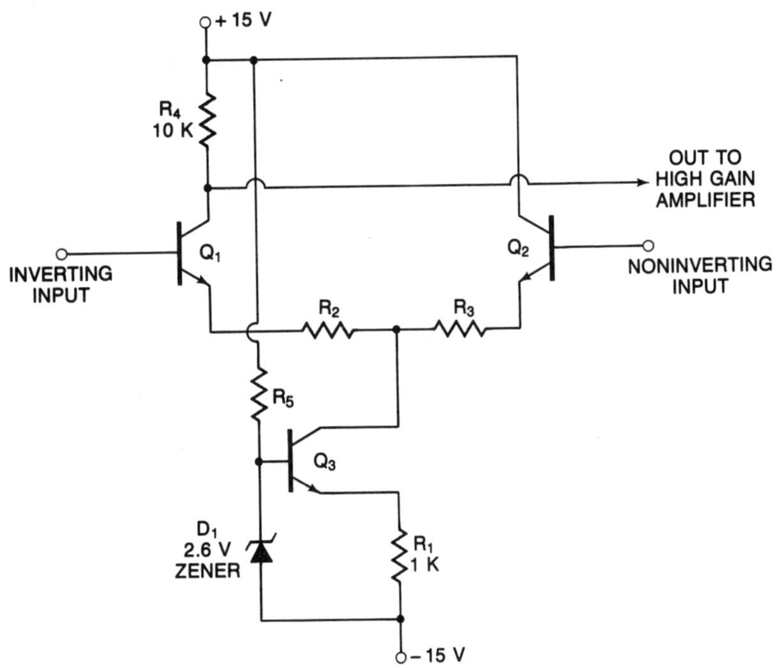


Figure 19–2

### Equal Signal to Both Inputs

If the same signal is applied to both the inverting and noninverting inputs, the output will not change.

Transistor $Q_3$ in Figure 19–2 is a constant-current source. When the same signal is applied to both $Q_1$ and $Q_2$, both transistors will try to change in the same direction, but they cannot. Transistor $Q_1$ can only increase in conduction by the amount that transistor $Q_2$ decreases and vice versa.

In an operational amplifier, the differential input is followed by amplifier stages with high gain, usually above 100,000 and higher. The amplifiers are adjusted so that 0 volts at the input provides an output of 0 volts. In most applications, the amplifier will be connected with some feedback network to control overall gain.

## *THE OPERATIONAL AMPLIFIER*

The standard operational amplifier (op amp) consists of a differential amplifier at the input followed by a high-gain dc amplifier. The symbol for the op amp is shown in Figure 19–3a.

The gain of the op amp is very high. As a rule, op amps are usually connected with some feedback network included. The exception to the rule is when the device is used simply to compare two signals.

Consider the gain of the amplifier as something from 50,000 to well above 200,000 for some units. Any small difference in input at the two input terminals will result in driving the amplifier to saturation in the positive or negative direction. Feedback is necessary in most cases if the amplifier is to be useful.

In standard operation, if the amplifier is not in saturation, both inputs will be at the same voltage level. This must be true, for the difference between the two input signals will be amplified by the high-gain amplifier. Consider the amplifier circuit in Figure 19–3b. Here the noninverting terminal of the amplifier is grounded. A signal of +2 volts is connected to the inverting input through a 1-kilohm resistor. The feedback resistor connected from the output back to the inverting input is 2 kilohms.

Because the input impedance at the + and − input terminals is very high, no current will flow into these terminals. If the amplifier is not in saturation, the voltage at both + and − input terminals must be the same. The + input terminal is at 0 volts. Therefore, the − terminal must be at 0 volts.

There must be 2 volts across the 1-kilohm input resistor $R_{in}$. This is shown in Figure 19–3c. The 2 volts across the input 1-kilohm resistor indicates a current of 2 ma flowing through the resistor.

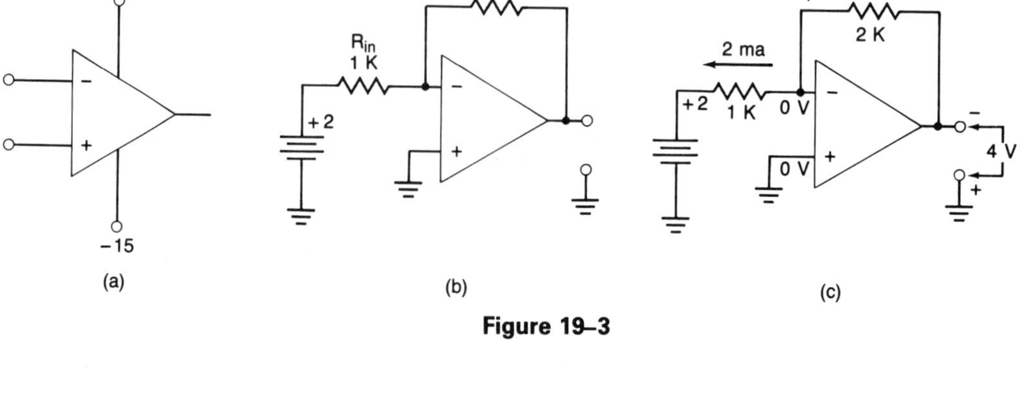


**Figure 19–3**

The 2 ma flowing from the − input of the op amp through the 1-kilohm resistor cannot be coming out of the amplifier because the op amp has a very high input impedance. This current must be coming through the external feedback resistor. The 2 ma flowing through the 2-kilohm feedback resistor must develop 4 volts across the resistor (Ohm's Law).

Because the current is flowing from the output of the amplifier through the feedback resistor back to the input, the 4 volts developed across the resistor are negative at the right and positive at the left. Current enters the negative terminal of a resistor and exits the positive terminal. The output terminal of the amplifier is at −4 volts as compared to the inverting input terminal, which is at 0 volts. The output of the amplifier is therefore at −4 volts with respect to ground. Ground is 0 volts reference.

It is obvious that a relation must exist between the ratio of input and feedback resistors and input and output voltage. The relationship may be expressed by the formula

$$V_{out} = -\left(V_{in}\frac{RF}{R_{in}}\right)$$

The formula may be important for solving problems quickly, but it is more important to remember the characteristics of this circuit operation.

1. Both input terminals must be at the same voltage level.
2. No current flows into or out of the input terminals.
3. Ohm's Law is true.

## THE SUMMING AMPLIFIER

A summing amplifier is used to combine two or more signals. Each signal may have a different amount of gain. Consider the circuit of Figure 19–4. The op amp is connected to provide summing-amplifier operation.

Output due to + 4 V input.

$$V_{out\,1} = -\left(V_{in}\frac{RF}{R_{in}}\right)$$

$$V_{out\,1} = -\left(4\,\frac{3}{1}\right)$$

$$V_{out\,1} = -12 \text{ volts}$$

Output due to −3 V input.

$$V_{out\,2} = -\left(V_{in}\frac{RF}{R_{in}}\right)$$

$$V_{out\,2} = -\left(-3\,\frac{3}{2}\right)$$

$$V_{out\,2} = +4.5 \text{ volts}$$

$$V_{out\,total} = V_{out\,1} + V_{out\,2}$$

$$VT = -12 + 4.5$$

$$VT = -7.5 \text{ volts}$$

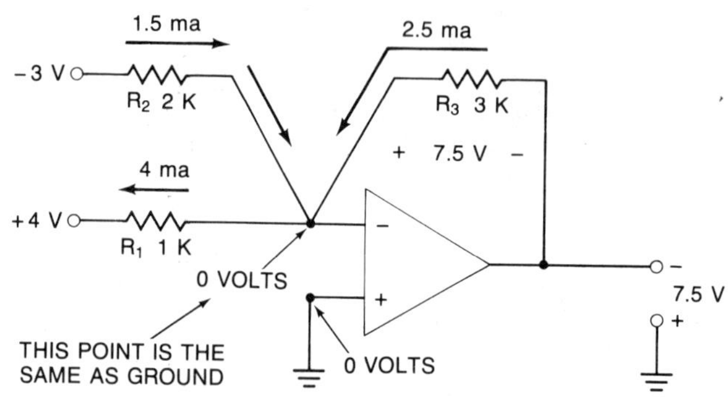


**Figure 19–4**

Check the problem using Ohm's Law and Kirchhoff's laws while recognizing the characteristics of op amp circuits.

First, check the currents at the junction at the inverting input.

The current through $R_1$ is:

$$I = \frac{V}{R} = \frac{4V}{1k} = 4 \text{ ma}$$

The current through $R_2$ is:

$$I = \frac{V}{R} = \frac{-3V}{2k} = -1.5 \text{ ma}$$

The sum of the currents through $R_1$ and $R_2$ indicates that 2.5 ma more are leaving the junction than are entering through $R_1$ and $R_2$. This extra 2.5 ma must be entering through $R_3$.

Because the noninverting input is at ground potential, the inverting input must be at ground potential.

The voltage across $R_3$ is 7.5 volts. The output terminal is 7.5 volts more negative than ground, or −7.5 V.

$$V = I \times R = -2.5 \text{ ma} \times 3 \text{ k} = -7.5 \text{ V}$$

## OP AMP FOLLOWER

The *follower circuit* is used where impedance levels must be changed from high to low levels. The input to an op amp is high impedance; the output is low impedance. Connection of a circuit as shown in Figure 19–5 provides a change from a high-impedance source to a low-impedance load. The gain of the stage is +1.

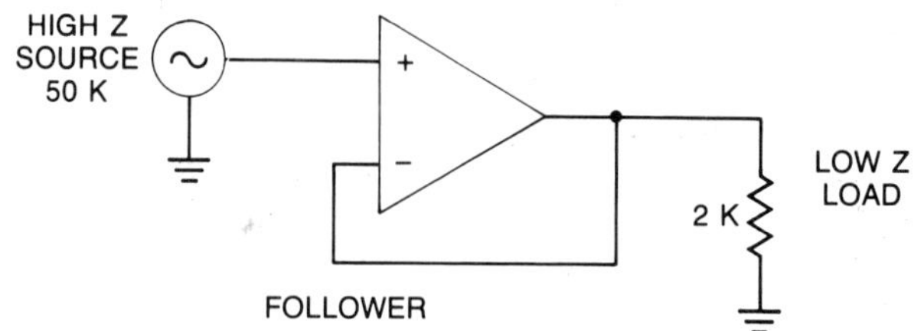


**Figure 19–5**

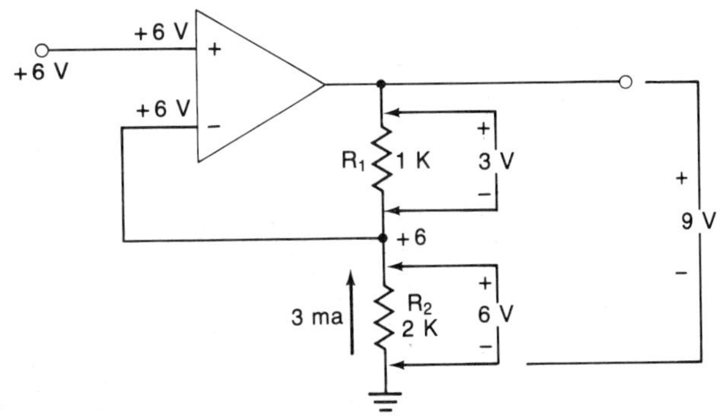


**Figure 19–6** Noninverting amplifier.

If the instantaneous value of the voltage from the source were +2 volts, the voltage across the load would be +2 volts. The output follows the input.

### Noninverting Amplifier

The noninverting amplifier functions using the + terminal for the input terminal. An example is shown in Figure 19–6.

The application of +6 volts to the + input terminal will cause the amplifier output to drive in the positive direction. The voltage level of the output will increase until +6 volts is available at the − input terminal. At this point, there will be +6 volts across $R_2$, 2 kilohms. There will be 3 ma flowing through $R_2$ and $R_1$ to the amplifier output. By Ohm's Law, the voltage across $R_1$ is 3 volts. The resistors $R_1$ and $R_2$ form a voltage divider. There must be +9 volts at the output.

The formula for calculating voltage output of a noninverting amplifier is

$$V_{out} = V_{in}\left(1 + \frac{R_1}{R_2}\right)$$

## *ADDER-SUBTRACTOR*

In certain circumstances, voltages may be connected to both the inverting and noninverting inputs of an operational amplifier. When this configuration is used, the op amp becomes an *adder-subtractor*. Figure 19–7 shows this configuration.

Determining the output voltage of the op amp circuit shown in Figure 19–7 is best accomplished using Ohm's Law rather than the complex formula required for such problems.

A. Determine the voltage at the noninverting input terminal. By using Millman's Theorem
   1. Place a short from + terminal to ground.
   2. Calculate the Millman current through $R_1$

$$I_m = \frac{V}{R} = \frac{-6}{1k} = -6 \text{ ma}$$

   3. Calculate the Millman current through $R_2$.

$$I_m = \frac{V}{R} = \frac{4}{2\text{ k}} = 2 \text{ ma}$$

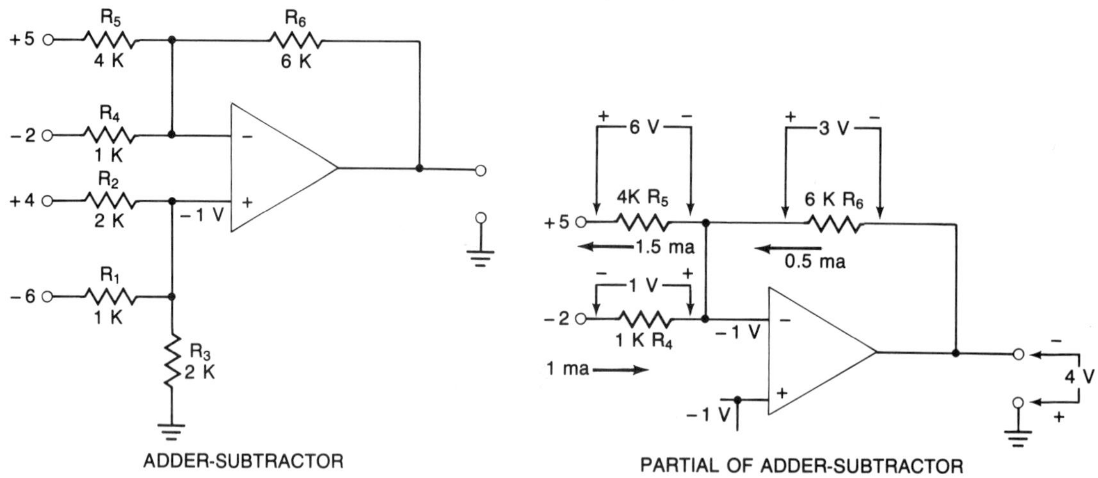


**Figure 19–7**

4. Both ends of $R_3$ are grounded. No Millman current flows through $R_3$.
5. Determine the Millman current through the short by adding currents.

$$-6 \text{ ma} + 2 \text{ ma} = -4 \text{ ma}$$

6. Calculate the Millman effective resistance of the combination.

$$R_m = \frac{1}{\frac{1}{R_1}\ \frac{1}{R_2}\ \frac{1}{R_3}} = \frac{1}{\frac{1}{2k}\ \frac{1}{1k}\ \frac{1}{2k}} = 0.5 \text{ k}$$

7. Determine the voltage at the noninverting input.

$$V = I \times R$$
$$V = -2 \text{ ma} \times 0.5 \text{ k}$$
$$V = -1 \text{ V}$$

The voltage at the noninverting input is $-1$ V. If the amplifier is not in saturation, the voltage at the inverting input must be $-1$ volt.

B. Determine the output of the amplifier shown in Figure 19–7 using Ohm's Law and Kirchhoff's Law.
1. The voltage across $R_4$ is the difference between $-1$ V and $-2$ V or $-1$ volt.
2. The current through $R_4$ by Ohm's Law is

$$I = \frac{V}{R} = \frac{-1\text{V}}{1\text{k}} = -1 \text{ ma}$$

3. The voltage across $R_5$ is the difference between $-1$ V and $+5$ V or $+6$ V.
4. The current through $R_5$ by Ohm's Law is:

$$I = \frac{V}{R} = \frac{+6}{4\text{k}} = +1.5 \text{ ma}$$

5. The current through $R_6$ by Kirchhoff's Law is

$$+1.5 \text{ ma} - 1 \text{ ma} = 0.5 \text{ ma}$$

6. By Ohm's Law, the voltage across $R_6$ is 0.5 ma × 6 kilohms = 3 volts.
7. The output terminal is 3 volts more negative than the inverting input terminal, which is at −1 V.
8. The output is −4 volts.

## *OP AMP INTEGRATOR*

An *integrator* is a circuit that could receive a voltage representing a rate of speed and produce a voltage representing the distance covered. An integrator measures the level of input voltage and the length of time the voltage is present. An integrator circuit is shown in Figure 19–8.

The formula given for integrators is

$$V_{out} = \frac{1}{RC} \int -V_{in}\, dt$$

Consider the circuit of Figure 19–8, an integrator circuit with a switch at the input. The switch moves from A to D, remaining for 2 minutes in positions A, B, and C. What is the voltage detail at the output of the integrator? Assume 0 volts output when the switch is in position A.

Position B

$$V_{out} = \frac{1}{2 \times 10^6 \times 10 \times 10^{-6}} \int -(-2 \times 120)$$

$$V_{out} = 0.05 \times 240$$

$$V_{out} = +12 \text{ V}$$

$V_{out}$ increases from 0 to +12 V.

Position C

$$V_{out} = \frac{1}{RC} \int -V_{in}\, dt$$

$$V_{out} = \frac{1}{20} \times \int -1 \times 120$$

$$V_{out} = 0.05 \times -120$$

$$V_{out} = -6 \text{ V}$$

$V_{out}$ decreases from +12 V to +6 V.

Position D, no change

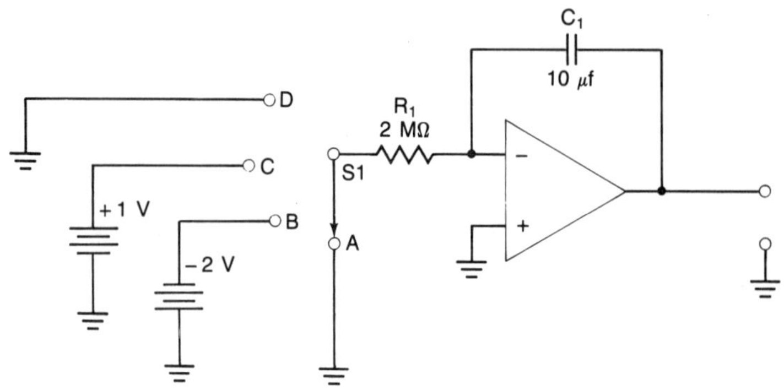


**Figure 19–8** Integrator.

A straight check of integrator operation may be made using the charging characteristics of a capacitor. In the circuit of Figure 19–8, the capacitor started with 0 volts across it. When the switch was placed in position B, the voltage across the input resistor was 2 volts. (+ terminal at 0 volts, − terminal must be at 0 volts.) With 2 volts across $R_1$, 2 megohms, a current of 1 μa will flow through $R_1$ into $C_1$.

One microampere is equal to 1 microcoulomb per second. One microcoulomb per second for 120 seconds equals 120 microcoulombs for total charge while the switch is in position B.

The formula for charge on a capacitor is:

$$Q = C\,V$$

Dividing both sides of the equation by C results in

$$V = \frac{Q}{C}$$

Substituting the values for Figure 19–9, switch in position B provides:

$$V = \frac{120 \text{ microcoulombs}}{10 \text{ microfarads}} = 12 \text{ V}$$

V = 12 V at the end of 120 seconds or 0.1 volt per second rise. This agrees with the results given by the integrator formula.

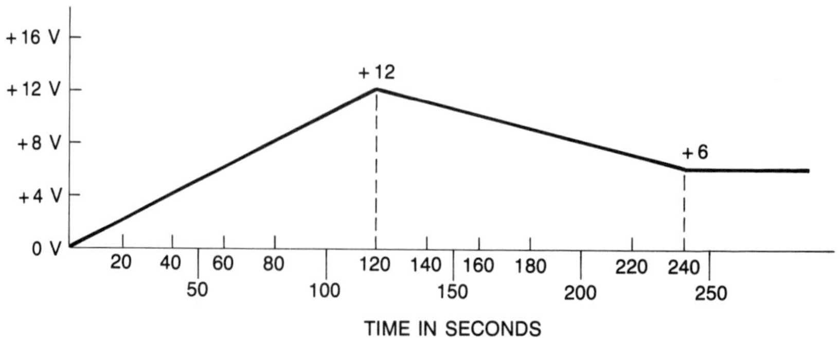


**Figure 19–9**

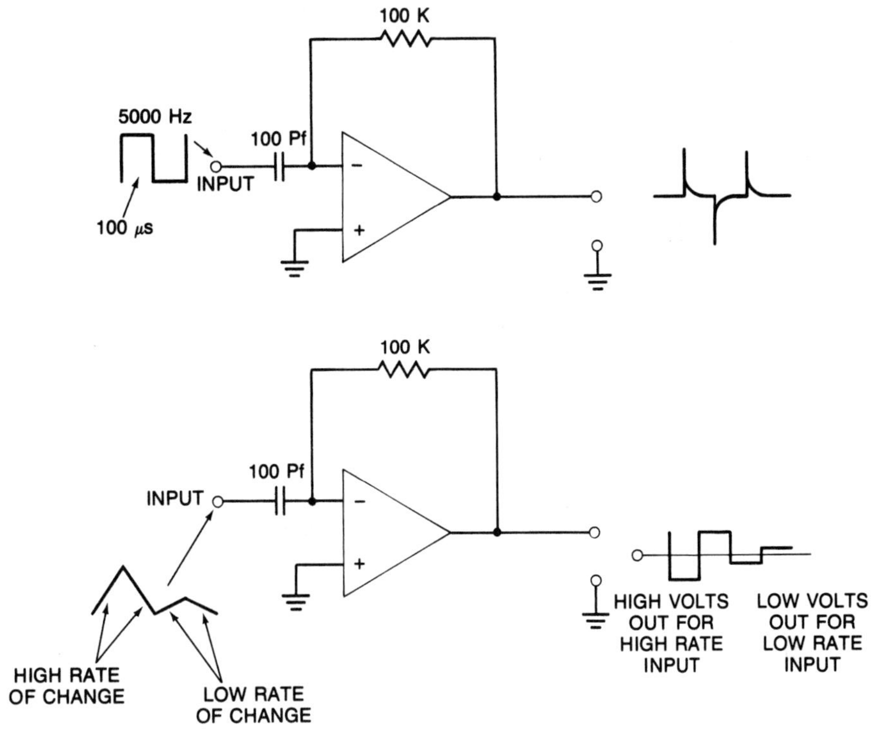


**Figure 19–10**

## *DIFFERENTIATOR*

A *differentiator* circuit provides an output that is proportional to the rate of change of the input voltage. When an op amp is included in the differentiator, the impedance level at the output is low. The circuit is shown in Figure 19–10. Examples of input and output signals are also provided.

## *FILTERS*

Passive filters were included in the units on RC circuits. The addition of an amplifier (op amp) to the circuit provides what is called an *active filter*. The addition of the amplifier provides for sharper cutoff characteristics of the filter and, as with op amps, low output impedance.

A number of different circuit configurations are usable for each type of filter: low pass, high pass, and band pass. Examples are provided in Figure 19–11 of a low-pass, band-pass, and high-pass active filter. The formulas provide the approximate cutoff and center frequencies.

**LOW PASS**

where $R_1 = R_2 = R_3$
$C_1 = C_2$

$$f = \frac{0.03\pi}{RC}$$

**BAND PASS**

where $C_1 = C_2$
$R_1 = R_2$
$R_3 = 2R_1$

$$f\ cf = \frac{1}{2\pi\ CR}$$

$$BW = \frac{\sqrt{2}}{CR}$$

**HIGH PASS**

$$f\ co = \frac{0.1\pi}{RC}$$

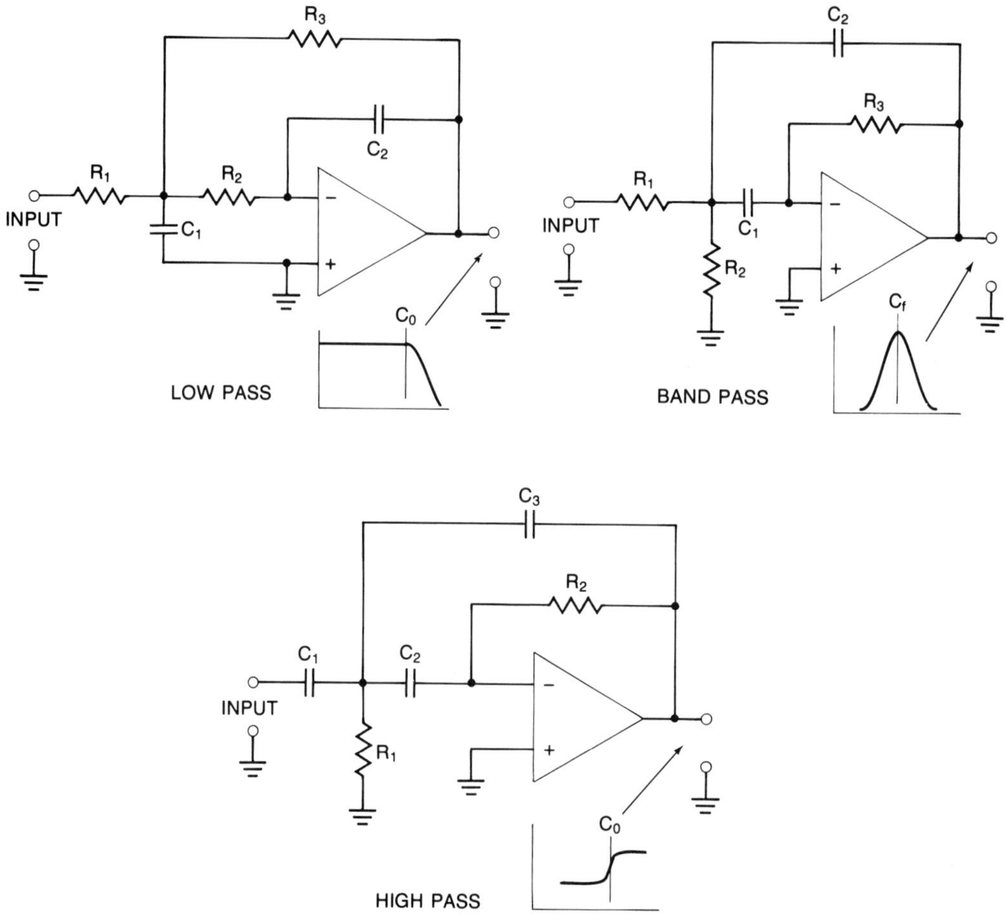


**Figure 19–11**

## *COMPARATORS*

An operational amplifier may be used as a simple *comparator*. The output of the op amp will be in either positive or negative saturation depending on the relation of the noninverting to the inverting input voltage. Observe Figure 19–12, which shows a simple comparator.

The voltage divider consisting of $R_1$ and $R_2$ provides +7.5 volts at the noninverting input. Potentiometer $P_1$ is at 0 volts. The output of the amplifier will be in positive saturation, approximately +14 V.

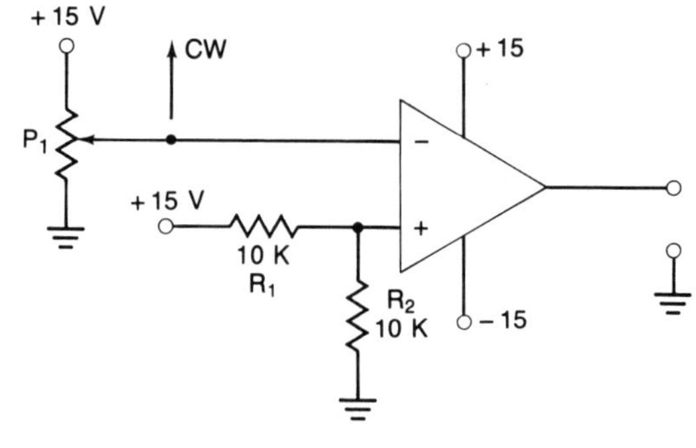


**Figure 19–12**

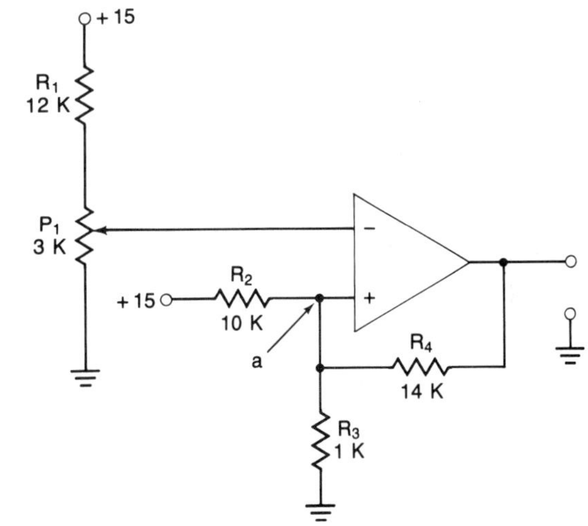


**Figure 19–13**

As potentiometer $P_1$ is rotated clockwise, the voltage at the inverting input becomes more positive. When the voltage at the inverting input reaches +7.5 volts, the output will switch to negative saturation, about −14 volts.

Comparators are used in many applications in industry. A comparator is used in every digital voltmeter. The comparator is used to determine when the digital counter has reached the level of the input voltage being measured.

### Regenerative Comparators

A *regenerative* op amp comparator uses the op amp output to determine the switchover voltage. An example of a regenerative comparator is shown in Figure 19–13. Assume the saturation voltage is +14 when positive and −14 when negative. These would be the approximate values considering 15-volt supplies.

Problem: Determine the input voltages for switchover. Assume $P_1$ input at 0 volts and op amp in positive saturation.

1. Use Millman's Theorem to determine the input at the + terminal.
   a. Assume a ground at a.
   b. Current through $R_2$ = 1.5 ma.
   c. Current through $R_4$ = 1 ma.
   d. Millman resistance = 854 ohms.
   e. Voltage at a = 2.5 ma × 854 ohms = 2.135 volts.

      The potentiometer $R_1$ would have to be adjusted to slightly over + 2.134 volts where the amplifier output would switch to negative saturation, −14 V.
2. Use Millman's Theorem to determine the new voltage at the noninverting input terminal. This is the voltage that potentiometer $R_1$ would have to be reduced to before the op amp would switch back to positive saturation.
   a. Assume a ground at a.
   b. Current through $R_2$ = 1.5 ma.
   c. Current through $R_4$ = −1 ma.
   d. Millman resistance = 854 ohms.
   e. Voltage at a = 0.5 ma × 854 ohms = 0.427 volts.

      The potentiometer $R_1$ would have to be adjusted to 0.427 volts where the output would switch back to +14 V.

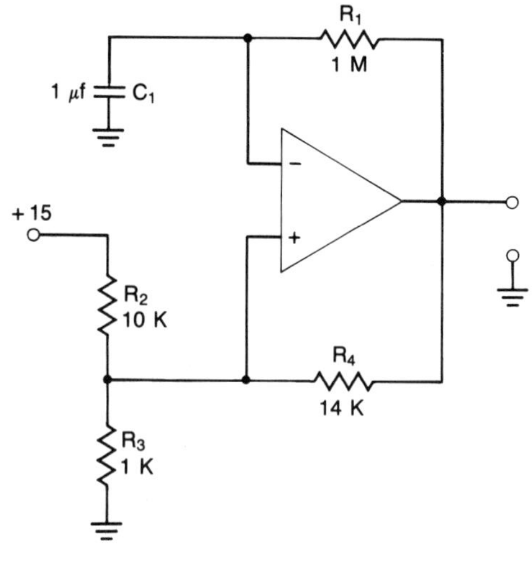


**Figure 19–14**

## *SQUARE-WAVE GENERATOR*

A regenerative comparator circuit may be used to produce square waves at the op amp output. The circuit needs a slight change, the addition of a feedback resistor, and a capacitor. Observe Figure 19–14. Note that the circuit of the noninverting input is the same as in Figure 19–13. Switchover voltages will be the same as in the previous problem.

The time constant of the circuit connected to the inverting input is 1 megohm times 1 MF = 1 second.

When power is applied to the circuit, the output will go to positive saturation because of the positive voltage at the noninverting input. Capacitor $C_1$, 1 MF, will start charging to the +14 output through resistor $R_1$, 1 megohm. When the capacitor voltage reaches +2.134 volts, the output will switch to negative saturation, −14V. The voltage at the noninverting input will reduce to 0.427 volts.

Capacitor $C_1$ will start charging in a negative direction from +2.134 V. When the voltage across the capacitor reaches 0.427, the output will switch back to +14 V and the process will repeat itself.

## REVIEW QUESTIONS

1. The gain of an operational amplifier is (high, low) ______.
2. The input impedance of an operational amplifier is (high, low) ______.
3. What will be the level of output of an operational amplifier if both the inverting and noninverting inputs are not at the same potential?
4. What will be the voltage gain of an operational amplifier circuit (adder) if the input resistor is 2 kilohms and the feedback resistor is 8 kilohms?
5. When using an operational amplifier as a follower, the input impedance is (high, low) ______.
6. The voltage-gain formula for a noninverting amplifier is ______.
7. When solving an adder-subtractor problem, the voltage at the (inverting, noninverting) ______ terminals must be determined first.

8. The formula for voltage output of an op amp integrator is ______.
9. An op amp differentiator provides an output that is proportional to the (rate of change, magnitude) ______ of the input voltage.
10. The use of an op amp in active-filter circuits provides for a (low, high) ______ output impedance filter.
11. In a simple op amp comparator circuit, if the inverting terminal is at +6 volts and the noninverting input is at +5 volts, the output will be at (positive, negative) ______ saturation.
12. In the regenerative comparator square-wave generator (Figure 19–14), if the capacitor is increased in value, the frequency of the square-wave output will (increase, decrease) ______.

# UNIT TWENTY

# DIGITAL SYSTEMS, GATES, AND DEVICES

## DIGITAL CIRCUITS

A digital circuit is one where the operation depends on counting. In analog circuits, operation is dependent on measurement or comparing of voltage levels.

There is an increasing use of digital circuitry in the control of electric and electronic devices. The following sections will cover some of the circuits used in simple digital devices.

## DIGITAL COUNTING SYSTEM

The basic digital counting system is called the *binary system.* It is a simple system using only two units, zeros and ones. A given 0 or 1 is weighted by its position in a line representing the total number. The weight of each 1 is as shown.

| 2048 | 1024 | 512 | 256 | 128 | 64 | 32 | 16 | 8 | 4 | 2 | 1 |
|---|---|---|---|---|---|---|---|---|---|---|---|
| 1 | 1 | 1 | 1 | 1 | 1 | 1 | 1 | 1 | 1 | 1 | 1 |

A 0 in any column indicates that weight is not present. For example, the digital number 1011 is equal to 11. The digital number indicates a 1 in the 1 column, a 1 in the 2 column, and a 1 in the 8 column. Adding 1 + 2 + 8 = 11.

| 8 | 4 | 2 | 1 |
|---|---|---|---|
| 1 | 0 | 1 | 1 |

### *BINARY ADDITION*

Binary numbers may be added just as simply as regular addition. The rules are

$0 + 0 = 0, 0 + 1 = 1, 1 + 0 = 1,$ and $1 + 1 = 0$ and a carry of 1

For example, add the following binary numbers:

| | | | | |
|---|---|---|---|---|
| | 1 | 0 | 1 | 1 |
| + | 1 | 1 | 1 | 0 |

| | | | | |
|---|---|---|---|---|
| carry | carry | | | |
| | | | carry | |
| 1 | 1 | 1 | | |
| | 1 | 0 | 1 | 1 |
| | 1 | 1 | 1 | 0 |
| 1 | 1 | 0 | 0 | 1 |

$16 + 8 + 0 + 0 + 1 = 25$

$1\ 0\ 1\ 1 = 11$

$1\ 1\ 1\ 0 = 14 \qquad 14 + 11 = 25$

### *BINARY SUBTRACTION*

Binary subtraction is also fairly simple. The rules are to change or invert all digits in the subtrahend and then add. The last carry is an end-around-carry, then add.

For example, subtract the binary number 0101 from the binary number 1110. $(14 - 5 = 9)$

| | | | | | | | | | | | | |
|---|---|---|---|---|---|---|---|---|---|---|---|---|
| | 1 | 1 | 1 | 0 | | | | 1 | 1 | 1 | 0 | |
| − | 0 | 1 | 0 | 1 | becomes | + | | 1 | 0 | 1 | 0 | |
| | | | | | | | 1 | 1 | 0 | 0 | 0 | |
| | | | | | | | | | | + | 1 | |
| | | | | | | | | 1 | 0 | 0 | 1 | = 9 |

In digital calculations, addition and subtraction are done using a system similar to the procedure shown as binary addition and subtraction. Multiplication and division are accomplished by repeated addition in multiplication and by repeated subtraction in division.

### *DIGITAL GATES*

There are three main types of gate circuits used in digital work. They are the AND gate, the OR gate, and the EXCLUSIVE OR gate. In order to understand their operation, look at their equivalent in switch circuits as shown in Figure 20–1.

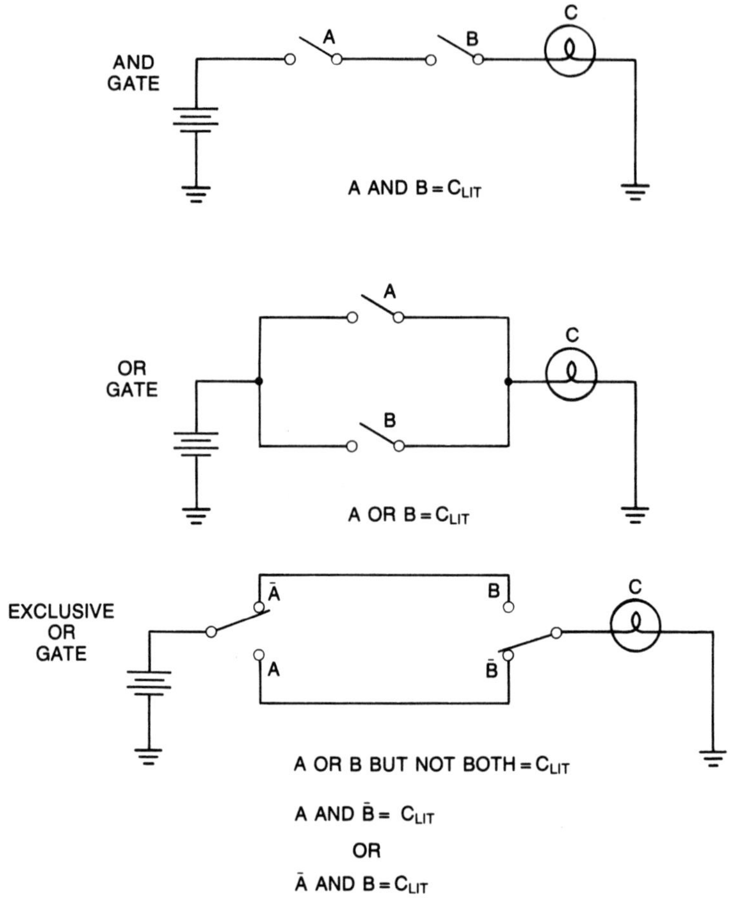


**Figure 20–1**

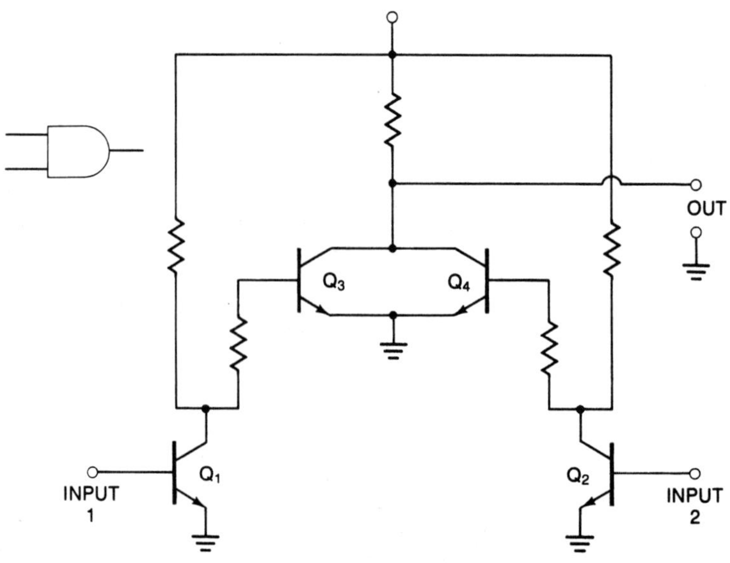


**Figure 20–2**

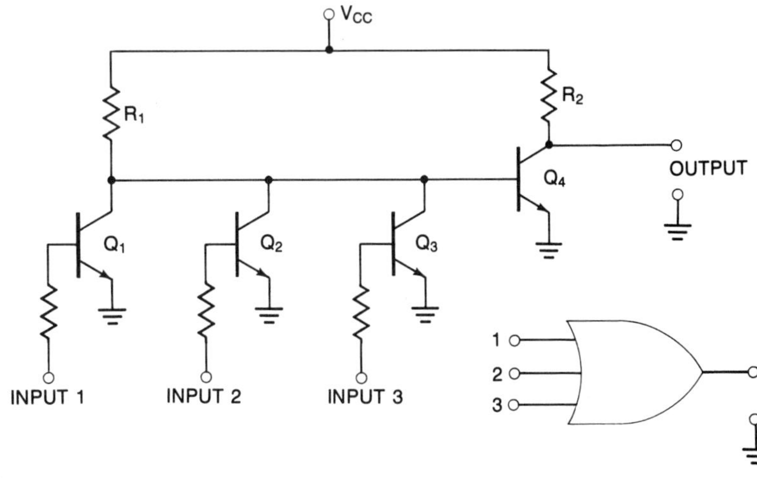


**Figure 20–3**

## AND Gate

The circuits used to provide the operating gates are rather simple. A sample circuit is shown in Figure 20–2. The symbol for the AND gate is also shown.

The output is taken from the common-collector resistor of the two transistors $Q_3$ and $Q_4$. In order to obtain a positive output, both transistors must be cut off. The base of $Q_3$ will go negative when $Q_1$ conducts. The base of $Q_4$ will go negative when $Q_2$ conducts. Both $Q_1$ and $Q_2$ inputs must be positive in order to get a positive output. (Note that inverters $Q_1$ and $Q_2$ are required in this operation.)

## OR Gate

A simple OR gate is shown in Figure 20–3. This is a three-input OR gate. Again note the requirement for an inverter $Q_4$ at the output in order to obtain a positive output for a positive input.

Transistors $Q_1$, $Q_2$, and $Q_3$ operate in the same fashion. If any input is positive, the transistor will turn on and current through the common-collector resistor will cause the collector to move negative. This negative-going signal will be felt at the base of $Q_4$, causing its collector to move positive. Any positive input causes a positive output.

## NAND and NOR Gates

In the circuit for the AND gate, inverters $Q_1$ and $Q_2$ were needed. In the OR gate circuit, inverter $Q_4$ was needed. Many standard gates are developed without the use of the inverters. These gates are called NAND and NOR gates. With a two-input NAND gate, two positive inputs cause a negative or zero output. In the NOR gate, a positive on any input terminal causes a negative or zero output. The gate symbol is changed by the addition of an inverter, shown at the output as in Figure 20–4.

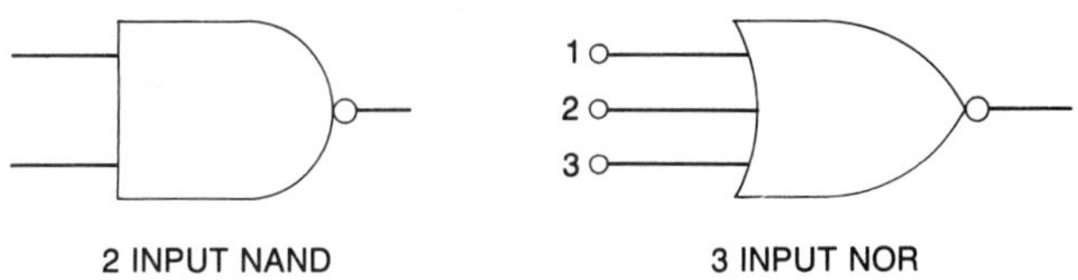


**Figure 20–4**

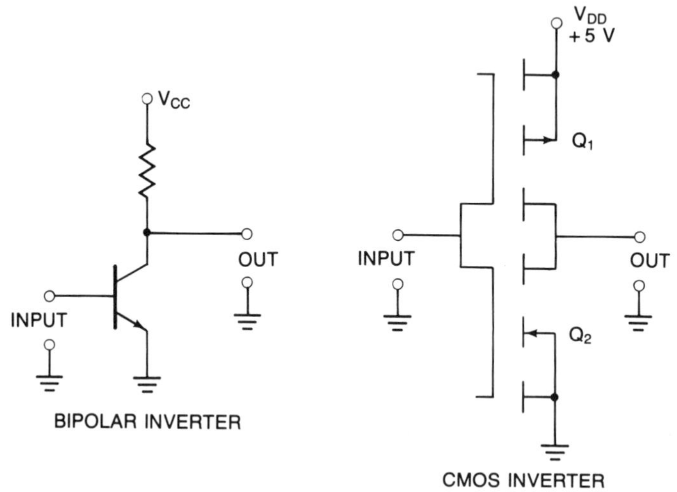


**Figure 20–5**

### Inverters

An inverter is another important circuit used in digital devices. Its purpose, as its title indicates, is to invert the signal of its input. There are a number of different circuits used as inverters. Two are shown in Figure 20–5.

The *bipolar inverter* is a straight transistor amplifier. A positive on the base turns the transistor on, and the collector is saturated resulting in low, close to zero output. A zero or low input to the base results in cutoff of the transistor and the collector goes to positive $V_{CC}$.

The MOSFET inverter uses two transistors. Transistor $Q_1$ is P-channel, enhancement mode, while $Q_2$ is N-channel, enhancement mode. A positive voltage at the common-gate connection turns $Q_1$ off and $Q_2$ on. The ground at $Q_2$ source is connected to the output. Positive-in provides zero-out. When a zero is provided at the input, $Q_1$ turns on and $Q_2$ turns off. The positive source voltage of $Q_1$ is connected to the output. Zero-in provides a positive output.

## *DIGITAL DEVICES*

Most electrical and electronic devices presently being manufactured are controlled by digital circuitry. The control is usually provided by a processor chip that in many cases includes a clock. The clock provides timing for the system operation as well as providing a means of automatic turn on and turn off.

Digital circuitry is highly developed at the present time. Systems that used to take up a full 6-by-2-by-2-foot rack are now housed in a single chip. The purpose of this section is not to explain the operation of digital circuitry step by step but to provide an overview of some system operations.

### Counters

A counting system is one of the main systems in digital devices. Counting systems are of various design, but in most every case they use flip-flops to provide the count. The basic flip-flop was covered in the unit on transistor circuits. In practice, there are a number of different types of flip-flops. The system covered here fits the general pattern of most.

Remember, a trigger at the input to a flip-flop causes the device to change

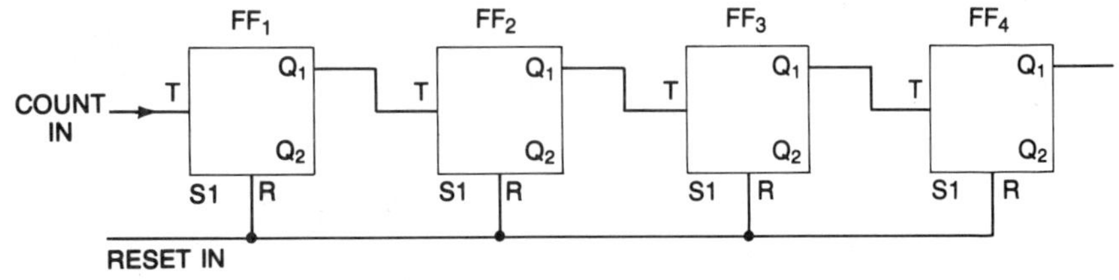


**Figure 20–6**

conduction states from $Q_1$ to $Q_2$ and back to $Q_1$, etc., as each trigger appears at the input. A series of four flip-flops is shown in Figure 20–6. This counter can count to 15. More flip-flops would provide a higher count.

The reset input is connected to the $Q_2$ transistor only. A positive reset pulse turns $Q_2$ of each FF on. This is equivalent to a zero count-out from each FF.

After a reset pulse, the system is ready to start counting. The digital count for the counter is 0000, or 0.

The leading edge of the first positive pulse will trigger $Q_1$, $FF_1$ on and $Q_2$, $FF_1$ off. $FF_1$ now records a 1 count. The output of $FF_1$ feeding $FF_2$ is the negative wavefront of $Q_1$, $FF_1$ turning on. This has no effect as only positive pulses trigger. The digital count is now 0001, or 1.

The next positive input pulse triggers $Q_1$, $FF_1$ off; $FF_1$ records a 0. The positive-going wavefront from $Q_1$, $FF_1$ is fed into $FF_2$ as the trigger pulse. $Q_1$, $FF_2$ triggers off; $Q_2$, $FF_2$ triggers on. $FF_2$ now indicates a 1 count. The digital count is now 0010, or 2.

The next positive input to $FF_1$ will trigger it into the 1 count. The digital count is now 0011, or 3. The fourth input count pulse triggers both $FF_1$ and $FF_2$ to 0. $FF_2$ going to 0 ($Q_1$ off) triggers $FF_4$ on. The digital count is now 0100, or 4. The count may continue as high as required dependent only on the number of flip-flops connected in the series.

## Memory

A second important use of flip-flops is as *memory*. The memory flip-flops, sometimes called a *register*, take in information and store it for a short time so that a counter can be used to count another series of operations. A simple transfer of information (or count) is shown in Figure 20–7.

The counter works the same as in the previous explanation. Assume the counter has the digital count of 1011, or 11. This number is to be stored in memory, so the counter may be used to receive another count. A memory reset is executed, which returns all memory flip-flops to the 0 condition, $Q_2$ off and $Q_1$ on. When the count 1011 is to be transferred to memory, a positive store pulse is fed to each AND gate, $A_1$, $A_2$, $A_3$, and $A_4$. The other input to the AND gates is coming from the counter flip-flops. If the counter flip-flop is a 1 (+), the output of the AND gate will be a positive during the store pulse. This will trigger the associated memory flip-flops to the correct count condition. After the store-positive pulse, the memory flip-flops will contain 1011.

The counter could then be reset and used to count another number. The first count is safely stored in the memory register.

## Digital Voltmeter

Another very common digital device is the *voltmeter*. A digital voltmeter may use a combination of analog and digital circuits. In Figure 20–8, a simplified digital voltmeter is shown. The voltmeter is capable of indicating voltages up to 63 volts. The

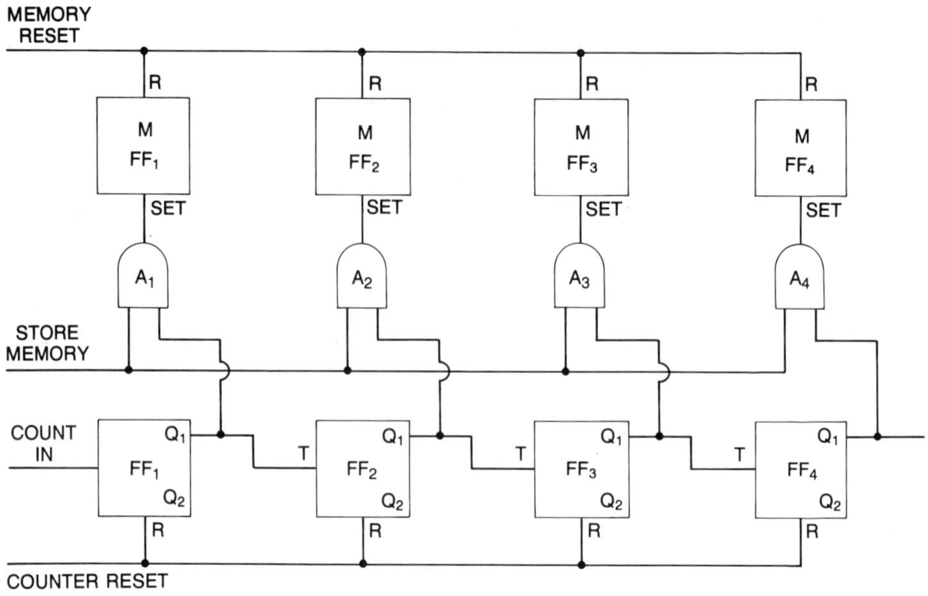


**Figure 20–7**

range of the meter could be increased by adding counter flip-flops or by changing the scaling of the analog amplifiers.

1. The counting circuit is capable of counting to 63 in increments of one.
2. The AND gate receives square-clock pulses at the rate of approximately 100 per second (enough to count through the full 63 before the next reset).
3. The amplifier $A_2$ is a *comparator*. Its output will be positive as long as the voltage to be measured input is more positive than the summed voltage from the counter.

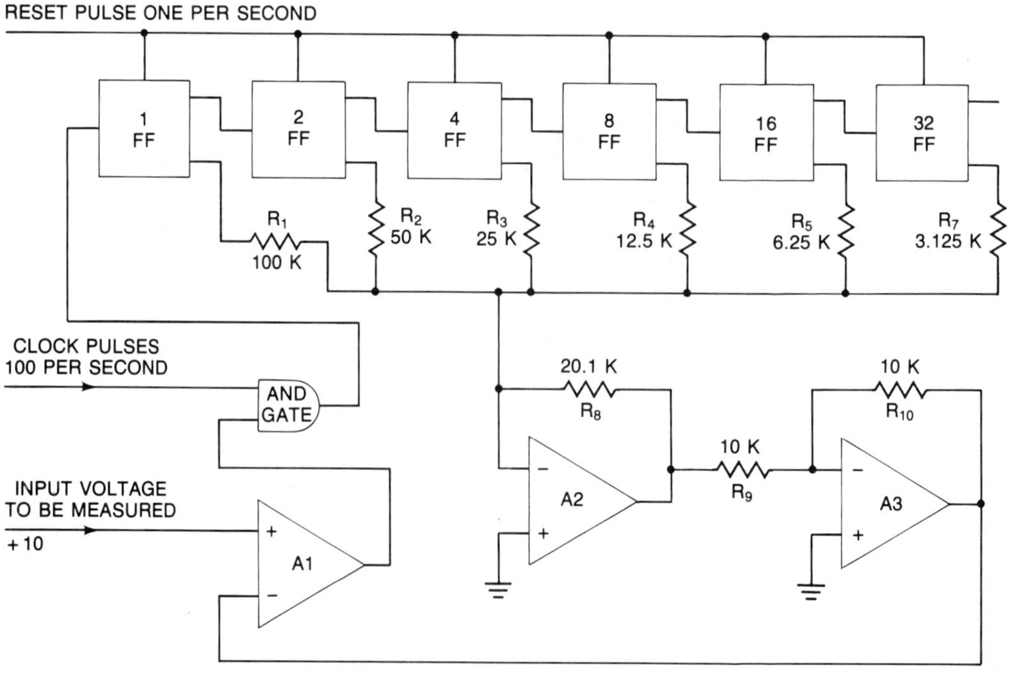


**Figure 20–8** Digital voltmeter.

4. $A_2$ received scaled voltage from each flip-flop of the counter. The output of an FF is +5 volts if that FF is in the 1-state.
5. $A_3$ is a simple inverter producing positive voltage at its output.

The voltmeter functions as the counter starts after a reset pulse. The counter will count up until the comparator $A_2$ switches over and produces a negative voltage at its output. The negative voltage at the output of $A_1$ is fed to the AND gate, which stops producing pulses at its output. Without pulses at the output of the AND gate, the counter cannot count further.

Assume +10 volts at the voltmeter input terminal. The positive input causes the comparator output to go to positive saturation. Positive out of the comparator and into the AND gate allows positive pulses through the AND gate and the counter starts counting up.

When the count reaches 9, $FF_8$ and $FF_1$will be in the 1-state. Input voltages to the amplifier $A_2$ will be +5 volts through each scaling resistor. The +5 volts input through $R_1$, 100k, will produce −1.005 volts at $A_2$ output.

The +5 volts input through $R_4$, 12.5k, will produce −8.04 volts at $A_2$ output. The sum of these two voltages is −9.045 volts. The inverter changes this to +9.045 volts, which is fed to the comparator. With +10 volts at its +input terminal and +9.045 volts at its −input terminal, the comparator will still produce +saturation output.

The counter will count one more step where $FF_1$ goes to 0 and $FF_2$ goes to 1. Now the input to $A_2$ changes. $FF_8$ still produces −8.04 volts at $A_2$ output, but now the input from $FF_2$ is producing −2.01 volts at $A_2$ output. The sum of −8.04 and −2.01 is −10.05. This voltage is inverted by $A_3$ and fed to the comparator.

The comparator now sees a slightly more positive voltage at the inverting input. The comparator output switches to negative saturation. This negative voltage fed to the AND gate shuts it down, stopping further pulses from getting to the counter. The counter stops at 10, indicating the input voltage to be measured.

## Automobile Third Stop Light

New automobiles now have a third stop light installed at the rear window. This light is at the optimum level for viewing by drivers approaching from the rear. In newer autos, the light is connected directly to the brake switch. In older models, with an add-on unit, the light is controlled from both rear brake lights.

Both brake lights must be considered to ensure that the light is not turned on by the directional signal. A simple circuit that will accomplish the required recognition of signals is shown in Figure 20–9.

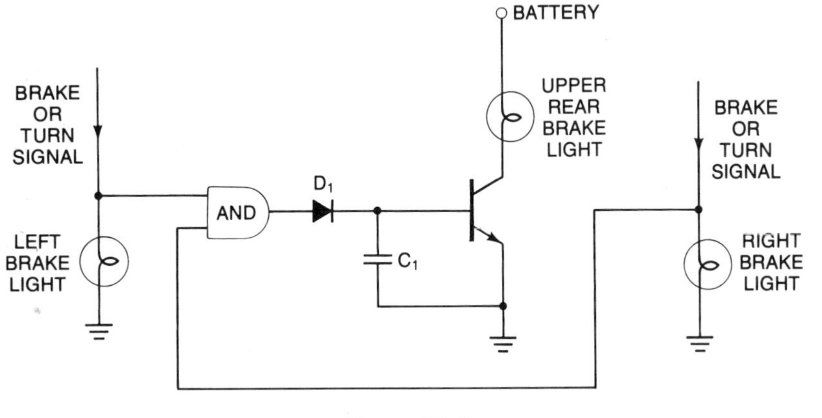


**Figure 20–9**

Three conditions must be considered that are associated with the lighting of a brake light. They are:

1. Brake pedal depressed alone.
2. Brake pedal depressed with turn signal engaged.
3. Turn signal engaged alone.

In conditions 1 and 2 above, the upper brake light should turn on. In condition 3, the upper light should not turn on.

*Condition 1.* When the brake pedal is depressed, both brake lights receive positive supply voltage directly through the brake switch. The AND gate will have + at both input terminals. It will produce + at the output. This turns $Q_1$ on, which will in turn light the upper rear brake light (Figure 20–9.)

*Condition 2.* When the brake pedal is depressed and a turn signal is on, one input to the AND gate will be steady + voltage while the other will be a pulsating on-off + voltage at a frequency usually above 1 pulse per second. The output of the AND gate would be positive pulses at the same frequency as the input. Capacitor $C_1$ would charge through diode $D_1$ to the pulse output peak amplitude. Capacitor $C_1$ would discharge through $R_1$ and the emitter-base junction of $Q_1$. The discharge time of $C_1$ must be greater than the pulse repetition time of the turn signal if the upper rear brake light is not to blink.

*Condition 3.* If a turn signal is activated alone, there will be a single input to the AND gate. There will not be an output from the AND gate. $Q_1$ and the upper rear brake light will remain off.

## REVIEW QUESTIONS

1. The difference between digital and analog circuits is that analog operation is by (measurement, counting) ______ whereas digital operation is by (measurement, counting) ______.
2. In the binary counting system, the number 11101 base 2 is equal to ______ base 10.
3. Add these digital numbers:

   $$\begin{array}{r} 11011 \\ +\ 10111 \\ \hline \end{array}$$

4. Subtract 1001 from 11001.
5. Draw the symbol for an AND gate.
6. Draw the symbol for an OR gate.
7. Draw the symbol for an inverter.
8. The standard circuit used for counting in a digital circuit is called a ______ ______.
9. The circuit used to store numbers in Figure 20–7 is a ______ ______.
10. In Figure 20–8, if +13 volts dc were the voltage being measured, which flip-flops would be providing +5-volt inputs to amplifier $A_2$?

# UNIT TWENTY-ONE

# AC METERS AND THE OSCILLOSCOPE

One of the most important instruments used to measure voltages is the oscilloscope. Steady-state dc voltages are easily measured using standard dc meters. Whenever a measurement must be made of a changing voltage, the oscilloscope is the proper instrument to use. Exceptions are the measurement of ac sine-wave voltages. If it is known that the voltage to be measured is in the form of a sine wave, a standard ac voltmeter may be used. (*Note*: ac meters will be covered later in this unit.) In situations where the shape of the voltage is uncertain, an oscilloscope must be used in order to obtain accurate amplitude and time information. Consider trying to use a voltmeter to obtain information about a waveform as shown in Figure 21–1, which is a varying positive dc voltage. The voltmeter would indicate something close to 8.5 volts positive. This is an accurate indication of the average voltage but it tells nothing about the variation in amplitude of the signal. If the voltage were connected to a device that had a maximum voltage rating of 10 volts, the device could be destroyed.

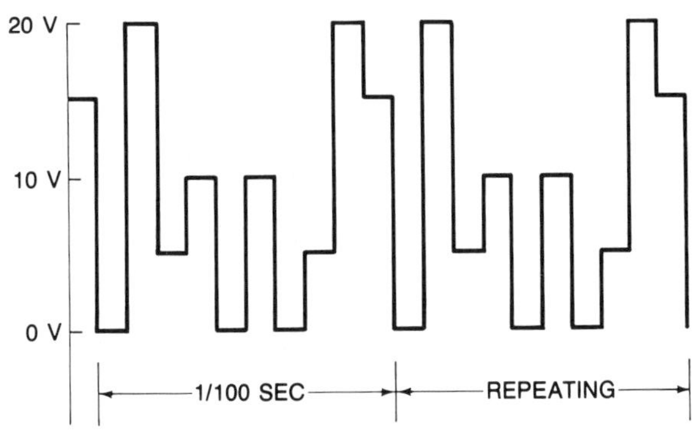


**Figure 21–1**

## THE CATHODE-RAY TUBE

The heart of the oscilloscope is the *cathode-ray tube* (CRT). The CRT is one of the few vacuum-tube holdouts from the vacuum-tube heyday (early 1900s to about 1965). Semiconductor devices now take the place of vacuum tubes in most applications. A simplified diagram of the cathode-ray tube is shown in Figure 21–2.

The *filament* is used to heat the cathode to a temperature where the cathode glows brightly. The cathode is coated with a material that gives up electrons readily when heated. Great numbers of electrons are boiled off the cathode.

### Control Grid

The *control grid* is in the form of a screen between the cathode and the focus anode. The control grid is used to fix the number of electrons entering the focus anode. This is accomplished by the application of a given potential to the grid. By controlling the number of electrons in the beam, the brightness or intensity of the cathode-ray tube presentation can be determined.

A negative potential is applied to the control grid during retrace time. The negative potential on the grid blocks the flow of electrons and thereby blanks out the presentation during this time period.

The *focus anode* is a cylinder closed at one end, with a single hole at the center of the closure. Electrons are drawn from the cathode, formed into a beam, and exit the focus anode through the small hole at the center of the closed end. (See Figure 21–2.)

The *accelerating anode*, shaped like the focus anode, has a higher positive potential applied to it. The electron beam exiting the focus anode is accelerated by the positive voltage. The high-speed electron beam passes through the accelerating anode to continue on to the face of the cathode-ray tube.

The heater (filament), cathode, focus anode, and accelerating anode are usually constructed as a unit and fitted into the remainder of the cathode-ray tube. As a unit, the heater, cathode, focus anode, and accelerating anode are called the *electron gun*.

Electrons leaving the electron gun next pass between the *vertical deflection plates*. The plates as shown in Figure 21–2 are set in the horizontal plane, but voltages applied to these plates will cause deflection to the electron beam in the vertical direction.

After passing through the vertical deflection plates, the electron beam passes

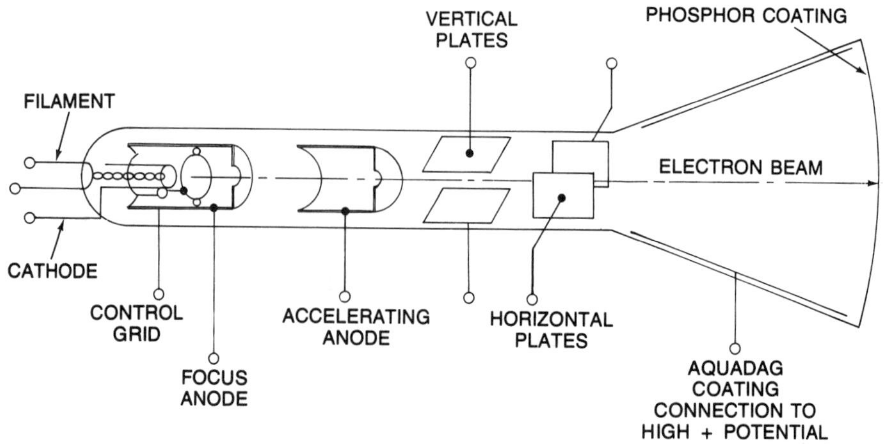


**Figure 21–2** Cathode-ray tube.

through the region of the *horizontal deflection plates*. Voltages supplied to these plates cause deflection across the face of the scope horizontally.

The electron beam continues on until it strikes the rear face of the CRT. The phosphor coating on the rear face glows brightly when impacted by high-speed electrons. If both the vertical and horizontal deflection plates are at ground potential, the electron beam should form a bright spot at the center of the CRT.

## DEFLECTION VOLTAGES

One of the reasons the CRT is important as a measurement device is that it can be operated to indicate things that are occurring at high speed.

Instruments that contain meter movements can only move so fast. Inertia limits the speed of movement. Of course, if a meter movement were moving very fast, the human eye would probably not determine the amount of movement. In the CRT, because electrons are very light, they can be deflected at very high speeds with almost no concern about inertia.

The deflected electron beam in an oscilloscope is displayed one presentation on top of another. The pattern of a repetitious waveform is shown as a steady display on the oscilloscope.

## DEFLECTION MEASUREMENT

A standard oscilloscope will have a *graticule* in front of the CRT face in order to determine the amount of deflection. The graticule is normally marked off in centimeters, as shown in Figure 21–3.

The graticule depicted in Figure 21–3 has 10 cm horizontally and 8 cm vertically. Deflection sensitivity on the front panel of the oscilloscope is usually given as volts-per-cm-deflection vertically and time-per-cm-deflection horizontally. Internally, of course, the deflection of the CRT electron beam is dependent on voltage, not time.

### Horizontal Deflection

Internally, the sensitivity of the horizontal plates is given in volts-per-centimeter of deflection. Assume the CRT shown in Figure 21–2 has a horizontal deflection sensitivity of 40 volts per cm. Now observe Figure 21–4.

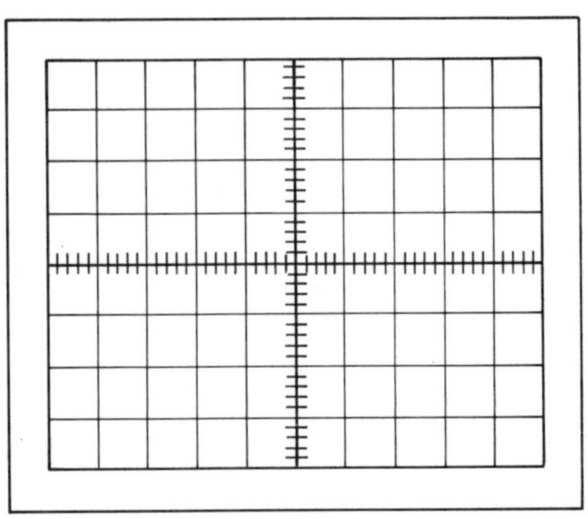

**Figure 21–3** Graticule markings.

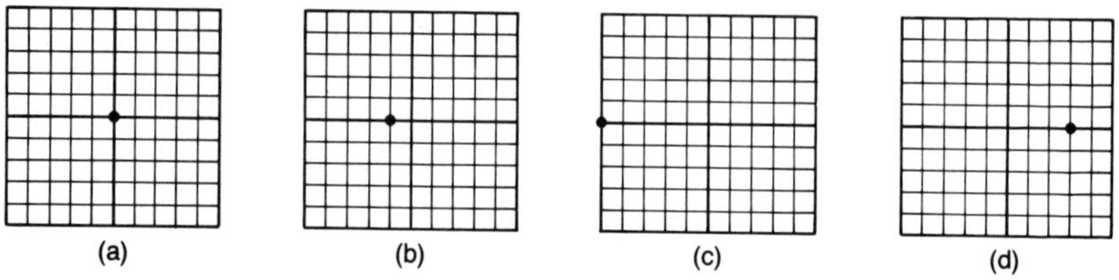


**Figure 21–4**

1. Figure 21–4a shows the electron beam centered. No voltage is applied to the horizontal deflection plates.
2. Figure 21–4b shows the electron beam deflected 1 cm left. A positive 20 volts is applied to the left horizontal plate while a negative 20 volts is applied to the right horizontal plate.
3. Figure 21–4c shows the electron beam deflected 5 cm to the left. A positive 100 volts is applied to the left horizontal plate and a negative 100 volts is applied to the right horizontal plate.
4. Figure 21–4d shows the electron beam deflected 3 cm to the right of center. A negative 60 volts is applied to the left horizontal plate, and a positive 60 volts is applied to the right horizontal plate.

In normal operation, the horizontal deflection in an oscilloscope is accomplished by using precision sawtooth voltage waveforms. The waveforms are precise in both amplitude and time. Now consider the two waveforms shown in Figure 21–5.

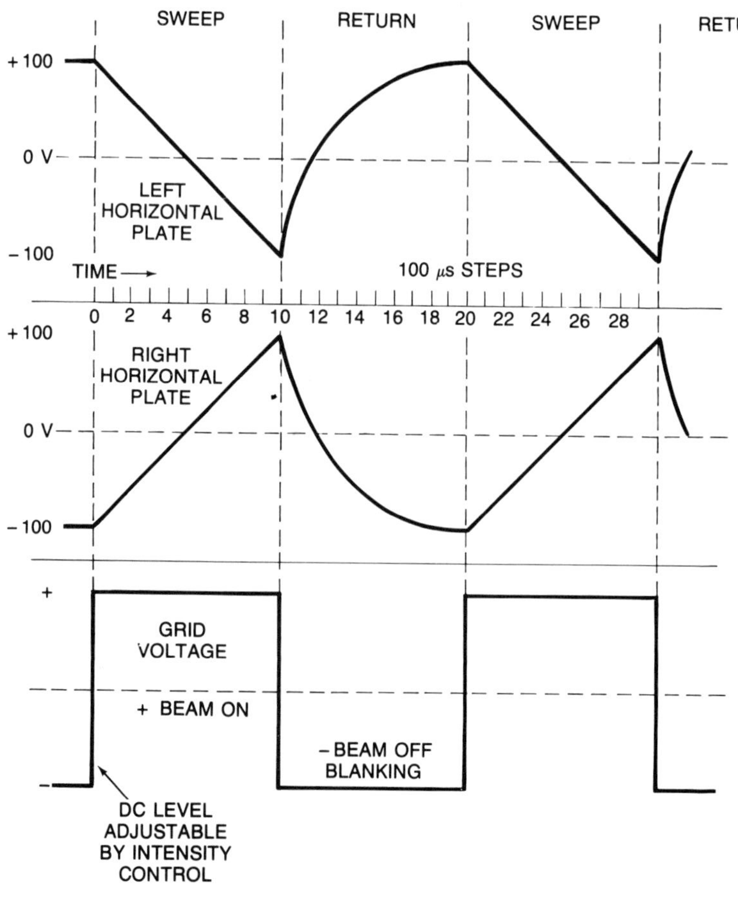


**Figure 21–5**

*Time Periods.* At the starting point, position 0, the left plate is at +100 and the right plate is at −100. The electron beam is deflected 5 cm toward the left side (200 volts applied, sensitivity 40 V/cm). The following list shows what happens at each time period.

1. At time period 1, 100 microseconds after 0, the voltage at the left plate has decreased to +80 volts and the voltage at the right plate is at −80 volts (−80 to +80 = 160 volts difference). The electron beam will move from the 5-cm position to the 4-cm position left of center. Because the voltage is decreasing linearly, the electron beam deflection will be linear.
2. At time period 2, the voltages are +60 volts left and −60 volts right for a total of 120 volts. The electron beam moves linearly to 3 cm left of center.
3. At time period 3, the voltages are +40 volts left and −40 volts right. The beam moves to 2 cm left of center.
4. At time period 4, the voltages are +20 volts left and −20 volts right. The beam moves to 1 cm left of center.
5. At time period 5, both plates have 0 volts. The beam moves to the center position.
6. At time period 6, the voltages are −20 volts left and +20 volts right. The beam moves to 1 cm right of center.
7. At time period 7, the voltages are −40 volts left and +40 volts right. The beam moves to 2 cm right of center.
8. At time period 8, the voltages are −60 volts left and +60 volts right. The beam moves to 3 cm right of center.
9. At time period 9, the voltages are −80 volts left and +80 volts right. The beam moves to 4 cm right of center.
10. At time period 10, the voltages are −100 volts left and +100 volts right. The beam moves to 5 cm right of center.
11. to 20. The aiming of the electron beam moves from full right to full left of center.

During time periods 1 through 10, the grid of the CRT is held positive, allowing electrons to flow to the CRT screen. During time periods 10 through 20, the grid of the CRT is held negative, cutting electron flow off and blanking the CRT screen. The presentation, without vertical deflection voltage, will be a straight line across the face of the CRT, as shown in Figure 21–6.

### Vertical Deflection

A deflection voltage may be applied to the vertical plate at the same time as the application of the horizontal deflection voltage, as shown in Figure 21–5.

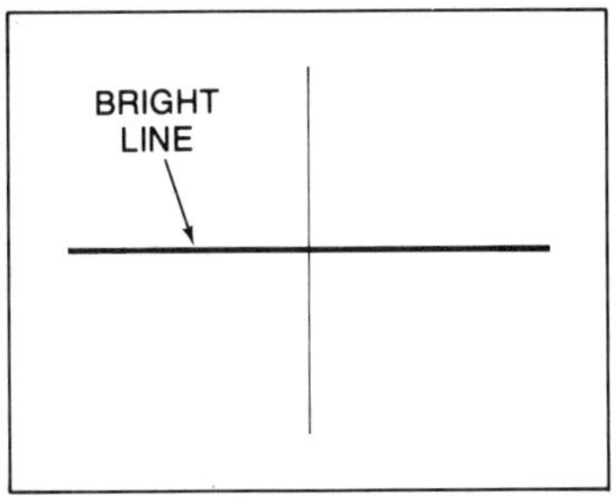


**Figure 21–6**

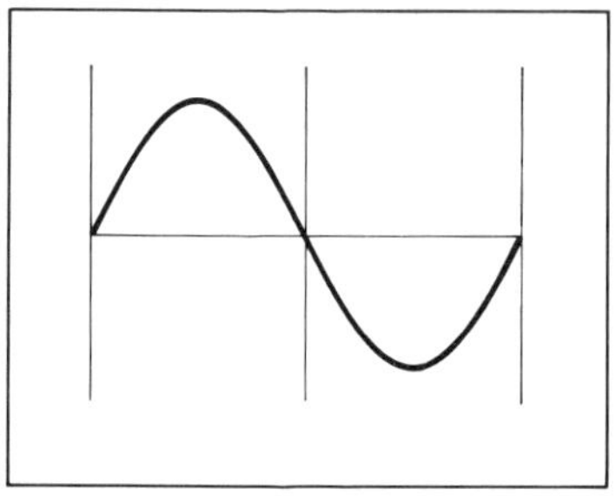

Figure 21–7

Assume a 1,000 Hz sine wave is applied to the vertical deflection plates. The vertical deflection plates are more sensitive than the horizontal plates because they are closer to the electron gun in the CRT. A deflection imparted to the electron beam acts over a longer distance. For this discussion, a sensitivity of 20 volts per cm may be assumed.

The 1,000-hertz signal impressed on the vertical plates causes deflection of the electron vertically at the same time the sawtooth voltages are deflecting the beam horizontally. The presentation on the oscilloscope will be as shown in Figure 21–7.

The presentation on the face of the CRT is one cycle of the 1,000-hertz signal. The second cycle is blanked out. The next cycle of the 1,000-hertz signal is presented directly over the first, thus providing a stable sine wave on the CRT face.

## *SYNCHRONIZATION*

In order for the presentation on the oscilloscope to appear stable, it is necessary that the triggering of the start of the horizontal sweep sawtooth be synchronized to some point on the signal presented to the vertical plates. A comparator circuit is included in the oscilloscope to accomplish this. A simplified version of a synchronization trigger circuit is shown in Figure 21–8.

Assume the start of the sawtooth waveform is triggered by a negative-going input. With the trigger level control set at 0 volts, the comparator output will be positive while the vertical input signal is negative. As soon as the vertical signal

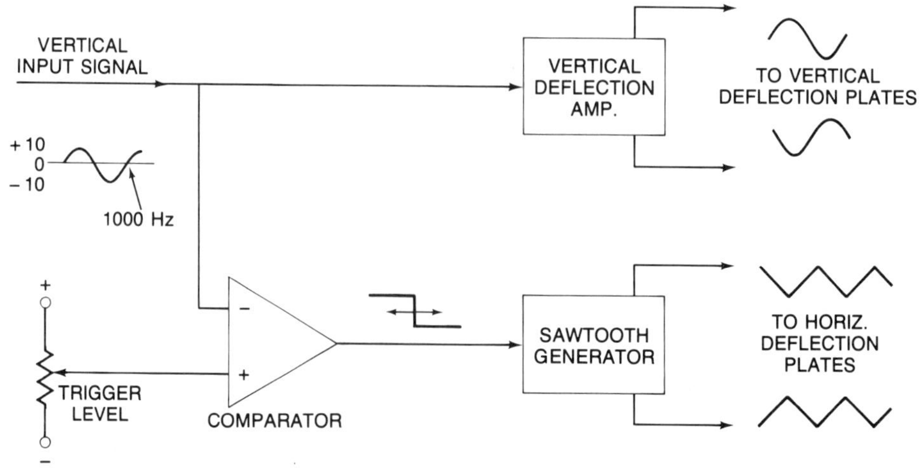


Figure 21–8

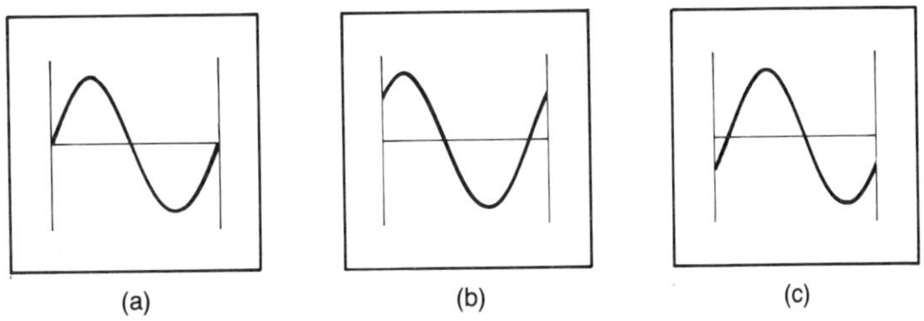


**Figure 21–9**

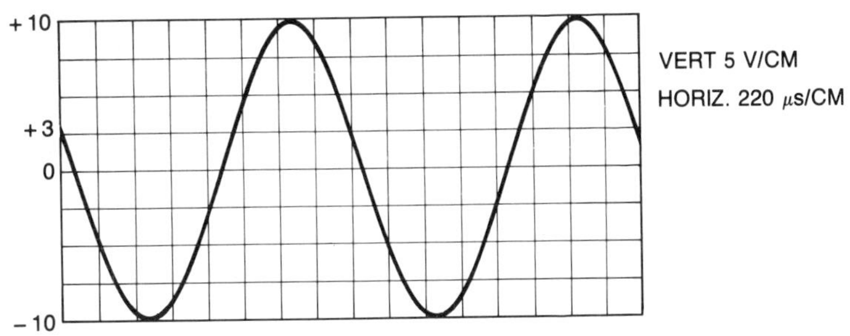


**Figure 21–10**

passes through 0, the comparator output will switch to negative. This will trigger the oscilloscope horizontal sweep circuit set at 100 μs/cm, providing a presentation as shown in Figure 21–9a. If the trigger level is set to +5 volts, the input vertical signal will have to rise past +5 volts before the comparator will switch to a negative output triggering the start of the sweep. The presentation will be as in Figure 21–9b.

If the trigger level is set to −3 volts, triggering will take place when the input vertical signal passes through −3 volts. The presentation will be as in Figure 21–9c.

Selection can usually be made for triggering during a rise in the positive direction, as in Figure 21–9, or for triggering during a rise in the negative direction, as shown in Figure 21–10. Here is a 10-volt peak 1,000 Hz sine wave fed to an oscilloscope vertical plate. The scope is set at 5 volts per cm vertical deflection sensitivity and 200 μsec per cm horizontal sweep time. Sweep triggering is set for approximately +3 volts level negative slope.

## *HORIZONTAL SWEEP SPEED*

Modern oscilloscopes usually provide for control of the horizontal sweep in time per centimeter of deflection. Sweep speeds from 1 sec/cm to 0.1 μsec/cm are relatively common. Specialized oscilloscopes are available with extremely high sweep speeds.

### Oscilloscope Operation

An oscilloscope should only be operated after reviewing the manufacturer's operating instruction manual. The operating instruction manual will include a list of front-panel controls and an explanation of their uses.

The specifications of the instrument are also included in the instruction manual. It is important to know the limits of the test equipment being used.

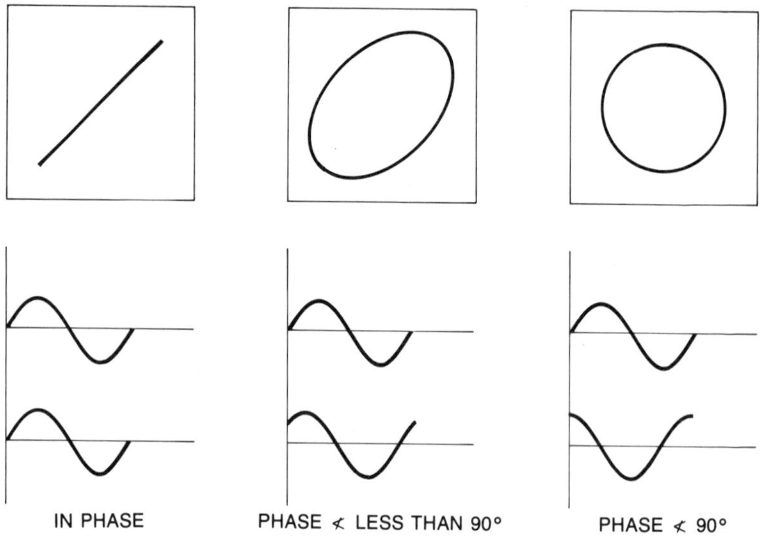


**Figure 21–11**

### Lissajous Figures

If the sweep circuits are disconnected from the horizontal plates and access is made to them directly, comparison of two signals may be made in an oscilloscope. A comparison of frequencies and/or phase angles may be accomplished using this technique. The presentation on the oscilloscope face is called a *lissajous figure.*

Suppose two signals of the same frequency are connected to the two deflection systems. The presentation on the scope face will be as shown in Figure 21–11.

If the frequency of the signals is a multiple of each other (an harmonic), the waveforms may appear as in Figure 21–12. The frequency ratio is equal to the number of times the electron beam extends to one horizontal plate as to the number of times it extends to one vertical plate.

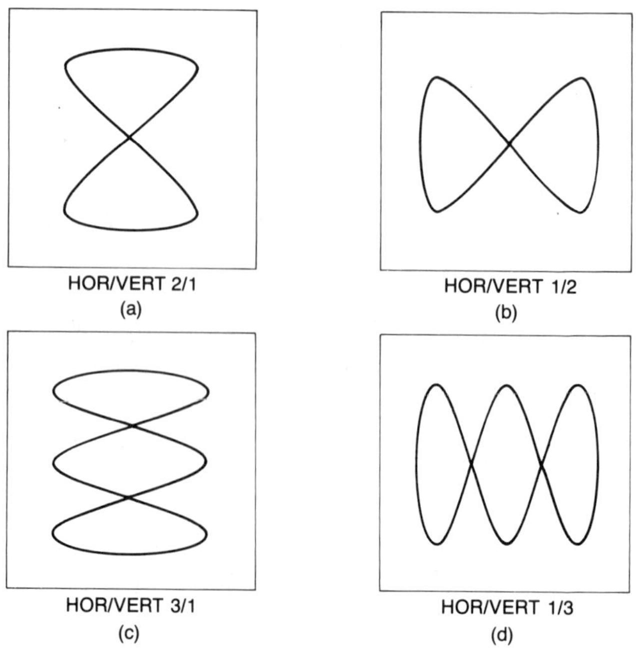


**Figure 21–12**

It is obvious that if signals of different wave shapes are combined in the oscilloscope, a variety of patterns will result.

## ALTERNATING CURRENT METERS

The instruments shown in Unit 6 of this text use permanent magnet meter movements. Meters of this type are capable of indicating dc levels. A standard 60-hertz signal connected to a permanent magnet meter movement would cause the movement to vibrate. It would not indicate voltage or current properly.

### Bridge Rectifier (ac)

A *bridge rectifier* is commonly used with permanent magnet meter movements to change ac signals to dc before application to the movement (see Figure 21–13). The meter responds to current flow through the meter in the same direction for both alternations of the input.

### Iron Vane Movement

A rugged meter movement that responds to low-frequency ac signals is the *iron vane meter*. The movement is essentially solenoid action against a spring tension. A simplified version of the instrument is shown in Figure 21–14.

As an increasing voltage is applied to the coil, thc iron vane is pulled farther into the coil. The spring inside the coil provides pressure pushing out. The pointer attached to the iron vane moves in response to the strength of the signal fed to the coil.

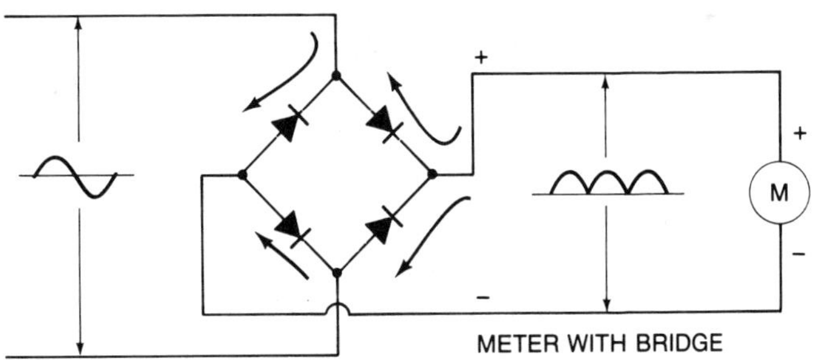


**Figure 21–13**

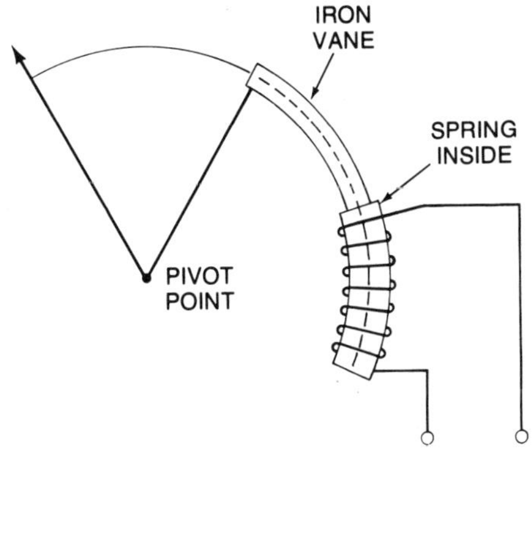


**Figure 21–14** Iron vane meter.

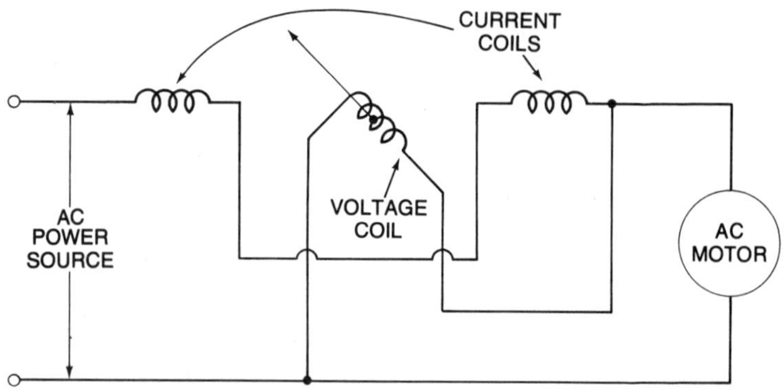


**Figure 21–15** Dynamometer used as a wattmeter.

## *ELECTRODYNAMOMETER*

If the permanent magnets of the meters shown in Unit 6 were replaced with electromagnets, an instrument suitable for use with ac signals would be available. This is called the *electrodynamometer*. The meter is not in common use in measuring voltage or current. Its present application is as a wattmeter. Shown in Figure 21–15 is an electrodynamometer connected as a wattmeter.

The meter responds to average power. The effect of phase angle on power is compensated in the meter as it continuously responds to instantaneous values of voltage and current and thus provides the average power output. When used in a 60-hertz circuit, the inertia of the meter movement smoothes out the reading.

## REVIEW QUESTIONS

1. The oscilloscope is set on 10 μsec/cm. What is the frequency of the waveform shown in Figure 21–16a?
2. The vertical deflection sensitivity is set at 10 volts per cm. What is the effective value of the voltage shown in Figure 21–16a ?
3. The horizontal sweep speed of the oscilloscope shown in Figure 21–16b is 50 μsec/cm. What is the frequency of the signal shown?
4. In Figure 21–16b, what is the amplitude of voltage shown 300 μsec after the waveform start?
5. In Figure 21–12a, if the frequency applied to the vertical plates is 200 Hz, what is the frequency of the horizontal signal?
6. In Figure 21–12d, if the frequency of the signal applied to the horizontal plate is 333 Hz, what is the frequency of the vertical signal?
7. What is the instantaneous value of voltage 75 sec after the start of the signal in Figure 21–16?
8. If the sweep speed were changed to 20 μsec/cm, how many cycles of the waveform shown in Figure 21–16b would appear on the oscilloscope face?

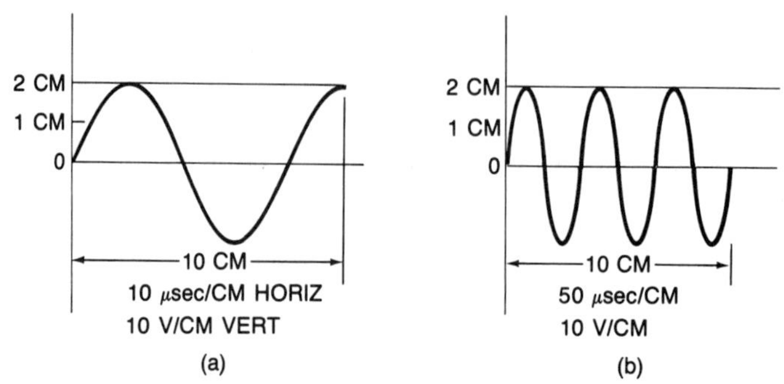


**Figure 21–16**

# APPENDIX A

# POWERS OF TEN

The use of the powers of ten provides for mathematical operations using very large or very small numbers. In electronics most calculations are held to three significant digits, that is, in the number 3,262,137,812, only the 326 and the position of the decimal would be important in most situations. Electronic components and measurements are usually only accurate to three significant digits.

Considering three-significant-digit use in electronics makes the powers of ten very important. The number 3,262,137,812 becomes $3.26 \times 10^9$. The decimal is simply moved to the point providing a single digit whole number. The number of places the decimal has been moved provides the power of ten. For numbers less than one, say, 0.0000000192, this becomes $1.92 \times 10^{-8}$.

Numbers with a power of ten may be added if the numbers have the same power of ten.

$$\begin{array}{r} 2.13 \times 10^4 \\ +1.33 \times 10^4 \\ \hline 3.46 \times 10^4 \end{array}$$

If the powers of ten are different, they must be made the same before adding.

---

**EXAMPLE 1**

Add $(2.14 \times 10^6)$ plus $(1.21 \times 10^5)$.

***Solution:*** Change $1.21 \times 10^5$ to $0.121 \times 10^6$.

$$\begin{array}{r} 0.121 \times 10^6 \\ +2.14\phantom{0} \times 10^6 \\ \hline 2.26\phantom{0} \times 10^6 \end{array}$$

---

The same conditions exist in subtraction.

---

**EXAMPLE 2**

Subtract $8.37 \times 10^5$ from $1.28 \times 10^6$.

***Solution:*** Change $1.28 \times 10^6$ to $12.80 \times 10^5$.

$$\begin{array}{r} 12.80 \times 10^5 \\ -8.37 \times 10^5 \\ \hline 4.43 \times 10^5 \end{array}$$

---

When powers of ten are used in multiplication, the numbers are multiplied and the powers of ten are added.

---

**EXAMPLE 3**

Multiply $(3.11 \times 10^4)$ times $(2.06 \times 10^2)$.

***Solution:***

$$\begin{array}{rl} 3.11 & \times 10^4 \\ \times 2.06 & \times 10^2 \\ \hline 6.4066 & \times 10^6 \end{array} \quad \text{or} \quad 6.41 \times 10^6$$

---

**EXAMPLE 4**

Multiply $(2.17 \times 10^{-4})$ times $(4.23 \times 10^3)$.

***Solution:***

$$\begin{array}{rl} 2.17 & \times 10^{-4} \\ \times 4.23 & \times 10^3 \\ \hline 9.1791 & \times 10^{-1} \end{array} \quad \text{or} \quad 9.18 \times 10^{-1}$$

---

In division, the number is divided by the divisor and the powers of ten are subtracted.

---

**EXAMPLE 5**

Divide $(6.11 \times 10^6)$ by $(2.07 \times 10^3)$.

***Solution:***

$$(6.11 \times 10^6) \div (2.07 \times 10^3) = 2.95 \times 10^3 \quad \text{or} \quad 2950$$

---

**EXAMPLE 6**

Divide $(3.78 \times 10^5)$ by $(2.16 \times 10^{-3})$.

***Solution:***

$$(3.78 \times 10^5) \div (2.16 \times 10^{-3}) = 1.75 \times 10^8$$

Powers of Ten

| | |
|---|---|
| $10^{10}$ | 10,000,000,000 |
| $10^9$ Giga | 1,000,000,000 |
| $10^8$ | 100,000,000 |
| $10^7$ | 10,000,000 |
| $10^6$ Mega | 1,000,000 |
| $10^5$ | 100,000 |
| $10^4$ | 10,000 |
| $10^3$ Kilo | 1,000 |
| $10^2$ | 100 |
| $10^1$ | 10 |
| $10^0$ | 1 |
| 0 | 0 |
| $10^{-1}$ | 0.1 |
| $10^{-2}$ | 0.01 |
| $10^{-3}$ Milli | 0.001 |
| $10^{-4}$ | 0.0001 |
| $10^{-5}$ | 0.00001 |
| $10^{-6}$ Micro | 0.000001 |
| $10^{-7}$ | 0.0000001 |
| $10^{-8}$ | 0.00000001 |
| $10^{-9}$ Nano | 0.000000001 |
| $10^{-10}$ | 0.0000000001 |

# APPENDIX B

## *PERIODIC TABLE*

# PERIODIC TABLE

| Group | I | II | III | IV | V | VI | VII | VIII | | | 0 |
|---|---|---|---|---|---|---|---|---|---|---|---|
| | Mono-Valent | Bi-Valent | Tri-Valent | Tetra-Valent | Penta-Valent | Hexa-Valent | Hepta-Valent | | | | Inert |
| 1 | 1<br>H<br>Hydrogen | | | | | | | | | | 2<br>He<br>Helium |
| 2 | 3<br>Li<br>Lithium | 4<br>Be<br>Beryllium | 5<br>B<br>Boron | 6<br>C<br>Carbon | 7<br>N<br>Nitrogen | 8<br>O<br>Oxygen | 9<br>F<br>Fluorine | | | | 10<br>Ne<br>Neon |
| 3 | 11<br>Na<br>Sodium | 12<br>Mg<br>Magnesium | 13<br>Al<br>Aluminum | 14<br>Si<br>Silicon | 15<br>P<br>Phosphorus | 16<br>S<br>Sulfur | 17<br>Cl<br>Chlorine | | | | 18<br>Ar<br>Argon |
| 4 A | 19<br>K<br>Potassium | 20<br>Ca<br>Calcium | 21<br>Sc<br>Scandium | 22<br>Ti<br>Titanium | 23<br>V<br>Vanadium | 24<br>Cr<br>Chromium | 25<br>Mn<br>Manganese | 26<br>Fe<br>Iron | 27<br>Co<br>Cobalt | 28<br>Ni<br>Nickel | |
| 4 B | 29<br>Cu<br>Copper | 30<br>Zn<br>Zinc | 31<br>Ga<br>Gallium | 32<br>Ge<br>Germanium | 33<br>As<br>Arsenic | 34<br>Se<br>Selenium | 35<br>Br<br>Bromine | | | | 36<br>Kr<br>Krypton |
| 5 A | 37<br>Rb<br>Rubidium | 38<br>Sr<br>Strontium | 39<br>Y<br>Yttrium | 40<br>Zr<br>Zirconium | 41<br>Nb<br>Niobium | 42<br>Mo<br>Molybdenum | 43<br>Tc<br>Technetium | 44<br>Ru<br>Ruthenium | 45<br>Rh<br>Rhodium | 46<br>Pd<br>Palladium | |
| 5 B | 47<br>Ag<br>Silver | 48<br>Cd<br>Cadmium | 49<br>In<br>Indium | 50<br>Sn<br>Tin | 51<br>Sb<br>Antimony | 52<br>Te<br>Tellurium | 53<br>I<br>Iodine | | | | 54<br>Xe<br>Xenon |
| 6 A | 55<br>Cs<br>Cesium | 56<br>Ba<br>Barium | 57-71*<br>La<br>Lanthanum | 72<br>Hf<br>Hafnium | 73<br>Ta<br>Tantalum | 74<br>W<br>Tungsten | 75<br>Re<br>Rhenium | 76<br>Os<br>Osmium | 77<br>Ir<br>Iridium | 78<br>Pt<br>Platinum | |
| 6 B | 79<br>Au<br>Gold | 80<br>Hg<br>Mercury | 81<br>Tl<br>Thallium | 82<br>Pb<br>Lead | 83<br>Bi<br>Bismuth | 84<br>Po<br>Polonium | 85<br>At<br>Astatine | | | | 86<br>Rn<br>Radon |
| 7 | 87<br>Fr<br>Francium | 88<br>Ra<br>Radium | 89<br>**Ac<br>Actinium | | | | | | | | |

* LANTHANIDES

| | | | | | | | | | |
|---|---|---|---|---|---|---|---|---|---|
| | 58<br>Ce<br>Cerium | 59<br>Pr<br>Praseodymium | 60<br>Nd<br>Neodymium | 61<br>Pm<br>Promethium | 62<br>Sm<br>Samarium | 63<br>Eu<br>Europium | 64<br>Gd<br>Gadolinium | 65<br>Tb<br>Terbium | 66<br>Dy<br>Dysprosium |
| | 67<br>Ho<br>Holmium | 68<br>Er<br>Erbium | 69<br>Tm<br>Thulium | 70<br>Yb<br>Ytterbium | 71<br>Lu<br>Lutetium | | | | |

**ACTINIDES

| | | | | | | | | | |
|---|---|---|---|---|---|---|---|---|---|
| | 90<br>Th<br>Thorium | 91<br>Pa<br>Protoactinium | 92<br>U<br>Uranium | 93<br>Np<br>Neptunium | 94<br>Pu<br>Plutonium | 95<br>Am<br>Americium | 96<br>Cm<br>Curium | 97<br>Bk<br>Berkelium | 98<br>Cf<br>Californium |
| | 99<br>Es<br>Einsteinium | 100<br>Fm<br>Fermium | 101<br>Md<br>Mendelevium | 102<br>No<br>Nobelium | | | | | |

# APPENDIX C

# *REACTANCE NOMOGRAPH*

The *nomograph of reactance* versus frequency, with indications of capacitance and inductance, is plotted on log paper, six cycles horizontal and five cycles vertical.

The nomograph may be used to provide approximate reactance of capacitors from 1 pf ($1 \times 10^{-12}$ farads) to 100 μf and inductors from 0.1 μH to 10 henrys. The reactance of the component may be determined for a frequency range of from 100 hertz to 100 megahertz if the reactance is between 1.0 ohms and 100,000 ohms.

The nomograph will also indicate compatible values for resonance at any of the included frequencies. Simply follow a vertical line at the concerned frequency. The line will intersect inductance and capacitance values providing resonance at that frequency. An interpolation scale for L and C is provided.

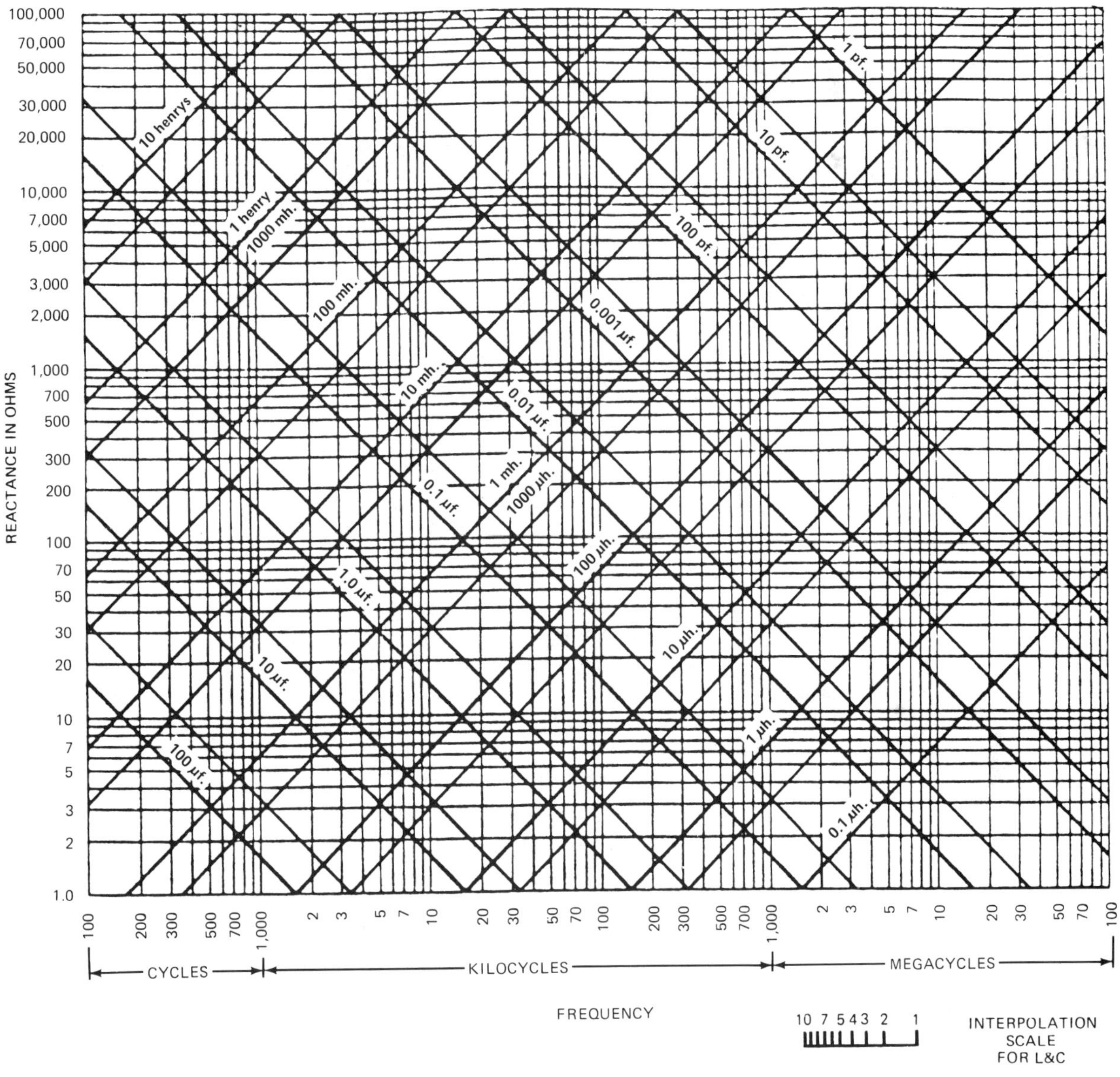
REACTANCE IN OHMS
100,000
70,000
50,000
30,000
20,000
10,000
7,000
5,000
3,000
2,000
1,000
700
500
300
200
100
70
50
30
20
10
7
5
3
2
1.0
10 henrys
1 henry
1000 mh.
100 mh.
10 mh.
1 mh.
1000 µh.
100 µh.
10 µh.
1 µh.
0.1 µh.
1 pf.
10 pf.
100 pf.
0.001 µf.
0.01 µf.
0.1 µf.
1.0 µf.
10 µf.
100 µf.
100 200 300 500 700 1,000 2 3 5 7 10 20 30 50 70 100 200 300 500 700 1,000 2 3 5 7 10 20 30 50 70 100
CYCLES
KILOCYCLES
MEGACYCLES
FREQUENCY
10 7 5 4 3 2 1
INTERPOLATION SCALE FOR L&C

**Figure C–1**

# APPENDIX D

# TRIGONOMETRIC FUNCTIONS

## Right-Triangle Trigonometry

A triangle has three sides and three angles. If any three factors, including at least one side, are known, the remaining sides and angles can be determined. This is true because, given a side and two angles or two sides and one angle, the size and shape of the triangle is fixed.

*Proportional Sides.* In triangle ABC of Figure D–1, the side AB is called the *hypotenuse*. It is the longest side.

The hypotenuse AB would be designated c because it is opposite the ∢ C.

In triangle ABC, the side AC is designated as the b side because it is opposite ∢ B.

In triangle ABC, the side BC is designated side a because it is opposite ∢ A.

If a line perpendicular to side b were drawn to the hypotenuse, a second triangle would be formed, as in Figure D–2.

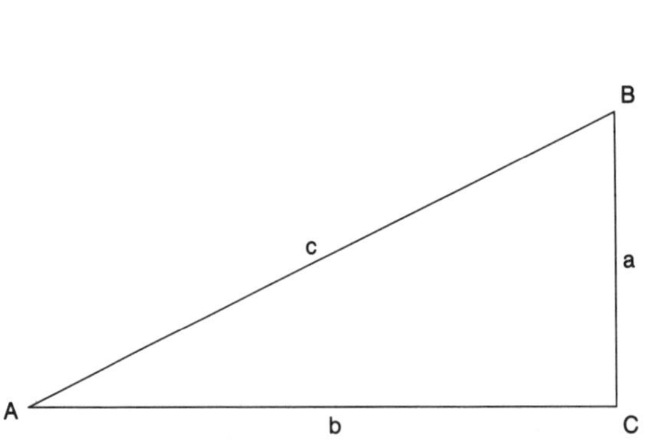


**Figure D–1**

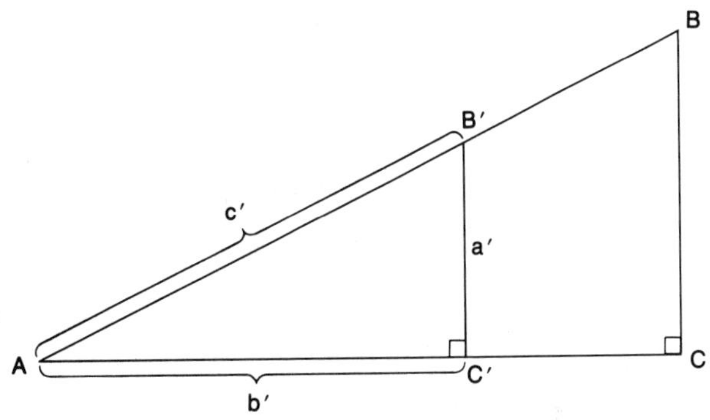


**Figure D–2**

In the triangle A B′C′, angle A is common to both triangle A B′C′ and triangle ABC. Angle C is a right angle as is angle C′. As all triangles contain 180° in the included angles, angle B′ must equal angle B.

In the triangle ABC, side a is proportional to side a′ in triangle A B′ C′. Side c is proportional to c′ as side b is proportional to side b′.

The ratio of side a to the hypotenuse c in ABC is the same as the ratio of side a′ to the hypotenuse c′ in A B′ C′.

In trigonometry, the ratio of side a to the hypotenuse is called the *sine of the angle*. The sine is usually abbreviated as *sin*.

$$\sin A = \frac{\text{side opposite}}{\text{hypotenuse}} = \frac{a}{c}$$

Another ratio is the side next to the angle to the hypotenuse. This is called the *cosine* and usually is abbreviated *cos*.

$$\cos A = \frac{\text{side adjacent}}{\text{hypotenuse}} = \frac{b}{c}$$

The third ratio is the side opposite the angle to the side adjacent to the angle. This is called the *tangent* and usually is abbreviated *tan*.

$$\tan A = \frac{\text{side opposite}}{\text{side adjacent}} = \frac{a}{b}$$

For angle B, the ratios are:

$$\sin B = \frac{b}{c}$$

$$\cos B = \frac{a}{c}$$

$$\tan B = \frac{b}{a}$$

The ratios of the sides of right triangles were discovered a long time ago. Tables of these ratios have been developed, and one can be found in this appendix. The tables may be used to find sides and angles of right triangles.

In Figure D-1, if angle A of triangle ABC is 25° and side c is 11 inches, how long are sides a and b?

$$\sin A = \frac{\text{opp}}{\text{hyp}} = \frac{a}{c} = \frac{a}{11}$$

$$0.4226 = \frac{a}{11}$$

$$a = 11 \times 0.4226$$

$$a = 4.649 \text{ in.}$$

$$\cos A = \frac{\text{adj}}{\text{hyp}} = \frac{b}{c} = \frac{b}{11}$$

$$0.9063 = \frac{b}{11}$$

$$b = 9.969 \text{ in.}$$

The tables may be used to determine angles of a triangle.

---

**EXAMPLE 1**

In Figure D–1, if side a is 2 feet long and side c is 4.5 feet long, what is angle A?

***Solution:***

$$\sin A = \frac{\text{opp}}{\text{hyp}} = \frac{a}{c} = \frac{2}{4.5}$$

$$\sin A = 0.4444$$

$$\measuredangle A = 26.4°$$

---

## TABLE OF TRIGONOMETRIC FUNCTIONS

| deg | sin | cos | tan | cot | | deg | sin | cos | tan | cot | | deg | sin | cos | tan | cot | |
|---|---|---|---|---|---|---|---|---|---|---|---|---|---|---|---|---|---|
| 0.0 | 0.0000 | 1.0000 | 0.0000 | ∞ | 90.0 | 5.0 | 0.0872 | 0.9962 | 0875 | 11.430 | 85.0 | 10.0 | 0.1736 | 0.9848 | 0.1763 | 5.671 | 80.0 |
| .1 | .0017 | 1.0000 | .0000 | 573.0 | .9 | .1 | .0889 | .9960 | .0892 | 11.205 | .9 | .1 | .1754 | .9845 | .1781 | 5.614 | .9 |
| .2 | .0035 | 1.0000 | .0033 | 286.5 | .8 | .2 | .0906 | .9959 | .0910 | 10.988 | .8 | .2 | .1771 | .9842 | .1799 | 5.558 | .8 |
| .3 | .0052 | 1.0000 | .0052 | 191.0 | .7 | .3 | .0924 | .9957 | .0928 | 10.780 | .7 | .3 | .1788 | .9839 | .1817 | 5.503 | .7 |
| .4 | .0070 | 1.0000 | .0070 | 143.24 | .6 | .4 | .0941 | .9956 | .0945 | 10.579 | .6 | .4 | .1805 | .9836 | .1835 | 5.449 | .6 |
| .5 | .0087 | 1.0000 | .0087 | 114.59 | .5 | .5 | .0958 | .9954 | .0968 | 10.385 | .5 | .5 | .1822 | .9833 | .1853 | 5.396 | .5 |
| .6 | .0105 | 0.9999 | .0105 | 95.49 | .4 | .6 | .0976 | .9952 | .0981 | 10.199 | .4 | .6 | .1840 | .9829 | .1871 | 5.343 | .4 |
| .7 | .0122 | .9999 | .0122 | 81.85 | .3 | .7 | .0993 | .9951 | .0998 | 10.019 | .3 | .7 | .1857 | .9826 | .1890 | 5.292 | .3 |
| .8 | .0140 | .9999 | .0140 | 71.62 | .2 | .8 | .1011 | .9949 | .1016 | 9.845 | .2 | .8 | .1874 | .9823 | .1908 | 5.242 | .2 |
| .9 | .0157 | .9999 | .0157 | 63.66 | .1 | .9 | .1028 | .9947 | .1033 | 9.677 | .1 | .9 | .1891 | .9820 | .1926 | 5.193 | .1 |
| 1.0 | 0.0175 | 0.9998 | 0.0175 | 57.29 | 89.0 | 6.0 | 0.1045 | 0.9945 | .1051 | 9.514 | 84.0 | 11.0 | 0.1908 | 0.9816 | 0.1944 | 5.145 | 79.0 |
| .1 | .0192 | .9998 | .0192 | 52.08 | .9 | .1 | .1068 | .9943 | .1069 | 9.357 | .9 | .1 | .1925 | .9813 | .1962 | 5.097 | .9 |
| .2 | .0209 | .9998 | .0209 | 47.74 | .8 | .2 | .1080 | .9942 | .1086 | 9.205 | .8 | .2 | .1942 | .9810 | .1980 | 5.050 | .8 |
| .3 | .0227 | .9997 | .0227 | 44.07 | .7 | .3 | .1097 | .9940 | .1104 | 9.058 | .7 | .3 | .1959 | .9806 | .1998 | 5.005 | .7 |
| .4 | .0244 | .9997 | .0244 | 40.92 | .6 | .4 | .1115 | .9938 | .1122 | 8.915 | .6 | .4 | .1977 | .9803 | .2016 | 4.959 | .6 |
| .5 | .0262 | .9997 | .0262 | 38.19 | .5 | .5 | .1132 | .9936 | .1139 | 8.777 | .5 | .5 | .1994 | .9799 | .2035 | 4.915 | .5 |
| .6 | .0279 | .9996 | .0279 | 35.80 | .4 | .6 | .1149 | .9934 | .1157 | 8.643 | .4 | .6 | .2011 | .9796 | .2053 | 4.872 | .4 |
| .7 | .0297 | .9996 | .0297 | 33.69 | .3 | .7 | .1168 | .9932 | .1175 | 8.513 | .3 | .7 | .2028 | .9792 | .2071 | 4.829 | .3 |
| .8 | .0314 | .9995 | .0314 | 31.82 | .2 | .8 | .1184 | .9930 | .1192 | 8.386 | .2 | .8 | .2045 | .9789 | .2089 | 4.787 | .2 |
| .9 | .0332 | .9995 | .0332 | 30.14 | .1 | .9 | .1201 | .9928 | .1210 | 8.264 | .1 | .9 | .2062 | .9785 | .2107 | 4.745 | .1 |
| 2.0 | 0.0349 | 0.9994 | 0.0349 | 28.64 | 88.0 | 7.0 | 0.1219 | 0.9925 | .1228 | 8.144 | 83.0 | 12.0 | 0.2079 | 0.9871 | 0.2126 | 4.705 | 78.0 |
| .1 | .0366 | .9993 | .0367 | 27.27 | .9 | .1 | .1236 | .9923 | .1245 | 8.028 | .9 | .1 | .2096 | .9778 | .2144 | 4.665 | .9 |
| .2 | .0384 | .9993 | .0384 | 26.03 | .8 | .2 | .1253 | .9921 | .1263 | 7.916 | .8 | .2 | .2113 | .9774 | .2162 | 4.625 | .8 |
| .3 | .0401 | .9992 | .0402 | 24.90 | .7 | .3 | .1271 | .9919 | .1281 | 7.806 | .7 | .3 | .2130 | .9770 | .2180 | 4.586 | .7 |
| .4 | .0419 | .9991 | .0419 | 23.86 | .6 | .4 | .1288 | .9917 | .1299 | 7.700 | .6 | .4 | .2147 | .9767 | .2199 | 4.548 | .6 |
| .5 | .0436 | .9990 | .0437 | 22.90 | .5 | .5 | .1305 | .9914 | .1317 | 7.596 | .5 | .5 | .2164 | .9763 | .2217 | 4.511 | .5 |
| .6 | .0454 | .9990 | .0454 | 22.02 | .4 | .6 | .1323 | .9912 | .1334 | 7.495 | .4 | .6 | .2181 | .9759 | .2235 | 4.474 | .4 |
| .7 | .0471 | .9989 | .0472 | 21.20 | .3 | .7 | .1340 | .9910 | .1352 | 7.396 | .3 | .7 | .2198 | .9755 | .2254 | 4.437 | .3 |
| .8 | .0488 | .9988 | .0489 | 20.45 | .2 | .8 | .1357 | .9907 | .1370 | 7.300 | .2 | .8 | .2215 | .9751 | .2272 | 4.402 | .2 |
| .9 | .0506 | .9987 | .0507 | 19.74 | .1 | .9 | .1374 | .9905 | .1388 | 7.207 | .1 | .9 | .2233 | .9748 | .2290 | 4.366 | .1 |
| 3.0 | 0.0523 | 0.9986 | .0524 | 19.081 | 87.0 | 8.0 | 0.1392 | 0.9903 | 0.1405 | 7.115 | 82.0 | 13.0 | 0.2250 | 0.9744 | 0.2309 | 4.331 | 77.0 |
| .1 | .0541 | .9985 | .0542 | 18.464 | .9 | .1 | .1409 | .9900 | .1423 | 7.026 | .9 | .1 | .2267 | .9740 | .2327 | 4.297 | .9 |
| .2 | .0558 | .9984 | .0559 | 17.886 | .8 | .2 | .1426 | .9898 | .1441 | 6.940 | .8 | .2 | .2284 | .9736 | .2345 | 4.264 | .8 |
| .3 | .0576 | .9983 | .0577 | 17.343 | .7 | .3 | .1444 | .9895 | .1459 | 6.855 | .7 | .3 | .2300 | .9732 | .2364 | 4.280 | .7 |
| .4 | .0593 | .9982 | .0594 | 16.832 | .6 | .4 | .1461 | .9893 | .1477 | 6.772 | .6 | .4 | .2317 | .9728 | .2382 | 4.198 | .6 |
| .5 | .0610 | .9981 | .0612 | 16.350 | .5 | .5 | .1478 | .9890 | .1495 | 6.691 | .5 | .5 | .2334 | .9724 | .2401 | 4.165 | .5 |
| .6 | .0628 | .9980 | .0629 | 15.895 | .4 | .6 | .1495 | .9888 | .1512 | 6.612 | .4 | .6 | .2351 | .9720 | .2419 | 4.134 | .4 |
| .7 | .0645 | .9979 | .0646 | 15.464 | .3 | .7 | .1513 | .9885 | .1530 | 6.535 | .3 | .7 | .2368 | .9715 | .2438 | 4.102 | .3 |
| .8 | .0663 | .9978 | .0664 | 15.056 | .2 | .8 | .1530 | .9882 | .1548 | 6.460 | .2 | .8 | .2385 | .9711 | .2456 | 4.071 | .2 |
| .9 | .0680 | .9977 | .0682 | 14.669 | .1 | .9 | .1547 | .9880 | .1566 | 6.386 | .1 | .9 | .2402 | .9707 | .2475 | 4.041 | .1 |
| 4.0 | 0.0698 | 0.9976 | .0699 | 14.301 | 86.0 | 9.0 | 0.1564 | 0.9877 | 0.1584 | 6.314 | 81.0 | 14.0 | 0.2419 | 0.9703 | 0.2493 | 4.011 | 76.0 |
| .1 | .0715 | .9974 | .0717 | 13.951 | .9 | .1 | .1582 | .9874 | .1602 | 6.243 | .9 | .1 | .2436 | .9699 | .2512 | 3.981 | .9 |
| .2 | .0732 | .9973 | .0734 | 13.617 | .8 | .2 | .1599 | .9871 | .1620 | 6.174 | .8 | .2 | .2453 | .9694 | .2530 | 3.952 | .8 |
| .3 | .0750 | .9972 | .0752 | 13.300 | .7 | .3 | .1616 | .9869 | .1638 | 6.107 | .7 | .3 | .2470 | .9690 | .2549 | 3.923 | .7 |
| .4 | .0767 | .9971 | .0769 | 12.996 | .6 | .4 | .1633 | .9866 | .1655 | 6.041 | .6 | .4 | .2487 | .9686 | .2568 | 3.895 | .6 |
| .5 | .0785 | .9969 | .0787 | 12.706 | .5 | .5 | .1650 | .9863 | .1673 | 5.976 | .5 | .5 | .2504 | .9681 | .2586 | 3.867 | .5 |
| .6 | .0802 | .9968 | .0805 | 12.429 | .4 | .6 | .1668 | .9860 | .1691 | 5.912 | .4 | .6 | .2521 | .9677 | .2605 | 3.839 | .4 |
| .7 | .0819 | .9966 | .0822 | 12.163 | .3 | .7 | .1685 | .9857 | .1709 | 5.850 | .3 | .7 | .2538 | .9673 | .2623 | 3.812 | .3 |
| .8 | .0839 | .9965 | .0843 | 11.909 | .2 | .8 | .1702 | .9854 | .1727 | 5.789 | .2 | .8 | .2554 | .9668 | .2642 | 3.785 | .2 |
| .9 | .0854 | .9963 | .0857 | 11.664 | .1 | .9 | .1719 | .9851 | .1741 | 5.730 | .1 | .9 | .2571 | .9664 | .2661 | 3.758 | .1 |
| | cos | sin | cot | tan | deg | | cos | sin | cot | tan | deg | | cos | sin | cot | tan | deg |

| deg | sin | cos | tan | cot | | deg | sin | cos | tan | cot | | deg | sin | cos | tan | cot | |
|---|---|---|---|---|---|---|---|---|---|---|---|---|---|---|---|---|---|
| 15.0 | 0.2588 | 0.9659 | 0.2679 | 3.732 | 75.0 | 20.0 | 0.3420 | 0.9397 | 0.3640 | 2.747 | 70.0 | 25.0 | 0.4226 | 0.9063 | 0.4663 | 2.145 | 65.0 |
| .1 | .2605 | .9655 | .2698 | 3.706 | .9 | .1 | .3437 | .9391 | .3659 | 2.733 | .9 | .1 | .4242 | .9056 | .4684 | 2.135 | .9 |
| .2 | .2622 | .9650 | .2717 | 3.681 | .8 | .2 | .3453 | .9385 | .3679 | 2.718 | .8 | .2 | .4258 | .9048 | .4706 | 2.125 | .8 |
| .3 | .2639 | .9646 | .2736 | 3.655 | .7 | .3 | .3469 | .9379 | .3699 | 2.703 | .7 | .3 | .4274 | .9041 | .4727 | 2.116 | .7 |
| .4 | .2656 | .9641 | .2754 | 3.630 | .6 | .4 | .3486 | .9373 | .3719 | 2.689 | .6 | .4 | .4289 | .9033 | .4748 | 2.106 | .6 |
| .5 | .2672 | .9636 | .2773 | 3.606 | .5 | .5 | .3502 | .9367 | .3739 | 2.675 | .5 | .5 | .4305 | .9026 | .4770 | 2.097 | .5 |
| .6 | .2689 | .9632 | .2792 | 3.582 | .4 | .6 | .3518 | .9361 | .3759 | 2.660 | .4 | .6 | .4321 | .9018 | .4791 | 2.087 | .4 |
| .7 | .2706 | .9627 | .2811 | 3.558 | .3 | .7 | .3535 | .9354 | .3779 | 2.646 | .3 | .7 | .4337 | .9011 | .4813 | 2.078 | .3 |
| .8 | .2723 | .9622 | .2830 | 3.534 | .2 | .8 | .3551 | .9348 | .3799 | 2.633 | .2 | .8 | .4352 | .9003 | .4834 | 2.069 | .2 |
| .9 | .2740 | .9617 | .2849 | 3.511 | .1 | .9 | .3567 | .9342 | .3819 | 2.619 | .1 | .9 | .4368 | .8996 | .4856 | 2.059 | .1 |
| 16.0 | 0.2756 | 0.9613 | 0.2867 | 3.487 | 74.0 | 21.0 | 0.3584 | 0.9336 | 0.3839 | 2.605 | 69.0 | 26.0 | 0.4384 | 0.8988 | 0.4877 | 2.050 | 64.0 |
| .1 | .2773 | .9608 | .2886 | 3.465 | .9 | .1 | .3600 | .9330 | .3859 | 2.592 | .9 | .1 | .4399 | .8980 | .4899 | 2.041 | .9 |
| .2 | .2790 | .9603 | .2905 | 3.442 | .8 | .2 | .3616 | .9323 | .3879 | 2.578 | .8 | .2 | .4415 | .8973 | .4921 | 2.032 | .8 |
| .3 | .2807 | .9598 | .2924 | 3.420 | .7 | .3 | .3633 | .9317 | .3899 | 2.565 | .7 | .3 | .4431 | .8965 | .4942 | 2.023 | .7 |
| .4 | .2823 | .9593 | .2943 | 3.398 | .6 | .4 | .3649 | .9311 | .3919 | 2.552 | .6 | .4 | .4446 | .8957 | .4964 | 2.014 | .6 |
| .5 | .2840 | .9588 | .2962 | 3.376 | .5 | .5 | .3665 | .9304 | .3939 | 2.539 | .5 | .5 | .4462 | .8949 | .4986 | 2.006 | .5 |
| .6 | .2857 | .9583 | .2981 | 3.354 | .4 | .6 | .3681 | .9298 | .3959 | 2.526 | .4 | .6 | .4478 | .8942 | .5008 | 1.997 | .4 |
| .7 | .2874 | .9578 | .3000 | 3.333 | .3 | .7 | .3697 | .9291 | .3979 | 2.513 | .3 | .7 | .4493 | .8934 | .5029 | 1.988 | .3 |
| .8 | .2890 | .9573 | .3019 | 3.312 | .2 | .8 | .3714 | .9285 | .4000 | 2.500 | .2 | .8 | .4509 | .8926 | .5051 | 1.980 | .2 |
| .9 | .2907 | .9568 | .3038 | 3.291 | .1 | .9 | .3730 | .9278 | .4020 | 2.488 | .1 | .9 | .4524 | .8918 | .5073 | 1.971 | .1 |
| 17.0 | 0.2924 | 0.9563 | 0.3057 | 3.271 | 73.0 | 22.0 | 0.3746 | 0.9272 | 0.4040 | 2.475 | 68.0 | 27.0 | 0.4540 | 0.8910 | 0.5095 | 1.963 | 63.0 |
| .1 | .2940 | .9558 | .3076 | 3.251 | .9 | .1 | .3762 | .9265 | .4061 | 2.463 | .9 | .1 | .4555 | .8902 | .5117 | 1.954 | .9 |
| .2 | .2957 | .9553 | .3096 | 3.230 | .8 | .2 | .3778 | .9259 | .4081 | 2.450 | .8 | .2 | .4571 | .8894 | .5139 | 1.946 | .8 |
| .3 | .2974 | .9548 | .3115 | 3.211 | .7 | .3 | .3795 | .9252 | .4101 | 2.438 | .7 | .3 | .4586 | .8886 | .5161 | 1.937 | .7 |
| .4 | .2990 | .9542 | .3134 | 3.191 | .6 | .4 | .3811 | .9245 | .4122 | 2.426 | .6 | .4 | .4602 | .8878 | .5184 | 1.929 | .6 |
| .5 | .3007 | .9537 | .3153 | 3.172 | .5 | .5 | .3827 | .9239 | .4142 | 2.414 | .5 | .5 | .4617 | .8870 | .5206 | 1.921 | .5 |
| .6 | .3024 | .9532 | .3172 | 3.152 | .4 | .6 | .3843 | .9232 | .4163 | 2.402 | .4 | .6 | .4633 | .8862 | .5228 | 1.913 | .4 |
| .7 | .3040 | .9527 | .3191 | 3.133 | .3 | .7 | .3859 | .9225 | .4183 | 2.391 | .3 | .7 | .4648 | .8854 | .5250 | 1.905 | .3 |
| .8 | .3057 | .9521 | .3211 | 3.115 | .2 | .8 | .3875 | .9219 | .4204 | 2.379 | .2 | .8 | .4664 | .8846 | .5272 | 1.897 | .2 |
| .9 | .3074 | .9516 | .3230 | 3.096 | .1 | .9 | .3891 | .9212 | .4224 | 2.367 | .1 | .9 | .4679 | .8838 | .5295 | 1.889 | .1 |
| 18.0 | 0.3090 | 0.9511 | 0.3249 | 3.078 | 72.0 | 23.0 | 0.3907 | 0.9205 | 0.4245 | 2.356 | 67.0 | 28.0 | 0.4695 | 0.8829 | 0.5317 | 1.881 | 62.0 |
| .1 | .3107 | .9505 | .3269 | 3.060 | .9 | .1 | .3923 | .9198 | .4265 | 2.344 | .9 | .1 | .4710 | .8821 | .5340 | 1.873 | .9 |
| .2 | .3123 | .9500 | .3288 | 3.042 | .8 | .2 | .3939 | .9191 | .4286 | 2.333 | .8 | .2 | .4726 | .8813 | .5362 | 1.865 | .8 |
| .3 | .3140 | .9494 | .3307 | 3.024 | .7 | .3 | .3955 | .9184 | .4307 | 2.322 | .7 | .3 | .4741 | .8805 | .5384 | 1.857 | .7 |
| .4 | .3156 | .9489 | .3327 | 3.006 | .6 | .4 | .3971 | .9178 | .4327 | 2.311 | .6 | .4 | .4756 | .8796 | .5407 | 1.849 | .6 |
| .5 | .3173 | .9483 | .3346 | 2.989 | .5 | .5 | .3987 | .9171 | .4348 | 2.300 | .5 | .5 | .4772 | .8788 | .5430 | 1.842 | .5 |
| .6 | .3190 | .9478 | .3365 | 2.971 | .4 | .6 | .4003 | .9164 | .4369 | 2.289 | .4 | .6 | .4787 | .8780 | .5452 | 1.834 | .4 |
| .7 | .3206 | .9472 | .3385 | 2.954 | .3 | .7 | .4019 | .9157 | .4390 | 2.278 | .3 | .7 | .4802 | .8771 | .5475 | 1.827 | .3 |
| .8 | .3223 | .9466 | .3404 | 2.937 | .2 | .8 | .4035 | .9150 | .4411 | 2.267 | .2 | .8 | .4818 | .8763 | .5498 | 1.819 | .2 |
| .9 | .3239 | .9461 | .3424 | 2.921 | .1 | .9 | .4051 | .9143 | .4431 | 2.257 | .1 | .9 | .4833 | .8755 | .5520 | 1.811 | .1 |
| 19.0 | 0.3256 | 0.9455 | 0.3443 | 2.904 | 71.0 | 24.0 | 0.4067 | 0.9135 | 0.4452 | 2.246 | 66.0 | 29.0 | 0.4848 | 0.8746 | 0.5543 | 1.804 | 61.0 |
| .1 | .3272 | .9449 | .3463 | 2.888 | .9 | .1 | .4083 | .9128 | .4473 | 2.236 | .9 | .1 | .4863 | .8738 | .5566 | 1.797 | .9 |
| .2 | .3289 | .9444 | .3482 | 2.872 | .8 | .2 | .4099 | .9121 | .4494 | 2.225 | .8 | .2 | .4879 | .8729 | .5589 | 1.789 | .8 |
| .3 | .3305 | .9438 | .3502 | 2.856 | .7 | .3 | .4115 | .9114 | .4515 | 2.215 | .7 | .3 | .4894 | .8721 | .5612 | 1.782 | .7 |
| .4 | .3322 | .9432 | .3522 | 2.840 | .6 | .4 | .4131 | .9107 | .4536 | 2.204 | .6 | .4 | .4909 | .8712 | .5635 | 1.775 | .6 |
| .5 | .3338 | .9426 | .3541 | 2.824 | .5 | .5 | .4147 | .9100 | .4557 | 2.194 | .5 | .5 | .4924 | .8704 | .5658 | 1.767 | .5 |
| .6 | .3355 | .9421 | .3561 | 2.808 | .4 | .6 | .4163 | .9092 | .4578 | 2.184 | .4 | .6 | .4939 | .8695 | .5681 | 1.760 | .4 |
| .7 | .3371 | .9415 | .3581 | 2.793 | .3 | .7 | .4179 | .9085 | .4599 | 2.174 | .3 | .7 | .4955 | .8686 | .5704 | 1.753 | .3 |
| .8 | .3387 | .9409 | .3600 | 2.778 | .2 | .8 | .4195 | .9078 | .4621 | 2.164 | .2 | .8 | .4970 | .8678 | .5727 | 1.746 | .2 |
| .9 | .3404 | .9403 | .3620 | 2.762 | .1 | .9 | .4210 | .9070 | .4642 | 2.154 | .1 | .9 | .4985 | .8669 | .5750 | 1.739 | .1 |
| | *cos* | *sin* | *cot* | *tan* | *deg* | | *cos* | *sin* | *cot* | *tan* | *deg* | | *cos* | *sin* | *cot* | *tan* | *deg* |

| *deg* | *sin* | *cos* | *tan* | *cot* | |
|---|---|---|---|---|---|
| 30.0 | 0.5000 | 0.8660 | 0.5774 | 1.7321 | 60.0 |
| .1 | .5015 | .8652 | .5797 | 1.7251 | .9 |
| .2 | .5030 | .8643 | .5820 | 1.7182 | .8 |
| .3 | .5045 | .8634 | .5844 | 1.7113 | .7 |
| .4 | .5060 | .8625 | .5867 | 1.7045 | .6 |
| .5 | .5075 | .8616 | .5890 | 1.6977 | .5 |
| .6 | .5090 | .8607 | .5914 | 1.6909 | .4 |
| .7 | .5105 | .8599 | .5938 | 1.6842 | .3 |
| .8 | .5120 | .8590 | .5961 | 1.6775 | .2 |
| .9 | .5135 | .8581 | .5985 | 1.6709 | .1 |
| 31.0 | 0.5150 | 0.8572 | 0.6009 | 1.6643 | 59.0 |
| .1 | .5165 | .8563 | .6032 | 1.6577 | .9 |
| .2 | .5180 | .8554 | .6056 | 1.6512 | .8 |
| .3 | .5195 | .8545 | .6080 | 1.6447 | .7 |
| .4 | .5210 | .8536 | .6104 | 1.6383 | .6 |
| .5 | .5225 | .8526 | .6128 | 1.6319 | .5 |
| .6 | .5240 | .8517 | .6152 | 1.6255 | .4 |
| .7 | .5255 | .8508 | .6176 | 1.6191 | .3 |
| .8 | .5270 | .8499 | .6200 | 1.6128 | .2 |
| .9 | .5284 | .8490 | .6224 | 1.6066 | .1 |
| 32.0 | 0.5299 | 0.8480 | 0.6249 | 1.6003 | 58.0 |
| .1 | .5314 | .8471 | .6273 | 1.5941 | .9 |
| .2 | .5329 | .8462 | .6297 | 1.5880 | .8 |
| .3 | .5344 | .8450 | .6322 | 1.5818 | .7 |
| .4 | .5358 | .8443 | .6346 | 1.5757 | .6 |
| .5 | .5373 | .8434 | .6371 | 1.5697 | .5 |
| .6 | .5388 | .8425 | .6395 | 1.5637 | .4 |
| .7 | .5402 | .8415 | .6420 | 1.5577 | .3 |
| .8 | .5417 | .8406 | .6445 | 1.5517 | .2 |
| .9 | .5432 | .8396 | .6469 | 1.5458 | .1 |
| 33.0 | 0.5446 | 0.8387 | 0.6494 | 1.5399 | 57.0 |
| .1 | .5461 | .8377 | .6519 | 1.5340 | .9 |
| .2 | .5476 | .8368 | .6544 | 1.5282 | .8 |
| .3 | .5490 | .8358 | .6569 | 1.5224 | .7 |
| .4 | .5505 | .8348 | .6594 | 1.5166 | .6 |
| .5 | .5519 | .8339 | .6619 | 1.5108 | .5 |
| .6 | .5534 | .8329 | .6644 | 1.5051 | .4 |
| .7 | .5548 | .8320 | .6669 | 1.4994 | .3 |
| .8 | .5563 | .8310 | .6694 | 1.4938 | .2 |
| .9 | .5577 | .8300 | .6720 | 1.4882 | .1 |
| 34.0 | 0.5592 | 0.8290 | 0.6745 | 1.4826 | 56.0 |
| .1 | .5606 | .8281 | .6771 | 1.4770 | .9 |
| .2 | .5621 | .8271 | .6796 | 1.4715 | .8 |
| .3 | .5635 | .8261 | .6822 | 1.4659 | .7 |
| .4 | .5650 | .8251 | .6847 | 1.4605 | .6 |
| .5 | .5664 | .8241 | .6873 | 1.4550 | .5 |
| .6 | .5678 | .8231 | .6899 | 1.4496 | .4 |
| .7 | .5693 | .8221 | .6924 | 1.4442 | .3 |
| .8 | .5707 | .8211 | .6950 | 1.4388 | .2 |
| .9 | .5721 | .8202 | .6976 | 1.4335 | .1 |
| 35.0 | 0.5736 | 0.8192 | 0.7002 | 1.4281 | 55.0 |
| .1 | .5750 | .8181 | .7028 | 1.4229 | .9 |
| | *cos* | *sin* | *cot* | *tan* | *deg* |

| *deg* | *sin* | *cos* | *tan* | *cot* | |
|---|---|---|---|---|---|
| 35.2 | .5764 | .8171 | .7054 | 1.4176 | .8 |
| .3 | .5779 | .8161 | .7080 | 1.4124 | .7 |
| .4 | .5793 | .8151 | .7107 | 1.4071 | .6 |
| .5 | .5807 | .8141 | .7133 | 1.4019 | .5 |
| .6 | .5821 | .8131 | .7159 | 1.3968 | .4 |
| .7 | .5835 | .8121 | .7186 | 1.3916 | .3 |
| .8 | .5850 | .8111 | .7212 | 1.3865 | .2 |
| .9 | .5864 | .8100 | .7239 | 1.3814 | .1 |
| 36.0 | 0.5878 | 0.8090 | 0.7265 | 1.3764 | 54.0 |
| .1 | .5892 | .8080 | .7292 | 1.3713 | .9 |
| 2 | .5906 | .8070 | .7319 | 1.3663 | .8 |
| .3 | .5920 | .8059 | .7346 | 1.3613 | .7 |
| .4 | .5934 | .8049 | .7373 | 1.3564 | .6 |
| .5 | .5948 | .8039 | .7400 | 1.3514 | .5 |
| .6 | .5962 | .8028 | .7427 | 1.3465 | .4 |
| .7 | .5976 | .8018 | .7454 | 1.3416 | .3 |
| .8 | .5990 | .8007 | .7481 | 1.3367 | .2 |
| .9 | .6004 | .7997 | .7508 | 1.3319 | .1 |
| 37.0 | 0.6018 | 0.7986 | 0.7536 | 1.3270 | 53.0 |
| .1 | .6032 | .7967 | .7563 | 1.3222 | 9. |
| .2 | .6046 | .7965 | .7590 | 1.3175 | .8 |
| .3 | .6060 | .7955 | .7618 | 1.3127 | .7 |
| .4 | .6074 | .7944 | .7646 | 1.3079 | .6 |
| .5 | .6088 | .7934 | .7673 | 1.3032 | .5 |
| .6 | .6101 | .7923 | .7701 | 1.2985 | .4 |
| .7 | .6115 | .7912 | .7729 | 1.2938 | .3 |
| .8 | .6129 | .7902 | .7757 | 1.2892 | .2 |
| .9 | .6143 | .7891 | .7785 | 1.2846 | .1 |
| 38.0 | 0.6157 | 0.7880 | 0.7813 | 1.2799 | 52.0 |
| .1 | .6170 | .7869 | .7841 | 1.2753 | .9 |
| .2 | .6184 | .7859 | .7869 | 1.2708 | .8 |
| .3 | .6198 | .7848 | .7898 | 1.2662 | .7 |
| .4 | .6211 | .7837 | .7926 | 1.2617 | .6 |
| .5 | .6225 | .7826 | .7954 | 1.2572 | .5 |
| .6 | .6239 | .7815 | .7983 | 1.2527 | .4 |
| .7 | .6252 | .7804 | .8012 | 1.2482 | .3 |
| .8 | .6266 | .7793 | .8040 | 1.2437 | .2 |
| .9 | .6280 | .7782 | .8069 | 1.2393 | .1 |
| 39.0 | 0.6293 | 0.7771 | 0.8098 | 1.2349 | 51.0 |
| .1 | .6307 | .7760 | .8127 | 1.2305 | .9 |
| .2 | .6320 | .7749 | .8156 | 1.2261 | .8 |
| .3 | .6334 | .7738 | .8185 | 1.2218 | .7 |
| .4 | .6347 | .7727 | .8214 | 1.2174 | .6 |
| .5 | .6361 | .7716 | .8243 | 1.2131 | .5 |
| .6 | .6374 | .7705 | .8273 | 1.2088 | .4 |
| .7 | .6388 | .7694 | .8302 | 1.2045 | .3 |
| .8 | .6401 | .7683 | .8332 | 1.2002 | .2 |
| .9 | .6414 | .7672 | .8361 | 1.1960 | .1 |
| 40.0 | 0.6428 | 0.7660 | 0.8391 | 1.1918 | 50.0 |
| .1 | .6441 | .7649 | .8421 | 1.1875 | .9 |
| .2 | .6455 | .7638 | .8451 | 1.1833 | .8 |
| | *cos* | *sin* | *cot* | *tan* | *deg* |

| *deg* | *sin* | *cos* | *tan* | *cot* | |
|---|---|---|---|---|---|
| 40.3 | .6468 | .7627 | .8481 | 1.1792 | .7 |
| .4 | .6481 | .7615 | .8511 | 1.1750 | .6 |
| .5 | .6494 | .7604 | .8541 | 1.1708 | .5 |
| .6 | .6508 | .7593 | .8571 | 1.1667 | .4 |
| .7 | .6521 | .7581 | .8601 | 1.1626 | .3 |
| .8 | .6534 | .7570 | .8632 | 1.1585 | .2 |
| .9 | .6547 | .7559 | .8662 | 1.1544 | .1 |
| 41.0 | 0.6561 | 0.7547 | 0.8693 | 1.1504 | 49.0 |
| .1 | .6574 | .7536 | .8724 | 1.1463 | .9 |
| .2 | .6587 | .7524 | .8754 | 1.1423 | .8 |
| .3 | .6600 | .7513 | .8785 | 1.1383 | .7 |
| .4 | .6613 | .7501 | .8816 | 1.1343 | .6 |
| .5 | .6626 | .7490 | .8847 | 1.1303 | .5 |
| .6 | .6639 | .7478 | .8878 | 1.1263 | .4 |
| .7 | .6652 | .7466 | .8910 | 1.1224 | .3 |
| .8 | .6665 | .7455 | .8941 | 1.1184 | .2 |
| .9 | .6678 | .7443 | .8972 | 1.1145 | .1 |
| 42.0 | 0.6691 | 0.7431 | 0.9004 | 1.1106 | 48.0 |
| .1 | .6704 | .7420 | .9036 | 1.1067 | .9 |
| .3 | .6717 | .7408 | .9067 | 1.1028 | .8 |
| .3 | .6730 | .7396 | .9099 | 1.0990 | .7 |
| .4 | .6743 | .7385 | .9131 | 1.0951 | .6 |
| .5 | .6756 | .7373 | .9163 | 1.0913 | .5 |
| .6 | .6769 | .7361 | .9195 | 1.0875 | .4 |
| .7 | .6782 | .7349 | .9228 | 1.0837 | .3 |
| .8 | .6794 | .7337 | .9260 | 1.0799 | .2 |
| .9 | .6807 | .7325 | .9293 | 1.0761 | .1 |
| 43.0 | 0.6820 | 0.7314 | 0.9325 | 1.0724 | 47.0 |
| .1 | .6833 | .7302 | .9358 | 1.0686 | .9 |
| .2 | .6845 | .7290 | .9391 | 1.0649 | .8 |
| .3 | .6858 | .7278 | .9424 | 1.0612 | .7 |
| .4 | .6871 | .7266 | .9457 | 1.0575 | .6 |
| .5 | .6884 | .7254 | .9490 | 1.0538 | .5 |
| .6 | .6896 | .7242 | .9523 | 1.0501 | .4 |
| .7 | .6909 | .7230 | .9556 | 1.0464 | .3 |
| .8 | .6921 | .7218 | .9590 | 1.0428 | .2 |
| .9 | .6934 | .7206 | .9623 | 1.0392 | .1 |
| 44.0 | 0.6947 | 0.7193 | 0.9657 | 1.0355 | 46.0 |
| .1 | .6959 | .7181 | .9691 | 1.0319 | .9 |
| .2 | .6972 | .7169 | .9725 | 1.0283 | .8 |
| .3 | .6934 | .7157 | .9759 | 1.0247 | .7 |
| .4 | .6997 | .7145 | .9793 | 1.0212 | .6 |
| .5 | .7009 | .7133 | .9827 | 1.0176 | .5 |
| .6 | .7022 | .7120 | .9861 | 1.0141 | .4 |
| .7 | .7034 | .7108 | .9896 | 1.0105 | .3 |
| .8 | .7046 | .7096 | .9930 | 1.0070 | .2 |
| .9 | .7059 | .7083 | .9965 | 1.0035 | .1 |
| 45.0 | 0.7071 | 0.7071 | 1.0000 | 1.0000 | 45.0 |
| | *cos* | *sin* | *cot* | *tan* | *deg* |

# INDEX

### *T*



### *U*



### *V*



### *W*



### *Z*